Applied Probability
Control
Economics
Information and Communication
Modeling and Identification
Numerical Techniques
Optimization

Applications of Mathematics

19

Edited by A. V. Balakrishnan

Applications of Mathematics

G. I. Marchuk
V. V. Shaidurov

Difference Methods
and Their Extrapolations

Springer-Verlag
New York Berlin Heidelberg Tokyo

G. I. Marchuk
Department of Numerical
  Mathematics of the
  USSR Academy of Sciences
Gorky 11
103905 Moscow
U.S.S.R.

V. V. Shaidurov
Computing Center of the
  Siberian Branch of the
  USSR Academy of Sciences
Mira 53
660049 Krasnoyarsk
U.S.S.R.

Managing Editor

A. V. Balakrishnan
Systems Science Department
University of California
Los Angeles, CA 90024
U.S.A.

AMS Subject Classifications: 41A25, 39A10, 65005

Library of Congress Cataloging in Publication Data
Marchuk, G. I. (Guriĭ Ivanovich), 1925–
  Difference methods and their extrapolations.
  (Applications of mathematics; 19)
  Translation of: Povyshenie tochnosti resheniĭ
raznostnykh skhem.
  Bibliography: p.
  Includes index.
  1. Difference equations. 2. Approximation theory.
I. Shaidurov, V. V. II. Title. III. Series.
QA431.M28613 1983   515'.625   82–19544

With 20 Figures.

Title of original Russian edition: Povyshenie Tochnosti resheniĭ Raznostnykh Skhem.
Nauka: Moscow, 1979.

Typeset by Composition House Ltd., Salisbury, England.

9 8 7 6 5 4 3 2 1

ISBN-13:978-1-4613-8226-3   e-ISBN-13:978-1-4613-8224-9
DOI: 10.1007/978-1-4613-8224-9

Preface

The stimulus for the present work is the growing need for more accurate numerical methods. The rapid advances in computer technology have not provided the resources for computations which make use of methods with low accuracy.

The computational speed of computers is continually increasing, while memory still remains a problem when one handles large arrays. More accurate numerical methods allow us to reduce the overall computation time by several orders of magnitude.

The problem of finding the most efficient methods for the numerical solution of equations, under the assumption of fixed array size, is therefore of paramount importance.

Advances in the applied sciences, such as aerodynamics, hydrodynamics, particle transport, and scattering, have increased the demands placed on numerical mathematics. New mathematical models, describing various physical phenomena in greater detail than ever before, create new demands on applied mathematics, and have acted as a major impetus to the development of computer science.

For example, when investigating the stability of a fluid flowing around an object one needs to solve the low viscosity form of certain hydrodynamic equations describing the fluid flow. The usual numerical methods for doing so require the introduction of a "computational viscosity," which usually exceeds the physical value; the results obtained thus present a distorted picture of the phenomena under study.

A similar situation arises in the study of behavior of the oceans, assuming weak turbulence. Many additional examples of this type can be given.

The natural way to solve such differential equations which arise in this type of problem is by the application of numerical methods which most closely approximate the original problem.

Several methods for increasing the accuracy of the standard numerical methods have been considered in the literature. Such methods include a decrease in the step size, use of multipoint difference schemes, and, most notably, Richardson extrapolation, which operates on a sequence of approximating nets. This monograph is devoted to the study of the Richardson extrapolation.

This method has several noteworthy features which make it both powerful and almost universally applicable. It makes use of the simplest form of difference approximation. It is a locally uniform method, that is, it uses a uniform step size on the various nets which are used, these nets entering with different approximation parameters. Finally, the algorithm as a whole is logically simple and easily applicable.

These advantages have caused this method to be widely used. Even though many papers have been devoted to describing this method and its various modifications, no coherent, general description of it exists in the literature. The purpose of this work is to provide such a description.

This work considers both the theoretical foundations and practical aspects of the method. For linear ordinary differential equations the method can usually be justified by the simple application of some abstract results. The situation is somewhat more complicated for partial differential equations; additional questions arise. For elliptic boundary-value problems the smoothness of the solution and its asymptotic behavior near cusps become important. For parabolic equations the question of smoothness is connected with the procedure for decomposing multidimensional problems into a series of one-dimensional problems. Some results on the application of the Richardson method to quasilinear equations and to eigenvalue problems are presented.

This work also discusses the application of the Richardson method to the solution of algebraic systems with singular coefficient matrices. The application of the method, at a small parameter, to boundary layer problems is also discussed.

The problem of finding more efficient algorithms is a very pressing one. This work collects and generalizes the research of the authors over the past ten years, which have focused on the application of various numerical methods to mathematical physics. The results presented in this book have been tested in practice on many occasions, with great success.

Some simple versions of these numerical methods formed the basis for a course given at the State Universities of Krasnoyarsk and Novosibirsk; an early version appeared as *Methods of Computational Mathematics* by G. I. Marchuk, Nauka Publishing House, 1977 (Russian).

The bibliography includes only those works containing results used in the text. Therefore, many interesting and important works, both theoretical and applied, have been omitted.

During the preparation of this work the authors had the opportunity to exchange ideas with many specialists in numerical mathematics. The authors wish to express their deep thanks to E. A. Volkov, Y. A. Kuznetsov, V. I. Lebedev, G.-L. Lions, V. I. Agoshkov, A. M. Matsokin, and many others, all of whom helped to shape this book through their most helpful discussions and comments.

The manuscript was read by G. V. Demidov, Y. N. Valitsky, V. A. Sapozhnikov, and V. A. Shchepanovsky. The authors are grateful for their valuable assistance and for several useful remarks which were incorporated into the final version. The authors also wish to thank A. V. Alexeev, T. E. Baburina, B. M. Bagaev, S. M. Bersenyov, B. S. Dobronets, and A. V. Zhukov who helped with the format of the final manuscript and who made numerical calculations.

G. I. MARCHUK
V. V. SHAIDUROV

Contents

Introduction

The widespread use of comupters in problem solving, in both science and technology, has stimulated the development of many numerical algorithms. Most of these algorithms are based on the reduction of an initial differential probelm, to the same problems in linear algebra. At present this method is the most widely used means for solving applied problems. It is natural that the dimensionality of the linear algebra problems obtained depends on the discretization parameter, the latter being generally the finite difference mesh-size. Thus, more accurate solutions require smaller discretization parameters. By decreasing the parameter we generally increase the number of linear equations, which usually results in an increase in the amount of computation time.

In spite of the growing efficiency of computer techniques, the requirements placed on the accuracy of the algorithms used grow so quickly that they cannot be properly met. This is what has prompted the search for a universal and economical approach to the creation of numerical algorithms, and to their implementation.

It is well known that the accuracy of the numerical solution is proportional to some power of the mesh parameter h. More economical schemes give solutions whose accuracy is proportional to the first or second power of h.

At the beginning of this century, Richardson formulated a completely new method for increasing the accuracy of numerical solutions of linear problems. He introduced a means for creating algorithms using numerical solutions with various different approximation parameters. A certain linear combination of these solutions seemed to give greater accuracy. This was the basic idea in constructing numerical algorithms for many problems connected with differential equations, which later was shaped into a universal approach whose importance for numerical and applied mathematics is hard to overestimate.

Unfortunately, Richardson's method, as with the corresponding Runge and Romberg methods had long been used heuristically, without a proper theoretical basis. In a number of cases this led to incorrect conclusions concerning the practicality and usefulness of the method. Recently interest in this method has been revived in the literature.

When developing economical numerical algorithms the behaviour of the solution is of great importance. The class of functions to which it belongs is often critical to the approximation. Indeed, let us assume that we know that the solution of a certain problem is a quadratic function. In order to find such a solution it is sufficient to determine its value at only three points. Then the solution at any point can be found from the given *a priori* assumption. This simple example emphasizes the fact that *a priori* information can radically decrease the dimensionality of the initial data. If one deals with differential equations the decisive *a priori* information will be the smoothness of the solution, its asymptotic behaviour at exceptional points, its dependence on initial data, and so on. In this case, the more complete the *a priori* information, the narrower the class of functions to which it will belong and, consequently, the easier it is to determine. It may happen that there will be so much information about the solution that the class of functions to which the solution can belong will have only one element. In most situations we deal with a certain set of useful data which usually determine the space of functions of the solution. The use of such *a priori* information sometimes is decisive in choosing the method of an approximation. Indeed, let us suppose a solution of a certain differential equation is sought using a numerical method with first-order accuracy with respect to the mesh size h. If this method is used with parameters h and $h/2$, then, generally speaking, one can claim that the second solution is twice as accurate as the first. If the parameter is chosen to be $h/3$, the solution is three times as accurate as the first, and so on. This formal process of the refinement leads us to conclude that the accuracy of the result under the *a priori* assumption of first-order accuracy is approximately proportional to the mesh-size. Therefore, if we need to obtain a result 100 times as accurate as the approximate solution with parameter h it is necessary to decrease the mesh-size by a factor of 100. Naturally such an approach cannot be always realized even with the most powerful computers. This is especially true if we deal with two- or three-dimensional problems, where the dimensions of the arrays will be squared and cubed, respectively.

Assume the solution of an initial-value problem has a greater degree of smoothness. Assume this allows us to conclude that the approximate solution has three derivatives which are bounded in (the parameter) h. Then, using a Taylor-series expansion in powers of h, one can prove that a linear combination of the three approximate solutions h, $h/2$, and $h/3$ causes a considerable increase in accuracy. In the given case the accuracy of the linear combination is $O(h^3)$. If the dimensionless parameter h is, for example, chosen to be $\frac{1}{10}$, the accuracy of the solution is of order 10^{-3}. We should note that without such a procedure we should have had to solve the approximate system with parameter 10^{-3}. In the one-dimensional case this leads to the number of difference nodes proportional to 10^3, while in the given procedure we have used only three solutions with the number of nodes proportional to 10, 20, and 30, respectively. For two- and three-dimensional problems the comparison is even more striking.

In the general case the situation is often the same as in the case when we find a quadratic by three points, with the difference that the "points" are the approximate solutions with the given parameters h, $h/2$, $h/3$, and the solution of the initial-value problem is not exact, but is an approximate solution with accuracy of order $O(h^3)$. It is natural that under the assumption of greater *a priori* smoothness we can obtain a higher order of accuracy. The application of a set of approximate solutions on a sequence of nets is the main idea of the Richardson extrapolation method for obtaining approximate solutions with greater accuracy.

The objective of this monograph is to develop some general approaches to using the Richardson method for a wide range of problems in mathematical physics, and to formulate requirements which will ensure the efficacy of its application. For linear problems it has been possible to formulate general requirements. The task of a researcher is then to give a constructive version of these requirements for a concrete problem. The reader is referred to Chapter 1, which presents some general convergence theorems based on these requirements.

The next four chapters are divided into sections according to the types of differential equation. We divide the material in this way for two reasons. For one thing, the numerical methods used are in many respects influenced by the properties of the problem; in many cases the type of differential equation influences the choice of the most economical numerical method. In addition, the theoretical results needed rely mainly on the smoothness of the data and not on the solution itself. We thus introduce some useful auxiliary results from the theory of differential equations on the relationship between the smoothness of the data and the solution. This information is also presented according to the type of problem.

The Richardson method allows us to obtain more accurate solutions. Any order of accuracy can be achieved, in principle, under suitable conditions of smoothness and consistency. These issues are dealt with in Chapters 3, 4, and 5. Equations with discontinuous coefficients and problems whose solutions have singularities at boundary points are discussed. It is known that it is particularly difficult to achieve greater accuracy when the differential equation is posed in a region with cusps. For example, this situation is studied in Chapter 4 for elliptic equations. In the neighborhood of a cusp the solution usually has singularities obeying a power law; the error near these points does not agree with the error far away from them. Therefore the Richardson extrapolation method cannot be applied directly. However, if we distinguish these singularities in an explicit way beforehand and represent the solution as a superposition of a singular part and a regular part, then one can use the Richardson method on the regular component. As a result, we can again solve this problem to a higher order of accuracy.

Applying the Richardson method to linear problems suggested that it was possible to use it for nonlinear problems. Of course, in this case the *a priori* restrictions on the differential operator, the initial data, and the solution turn out to be more strict than in the linear case. Nevertheless the class of possible applications of the method appears to be rather broad.

The Richardson method can be used for nonstationary problems in mathematical physics, primarily for parabolic equations. In recent years many algorithms have been constructed for the approximate solution of such problems. In particular, let us mention the "splitting-up" method, which splits complex multidimensional problems into systems of one-dimensional problems. At first it might seem that approximation errors arising as a consequence of using this method would exclude the Richardson method. However, the Richardson method has proved to be effective in this case; no modifications of the algorithm are in fact needed. The authors were able to apply the Richardson extrapolation to one splitting-up scheme which has been widely used. Moreover, we have reason to suppose that the class of such splitting-up schemes will grow.

The idea behind the Richardson method has proved to be fruitful for problems other than differential equations. It can be used, for example, to construct algorithms for solving singular linear algebraic equations. This book discusses such an algorithm due to A. N. Tikhonov. The basic idea is as follows. The original algebraic system is replaced by another similar, but nonsingular system, with parameter ε which we let go to zero. We first choose this parameter so as to obtain a well-conditioned system of equations and so that the solution is not too inaccurate. Then we solve the problems with regularization parameter $\varepsilon/2$, then $\varepsilon/3$, etc. A weighted linear combination of these solutions leads to a normal solution with accuracy $O(\varepsilon^n)$ where n is the number of auxiliary problems with parameters $\varepsilon, \varepsilon/2, \ldots, \varepsilon/n$. This approach can be applied to many problems of linear algebra, to differential equations with a small parameter in the higher derivatives, and to other types of problems. Such problems arise, for example, in the physics of the atmosphere and ocean, in conditions of weak turbulence, in hydrodynamic flows with large Reynolds' numbers, and so on. At the end of Chapter 6 we illustrate the application of the Richardson extrapolation to integral equations.

Chapter 7 contains some auxiliary results used repeatedly in our theoretical developments.

In all the experimental data given the effect of round-off error, and computational error does not exceed the last significant digit in the results presented. This allows us to evaluate the effect of extrapolation, irrespective of the type of computer used, compilers, or the skill of the programmer.

The only exception to this is the example of §2.1, which is devoted to the effect of the round-off error on the results of the extrapolation.

In conclusion, let us once again stress the universality of the Richardson method. The repeated solution of finite difference problems aimed at obtaining results with a higher order of accuracy does not change the structure of finite difference scheme approximating the original problem. Only the discretization parameter and the number of linear equations approximating the problem change. The invariant nature of the algorithm is essential when one constructs automated versions for problem solving.

CHAPTER 1
General Properties

We begin this chapter with a concrete example. In this example we use the solutions of a difference scheme of rather low accuracy and achieve a more accurate approximate solution using the Richardson extrapolation with higher-order differences. We then extend these results to abstract problems: we first, formulate a sufficient condition for the existence of an expansion of the approximate solution in powers of the difference net mesh-size; then we prove that this expansion allows us to use either the Richardson extrapolation or higher-order difference corrections.

1.1. The Simplest Example

In this section we consider the solution of first-order linear differential equations. In this archetypical example we use the Richardson extrapolation on a set of solutions of the difference problem with rather low accuracy. We present a method for justifying the increased accuracy of the extrapolated solution. Then we consider a simple version of this method using higher-order differences and present numerical results illustrating the effectiveness of algorithms.

1.1.1. Local and Global Richardson Extrapolation

Let us consider the differential equation

$$u' + u = f \quad \text{on } (0, 1), \tag{1.1}$$

with initial condition

$$u(0) = u_0. \tag{1.2}$$

Assume that the function f is infinitely differentiable on the interval $[0, 1]$.

To solve this problem numerically we construct the regular net

$$\bar{\omega}_\tau = \{t_j = j\tau; j = 0, 1, \ldots, M\} \tag{1.3}$$

with mesh-size $\tau = 1/M$ ($M \geq 2$ is an integer) and introduce the mid-points

$$\breve{\omega}_\tau = \{t_{j+1/2} = (j + 1/2)\tau; j = 0, 1, \ldots, M - 1\}. \tag{1.4}$$

In the Crank–Nicholson scheme we replace equation (1.1) at the mid-points by the system of algebraic equations

$$u_{\dot{t}}^\tau + u_{\hat{t}}^\tau = f \quad \text{on } \breve{\omega}_\tau. \tag{1.5}$$

Here

$$u_{\dot{t}}(t) = \{u(t + \tau/2) - u(t - \tau/2)\}/\tau, \qquad u_{\hat{t}}(t) = \{u(t + \tau/2) + u(t - \tau/2)\}/2.$$

Augment (1.5) with the condition

$$u^\tau(0) = u_0, \tag{1.6}$$

from (1.2). The solution of the problem we obtain is a net function u^τ which approximates the function u with second-order accuracy (see [11]) at all points of the net $\bar{\omega}_\tau$:

$$\|u^\tau - u\|_{C,\tau} \leq c_1 \tau^2. \tag{1.7}$$

We denote the constants which are independent of τ, t, x, h by c_i (where the index i is an integer).

The error is taken to be the difference between the approximate (difference) solution u^τ and the exact solution u at the net points $t \in \bar{\omega}_\tau$. We show that when $\tau \to 0$ the relation

$$u^\tau - u = \tau^2 v + \eta^\tau \quad \text{on } \bar{\omega}_\tau \tag{1.8}$$

holds, where v is a certain smooth function defined on $[0, 1]$ and independent of τ; η^τ is a net function defined on $\bar{\omega}_\tau$ with values of order $O(\tau^4)$. The notation $O(\tau^4)$ implies that for an arbitrary net function φ the equality $\varphi = O(\tau^k)$ on the set Ω is equivalent to the existence of a constant $c \in [0, \infty)$ for which the inequality

$$|\varphi| \leq c\tau^k \quad \text{on } \Omega$$

holds.

We first suppose that the expansion (1.8) is valid. Then

$$u^\tau = u + \tau^2 v + \eta^\tau \quad \text{on } \bar{\omega}_\tau.$$

Using these equations we substitute the values u^τ into (1.5), (1.6) and obtain

$$u_{\dot{t}} + \tau^2 v_{\dot{t}} + \eta_{\dot{t}}^\tau + u_{\hat{t}} + \tau^2 v_{\hat{t}} + \eta_{\hat{t}}^\tau = f \quad \text{on } \breve{\omega}_\tau, \tag{1.9}$$

$$u(0) + \tau^2 v(0) + \eta^\tau(0) = u_0. \tag{1.10}$$

Taking the Taylor-series expansion of the functions u and v around t in (1.9) we have

$$u' + u + \tau^2(u'''/24 + u''/8 + v' + v) + O(\tau^4) + \eta_{\bar{t}}^{\tau} + \eta_{\hat{t}}^{\tau} = f.$$

Since the coefficients of τ^0 and τ^2 are independent of τ and the other terms on the right-hand side are assumed to be of higher order, then because τ is arbitrary it follows that the relations

$$u' + u = f, \tag{1.11}$$

$$u'''/24 + u''/8 + v' + v = 0 \tag{1.12}$$

hold at all points.

Similar reasoning for (1.10) leads to

$$u(0) = u_0, \tag{1.13}$$

$$v(0) = 0. \tag{1.14}$$

It is natural to treat the first two relations as equations for the functions u and v and the last two equations as the initial conditions. These conditions hold automatically for u, since it is the solution of (1.1), (1.2). We have obtained an important fact about v which we take advantage of in what follows.

Let us note that the function v is solution of the equation

$$v' + v = -u'''/24 - u''/8, \qquad t \in (0, 1), \tag{1.15}$$

with initial condition

$$v(0) = 0. \tag{1.16}$$

Thus the system (1.15), (1.16) is solvable and has unique solution

$$v(t) = -\frac{1}{24} \int_0^t e^{x-t}(u'''(x) + 3u''(x)) \, dx \tag{1.17}$$

which is infinitely differentiable on the interval $[0, 1]$. Note that this function v is independent of the parameter τ.

Let us show that the net function η^{τ}, which is defined at the points of $\bar{\omega}_{\tau}$ by

$$\eta^{\tau} = u^{\tau} - u - \tau^2 v \tag{1.18}$$

is uniformly bounded. Consider the expression

$$\eta_{\bar{t}}^{\tau} + \eta_{\hat{t}}^{\tau} = (1/\tau + 1/2)\eta^{\tau}(t + \tau/2) - (1/\tau - 1/2)\eta^{\tau}(t - \tau/2).$$

Keeping in mind (1.18) we obtain

$$\eta_{\bar{t}}^{\tau} + \eta_{\hat{t}}^{\tau} = u_{\bar{t}}^{\tau} + u_{\hat{t}}^{\tau} - (u_{\bar{t}} + u_{\hat{t}}) - \tau^2(v_{\bar{t}} + v_{\hat{t}}) \quad \text{on } \bar{\omega}_{\tau}.$$

Fix any $t \in \bar{\omega}_{\tau}$ and rewrite the right-hand side of the equality. By virtue of (1.5) the sum of the first two terms is equal to f. In the other terms we replace the functions by their values at the point t, using the Taylor-series expansion with the Lagrange form for the remainder (Lemma 1.1 of §7.1). As a result we have

$$\eta_{\bar{t}}^\tau + \eta_{\bar{t}}^\tau = (f - u' - u) - \tau^2(u'''/24 + u''/8 + v' + v) + \xi,$$

where

$$|\xi| \le c_2 \tau^4 \quad \forall t \in \breve{\omega}_\tau.$$

Using equations (1.1) and 1.15) we have

$$\eta_{\bar{t}}^\tau + \eta_{\bar{t}}^\tau = \xi \quad \text{on } \breve{\omega}_\tau. \qquad (1.19)$$

Hence we obtain

$$|\eta^\tau(t + \tau/2)| \le |\eta^\tau(t - \tau/2)| + \tau|\xi(t)|.$$

It is easy to see that by virtue of (1.2), (1.6), and (1.16), $\eta^\tau(0) = 0$. Therefore, using induction we obtain

$$\|\eta^\tau\|_{c,\tau} \le \sum_{\breve{\omega}_\tau} |\xi| \tau \quad \text{or} \quad \|\eta^\tau\|_{c,\tau} \le c_2 \tau^4. \qquad (1.20)$$

Note that from (1.19) and (1.20) we set an estimate for the difference derivative

$$\max_{\breve{\omega}_\tau} |\eta_{\bar{t}}^\tau| \le \max_{\breve{\omega}_\tau} |\eta_{\bar{t}}^\tau| + O(\tau^4) = O(\tau^4). \qquad (1.21)$$

Now we present two methods for increasing accuracy based on this expansion. We construct nets (1.4) with mesh-sizes τ and $\tau/2$ solve approximation (1.5), (1.6) for both. Let u^τ and $u^{\tau/2}$ be the solutions of these problems (the accuracy of each solution being of order τ^2). Form the linear combination

$$V = \tfrac{4}{3} u^{\tau/2} - \tfrac{1}{3} u^\tau \quad \text{on } \bar{\omega}_\tau; \qquad (1.22)$$

we will show that the function V approximates u with accuracy of order τ^4.

At each point of $\bar{\omega}_\tau$ the expansions

$$u^\tau = u + \tau^2 v + O(\tau^4),$$

$$u^{\tau/2} = u + (\tau^2/4)v + O(\tau^4)$$

are valid. Hence

$$V = u + O(\tau^4). \qquad (1.23)$$

Thus the corrected solution (1.22) using a linear conbination of the approximate solutions of (1.5), (1.6) (with accuracy of order τ^2) approximates the exact solution u at the points of $\bar{\omega}_\tau$ with order τ^4.

Richardson called this method of obtaining more accurate approximate solutions from solutions of lesser accuracy (see [107]) a limit method. Here the limit is understood as being

$$u(t) = \lim_{\tau \to 0} u^\tau(t).$$

Later this approach was referred to as "Richardson extrapolation". This method is said to be *global* because the extrapolation is used mainly for several approximate solutions which have already been found. Let us consider another method which can be said to be *local*. In this method extrapolation is carried out for each mesh-size; the result is then used as the initial approximation for the next step. Assume the approximate solution w^τ is already known at a point x. Assume

$$v^\tau(x) = v^{\tau/2}(x) = w^\tau(x). \tag{1.24}$$

For v^τ we take one step in a Crank–Nicholson scheme of length τ, while for $v^{\tau/2}$ we take two steps of length $\tau/2$. We have

$$v_{\bar t}^\tau(t) + v_{\bar t}^\tau(t) = f(t), \qquad t = x + \tau/2,$$
$$v_{\bar t}^{\tau/2}(t) + v_{\bar t}^{\tau/2}(t) = f(t), \qquad t = x + \tau/4, x + 3\tau/4.$$

We add the solutions thus obtained using the same weights as in the global method:

$$w^\tau(x + \tau) = \tfrac{4}{3}v^{\tau/2}(x + \tau) - \tfrac{1}{3}v^\tau(x + \tau). \tag{1.25}$$

Using the initial approximation $w^\tau(0) = u_0$ and (1.24), (1.25) one can find the approximate solution w^τ at the points of $\bar\omega_\tau$.

Although the algorithms presented above appear to be similar, they have a vastly different theoretical structure. The first algorithm does not alter the stability of the difference schemes used initially. The second algorithm actually uses the Runge principle to achieve a higher order of approximation. Thus the algorithm leads to another qualitatively different Runge–Kutta-type method. The stability of this method does not follow automatically from the stability of the difference scheme used initially. The example presented above can be used to illustrate this fact. To solve (1.1) we used the Crank–Nicholson scheme, which is stable for any $\tau > 0$. At the same time (1.24), (1.25) (see [28]) is unstable for all τ greater than some τ_0. This fact comes up in [123], where a modification of the local Richardson extrapolation method free of this defect is described.

Very often one does not distinguish these methods, and therefore calculations sometimes give contradictory results.

From now on we will only consider global extrapolation and drop the term "global" for brevity.

1.1.2. Correction by Higher-Order Differences

Now we consider another method for increasing the accuracy of difference solutions of (1.5), (1.6). We use expansion (1.8) to justify it. In the literature this is referred to as the method of correction by higher-order differences [135, 136, 138, 139] or the method of difference corrections [104, 105].

Before we describe this method let us alter the difference equation (1.5) in such a way that it attains a higher order of approximation

$$L^\tau v^\tau \equiv v_{\bar{t}}^\tau + v_{\hat{t}}^\tau - \tau^2(v_{\bar{t}\hat{t}\hat{t}}^\tau/24 + v_{\bar{t}\hat{t}}^\tau) = f \quad \forall t \in \bar{\omega}_\tau \backslash \{\tau/2, 1 - \tau/2\}. \quad (1.26)$$

Here

$$v_{\hat{t}\hat{t}}(t) = (v_{\hat{t}}(t))_{\hat{t}} = \{v_{\hat{t}}(t + \tau/2) - v_{\hat{t}}(t - \tau/2)\}/\tau$$
$$= \{v(t + \tau) - 2v(t) + v(t - \tau)\}/\tau^2,$$
$$v_{\hat{t}\hat{t}\hat{t}}(t) = \{v_{\hat{t}\hat{t}}(t + \tau/2) + v_{\hat{t}\hat{t}}(t - \tau/2)\}/2$$
$$= \{v(t + 3\tau/2) - v(t + \tau/2) - v(t - \tau/2) + v(t + 3\tau/2)\}/(2\tau^2), \text{ etc.}$$

It is easy to check using Taylor-series that the difference operator L^τ on the left-hand side of (1.26) approximates the differential operator of (1.1) with fourth-order accuracy

$$L^\tau u = u' + u + O(\tau^4) \quad \forall t \in \bar{\omega}_\tau \backslash \{\tau/2, 1 - \tau/2\}. \quad (1.27)$$

Let us note that the four-point equation (1.27) is not suitable at the points $\tau/2$ and $1 - \tau/2$ since it contains the values of v^τ at the points $-\tau, 1 + \tau$ which are outside the interval $[0, 1]$. At these points we use the following difference equations:

$$L^\tau v^\tau(\tau/2) \equiv v_{\bar{t}}^\tau(\tau/2) + v_{\hat{t}}^\tau(\tau/2) - \frac{\tau^2}{24} v_{\bar{t}\hat{t}\hat{t}}^\tau(\tau/2)$$

$$- \frac{\tau^2}{8} v_{\hat{t}\hat{t}\hat{t}}^\tau(\tau/2) + \frac{\tau^3}{48}(3\tau - 2)v_{\hat{t}\hat{t}\hat{t}\hat{t}}^\tau(\tau) = f(\tau/2), \quad (1.28)$$

$$L^\tau v^\tau(1 - \tau/2) \equiv v_{\bar{t}}^\tau(1 - \tau/2) + v_{\hat{t}}^\tau(1 - \tau/2) - \frac{\tau^2}{24} v_{\bar{t}\hat{t}\hat{t}}^\tau(1 - \tau/2)$$

$$- \frac{\tau^2}{8} v_{\hat{t}\hat{t}\hat{t}}^\tau(1 - \tau/2) + \frac{\tau^3}{48}(3\tau + 2)v_{\hat{t}\hat{t}\hat{t}\hat{t}}^\tau(1 - \tau) = f(1 - \tau/2).$$

$$(1.29)$$

It is easy to check that the coefficients of $v^\tau(-\tau)$ and $v^\tau(1 + \tau)$ are equal to zero in these equations and that the error is of order $O(\tau^3)$:

$$L^\tau u(t) = u'(t) + u(t) + O(\tau^3), \quad t = \tau/2, 1 - \tau/2. \quad (1.30)$$

We augment equations (1.26), (1.28), and (1.29) with the initial condition which follows from (1.2). We get the system of linear algebraic equations

$$v^\tau(0) = u_0,$$
$$L^\tau v^\tau = f \quad \text{on } \bar{\omega}_\tau. \quad (1.31)$$

The matrix of this system cannot be put in triangular form at any renumbering of the unknowns; this fact complicates its solution. Proving the stability of the system requires some special justification.

We will now consider another modification of (1.31) which is free of these two defects. First we solve (1.5), (1.6) and then from the known function u^τ we find w^τ, which solves

$$w^\tau(0) = u_0, \tag{1.32}$$

$$w^\tau_{\hat{t}} + w^\tau_{\check{t}} = u^\tau_{\hat{t}} + u^\tau_{\check{t}} - L^\tau u^\tau + f \quad \text{on } \check{\omega}_\tau. \tag{1.33}$$

These difference problems differ only in their right-hand sides, and can be solved by a simple stable algorithm. Let us show that the solution w^τ obeys the estimate

$$\|w^\tau - u\|_{C,\tau} \le c_3 \tau^4. \tag{1.34}$$

Indeed, the function u^τ can be expanded via (1.8). Substitute this on the right-hand side of (1.33). For the points $\omega_\tau \backslash \{\tau/2, 1 - \tau/2\}$ we have

$$
\begin{aligned}
w^\tau_{\hat{t}} + w^\tau_{\check{t}} &= \tau^2 u^\tau_{\hat{t}\hat{t}\hat{t}}/24 + \tau^2 u^\tau_{\check{t}\check{t}\check{t}}/8 \\
&= \tau^2 u_{\hat{t}\hat{t}\hat{t}}/24 + \tau^2 u_{\check{t}\check{t}\check{t}}/8 + \tau^4 v_{\hat{t}\hat{t}\hat{t}}/24 + \tau^4 v_{\check{t}\check{t}\check{t}}/8 + \tau^2 \eta^\tau_{\hat{t}\hat{t}\hat{t}}/24 \\
&\quad + \tau^2 \eta^\tau_{\check{t}\check{t}\check{t}}/8 + f.
\end{aligned}
\tag{1.35}
$$

In order to estimate the right-hand we use a Taylor-series expansion. We have

$$u_{\hat{t}\hat{t}\hat{t}} = u''' + O(\tau^2), \qquad u_{\check{t}\check{t}\check{t}} = u'' + O(\tau^2),$$

$$|v_{\hat{t}\hat{t}\hat{t}}| \le \tfrac{5}{4}\max_{[0,1]}|v'''| = O(1), \qquad |v_{\check{t}\check{t}\check{t}}| \le \tfrac{5}{4}\max_{[0,1]}|v''| = O(1). \tag{1.36}$$

Now we use (1.20) and (1.21):

$$|\eta^\tau_{\hat{t}\hat{t}\hat{t}}| \le \frac{4}{\tau^2} \max_{\check{\omega}_\tau} |\eta^\tau_{\hat{t}}| = O(\tau^2),$$

$$|\eta^\tau_{\check{t}\check{t}\check{t}}| \le \frac{4}{\tau^2} \max_{\check{\omega}_\tau} |\eta^\tau_{\check{t}}| = O(\tau^2). \tag{1.37}$$

Using the inequalities (1.36), (1.37) we rewrite (1.35):

$$w^\tau_{\hat{t}} + w^\tau_{\check{t}} = \tau^2 u'''/24 + \tau^2 u''/8 + f + \zeta_1, \tag{1.38}$$

where

$$|\zeta_1| \le c_4 \tau^4 \quad \forall\, t \in \check{\omega}_\tau \backslash \{\tau/2, 1 - \tau/2\}. \tag{1.39}$$

A similar expansion (1.38) can be proved for $\tau/2$ and $1 - \tau/2$. In this case the estimate for the remainder is

$$|\zeta_1| \le c_5 \tau^3, \qquad t = \tau/2, 1 - \tau/2. \tag{1.40}$$

In the previous section we used the expansion for u

$$u_{\hat{t}} + u_{\check{t}} = u' + u + \tau^2 u'''/24 + \tau^2 u''/8 + \zeta_2,$$

where

$$|\zeta_2| \le c_6 \tau^4 \quad \forall t \in \bar{\omega}_\tau. \tag{1.41}$$

Rewrite it, using equation (1.1). We have

$$u_{\bar{t}} + u_{\hat{t}} = f + \tau^2 u'''/24 + \tau^2 u''/8 + \zeta_2.$$

Subtract this relation from (1.38) and introduce the notation $\varepsilon^\tau = w^\tau - u$. We have

$$\varepsilon_{\bar{t}}^\tau + \varepsilon_{\hat{t}}^\tau = \zeta_1 - \zeta_2 \quad \forall t \in \bar{\omega}_\tau. \tag{1.42}$$

From the initial conditions (1.2), and (1.3) it follows that

$$\varepsilon^\tau(0) = 0. \tag{1.43}$$

Therefore the solution of (1.42), (1.43) obeys an estimate similar to (1.20):

$$\|\varepsilon^\tau\|_{C,\tau} \le \sum_{\bar{\omega}_\tau} |\zeta_1 - \zeta_2|\tau.$$

Keeping in mind (1.39)–(1.41) we have

$$\|\varepsilon^\tau\|_{C,\tau} \le (c_4 + 2c_5 + c_6)\tau^4$$

which is equivalent to (1.34).

Thus we have proved that the solution of problem (1.32), (1.33) converges to the exact solution with fourth-order accuracy.

This method of correction by higher-order differences for unstable schemes is quite useful in obtaining higher orders of accuracy (see also [122]). In the following sections we will present a general formulation of this method.

Note that in the latter method, as opposed to the Richardson extrapolation method, both difference systems live on a single difference net. This property is independent of the differential equation and gives us a certain advantage in some cases. The structure of the right-hand side of equation (1.37) is much more complicated than that of equation (1.5) however. In many problems this can lead to considerably more complicated computer programs.

1.1.3. Some Numerical Examples

Consider the problem

$$u' + u = t(1 + t)^{-2} \quad \text{on } (0, 1),$$
$$u(0) = 1. \tag{1.44}$$

Its solution is $u(t) = (1 + t)^{-1}$.

Consider the five values $M_k = 5 \cdot 2^{k-1}$; construct the nets $\bar{\omega}_{\tau_k}$ and solve the difference problem (1.5), (1.6) for each. These solutions have maximal values

$$\zeta(M_k) = \|u^{\tau_k} - u\|_{C, \tau_k}.$$

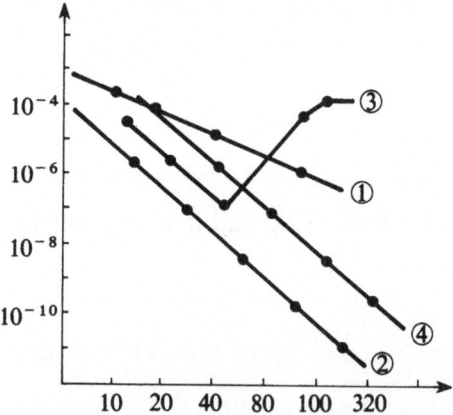

Figure 1.1. Errors of the approximate solutions for (1.48).

(1) Graph of the maximal error of the solution of the Crank–Nicholson scheme (1.5), (1.6); (2) graph of the maximal error of the extrapolated solution (1.22); (3) graph of the maximal error of the solution of the "doubtful" system (1.31); (4) graph of the maximal error of the solution of (1.32), (1.33) by the method of correction by higher-order differences.

The graphs of the values are shown in logarithmic coordinates in Figure 1.1. On each net of mesh-size τ_k form the linear combination:

$$V^{\tau_k} = \tfrac{4}{3}u^{\tau_k/2} - \tfrac{1}{3}u^{\tau_k} \quad \text{on } \bar{\omega}_{\tau_k};$$

the maximal value is

$$\varkappa_k = \| V^{\tau_k} - u \|_{C,\,\tau_k}. \tag{1.45}$$

To get an idea of the numerical effectiveness of the algorithm one should graph (1.49) versus $M_k + 2M_k$, since the number of the computer operations for the twofold solution of (1.5), (1.6) is proportional to $3M_k$. This number also characterizes the number of computations needed for the right-hand side (an additional criterion for comparing difference methods).

To illustrate the method of correction by higher-order differences we first use Gauss' elimination to find the solutions v^{τ_k} of the "doubtful" system (1.31) for the five values of M_k: the maximal error is

$$\rho_k = \| v^{\tau_k} - u \|_{C,\,\tau_k}.$$

Taking into account the number of calculations needed for the right-hand side in (1.31) we present a graph of this quantity in Figure 1.1 versus the argument M_k. For each of the five values of M_k the solution w^{τ_k} of (1.32), (1.33) is next found and the error is calculated

$$\xi_k = \| w^{\tau_k} - u \|_{C,\,\tau_k}.$$

Since to find w^τ one first needs to solve the difference system (1.5), (1.6) we graph the value ζ_k in Figure 1.1 with argument $2M_k$.

To illustrate pointwise convergence, the errors of the approximate solutions at several net points are given in Table 1.1. When the computations

Table 1.1.

t	Exact solution $u(t)$	Error of solution to (1.5), (1.6) $\tau = 1/80$	Error of solution to (1.5), (1.6) $\tau = 1/160$	Error of extrapolated solution (1.22) for $\tau = 1/80$	Error of solution of "doubtful" system (1.31) for $\tau = 1/80$	Error of solution of (1.32), (1.33) for $\tau = 1/80$
0.1	0.909090	1.46×10^{-7}	3.66×10^{-8}	2.75×10^{-11}	3.29×10^{-6}	2.07×10^{-8}
0.2	0.833333	4.48×10^{-7}	1.20×10^{-7}	4.12×10^{-11}	6.93×10^{-6}	1.88×10^{-8}
0.3	0.769230	7.84×10^{-7}	1.96×10^{-7}	4.77×10^{-11}	9.29×10^{-6}	1.72×10^{-8}
0.4	0.714285	1.10×10^{-6}	2.75×10^{-7}	5.03×10^{-11}	1.18×10^{-5}	1.56×10^{-8}
0.5	0.666666	1.37×10^{-6}	3.43×10^{-7}	5.04×10^{-11}	1.33×10^{-5}	1.42×10^{-8}
0.6	0.625000	1.60×10^{-6}	3.99×10^{-7}	4.93×10^{-11}	1.49×10^{-5}	1.29×10^{-8}
0.7	0.588235	1.77×10^{-6}	4.43×10^{-7}	4.73×10^{-11}	1.62×10^{-5}	1.17×10^{-8}
0.8	0.555555	1.90×10^{-6}	4.75×10^{-7}	4.49×10^{-11}	1.69×10^{-5}	1.07×10^{-8}
0.9	0.526315	1.99×10^{-6}	4.97×10^{-7}	4.22×10^{-11}	1.77×10^{-5}	9.68×10^{-9}
1.0	0.500000	2.04×10^{-6}	5.11×10^{-7}	3.94×10^{-11}	1.79×10^{-5}	7.98×10^{-9}

were carried out on a computer the relative level of round-off errors was 10^{-15}.

An analysis of the data confirms the theoretical estimates obtained in the previous sections. The solutions v^τ of the "doubtful" system (1.31), having a low level of accuracy is caused by the ill-conditioned matrix of this system.

1.2. Expansion Theorems

In §1.1 an example of Richardson extrapolation was presented illustrating the resulting increase in accuracy thereby. This method made use of the expansion

$$u^\tau = u + \tau^2 v + \eta^\tau \quad \text{on } \bar{\omega}_\tau, \tag{2.1}$$

with v independent of τ. In this section we will establish certain abstract sufficient conditions for the existence of an expansion of type (2.1) with an arbitrary number of terms on the right-hand side for a wide class of linear systems.

Let Ω be a bounded region in n-dimensional space R^n $(n \geq 1)$. Denote its closure by $\bar{\Omega}$. Assume that one wants to solve the following problem (typical in mathematical physics):

$$\begin{aligned} Lu &= f \quad \text{in } \Omega, \\ lu &= g \quad \text{on } D. \end{aligned} \tag{2.2}$$

Here L and l are linear differential operators, D is either the boundary of the region Ω or some subset of it. The functions f, g, u are defined in Ω, on D and on $\bar{\Omega}$, respectively.

In the linear spaces of functions defined on these sets we will introduce certain function classes $M_k(\Omega)$, $N_k(D)$ and $P_k(\bar{\Omega})$ which will depend on an integer parameter $k \geq 0$.

As a rule these classes characterize the smoothness of the right-hand side of the equation, of the boundary values, and of the solution itself. How they are defined for a concrete problem is dictated by known results on the existence of solutions. For example, for elliptic equations, one naturally chooses M_k, N_k, P_k to be the Hölder spaces $C^{k+\alpha}(\bar{\Omega})$, $C^{k+2+\alpha}(\Gamma)$ and $C^{k+2+\alpha}(\bar{\Omega})$. The problem of the existence of solutions for the ordinary differential equations and their differentiability have been fully investigated for the spaces $C^m[0, 1]$; therefore, it is convenient to introduce the classes M_k and P_k in terms of these spaces. In this case the set D will consist of one or two points, so that N_k will be R or R^2.

The following condition will thus involve the domain of definition of the equation and its coefficients in an implicit way and will characterize equations with smooth right-hand sides.

Condition A. For any integer $k, 0 \le k \le m$, and any pair of functions $f \in M_k(\Omega)$, $g \in N_k(D)$ there exists a unique solution $u \in P_k(\overline{\Omega})$ of (2.2).

To find a numerical solution of (2.2) we introduce the difference net $\overline{\Omega}_h \subset \overline{\Omega}$ with variable parameter h, which can be as small as we want. We will search for an approximate solution in the space of net functions defined at the knots of this net. We replace the system of differential equations by a system of finite-difference (algebraic) equations defined at the knots of certain finite subsets $\check{\Omega}_h \subset \Omega$ and $D_h \subset D$. The subsets $\overline{\Omega}_h$, $\check{\Omega}_h$ and D_h are the discrete analogs of the sets $\overline{\Omega}$, Ω and, D, respectively. We have

$$L_h u^h = f \quad \text{on } \check{\Omega}_h,$$
$$l_h u^h = g \quad \text{on } D_h. \tag{2.3}$$

Here L_h and l_h are linear algebraic operators and u^h is the net function approximating the solution u of the initial differential equation on $\overline{\Omega}_h$. In the linear spaces of net functions defined on $\overline{\Omega}_h$, $\check{\Omega}_h$, D_h we introduce norms $\| \cdot \|_{\overline{\Omega}_h}$, $\| \cdot \|_{\check{\Omega}_h}$, $\| \cdot \|_{D_h}$, respectively.

We will formulate the condition in terms of these norms characterizing the solvability of (2.3) and the stability of its solution.

Condition B. If the net function ψ^h is defined on $\overline{\Omega}_h$ and is a solution of the problem

$$L_h \psi^h = f^h \quad \text{on } \check{\Omega}_h,$$
$$l_h \psi^h = g^h \quad \text{on } D_h. \tag{2.4}$$

where f^h, g^h are net functions with the domains of definition $\overline{\Omega}_h, D_h$, respectively, then the estimate

$$\|\psi^h\|_{\overline{\Omega}_h} \le c(\|f^h\|_{\check{\Omega}_h} + \|g\|_{D_h}) \tag{2.5}$$

holds.

Note that the existence of a unique solution of (2.3) follows from (2.5). Indeed, this problem has a unique solution for any right-hand side if the corresponding homogeneous problem has only the trivial solution. But the uniform problem

$$L_h \xi^h = 0 \quad \text{on } \check{\Omega}_h,$$
$$l_h \xi^h = 0 \quad \text{on } D_h$$

cannot have any other solutions, since from (2.5) it follows that $\|\xi^h\|_{\overline{\Omega}_h} = 0$. Hence $\xi^h = 0$ on $\overline{\Omega}_h$.

One additional condition is directly related to the approximation of differential operators by difference operators.

Condition C. The expansions†

$$L_h \varphi = L\varphi + \sum_{j=1}^{k} h^j a_j + \sigma^h \quad \text{on } \check{\Omega}_h,$$

$$l_h \varphi = l\varphi + \sum_{j=1}^{k} h^j b_j + \rho^h \quad \text{on } D_h \tag{2.6}$$

are valid for any function $\varphi \in P_k(\overline{\Omega})$, where $0 \le k \le m$. The functions a_j, b_j are independent of h, $a_j \in M_{k-j}(\Omega)$, $b_j \in N_{k-j}(D)$ and the remainders σ^h, ρ^h obey the estimates

$$\|\sigma^h\|_{\check{\Omega}_h} \le c_1 h^{k+\beta}, \qquad \|\rho^h\|_{D_h} \le c_2 h^{k+\beta}, \tag{2.7}$$

where the constants c_1, c_2 are independent of h, and β is independent of h, k, and φ.

These conditions allow us to obtain expansions similar to (2.6) for the difference solution u^h of (2.3).

Theorem 2.1. *Assume Conditions A, B, and C hold for (2.2), and (2.3) and $f \in M_m(\Omega)$, $g \in N_m(D)$. Then the difference solution u^h has the expansion*

$$u^h = u + \sum_{j=1}^{m} h^j v_j + \eta^h \quad \text{on } \overline{\Omega}_h. \tag{2.8}$$

Here the functions v_j are independent of h, $v_j \in P_{m-j}(\overline{\Omega})$, and the remainder η^h obeys the estimate

$$\|\eta^h\|_{\overline{\Omega}_h} \le c_3 h^{m+\beta}, \tag{2.9}$$

where the constant c_3 is independent of h.

PROOF. Let us consider an arbitrary set of functions $v_j \in P_{m-j}(\overline{\Omega})$ which are independent of h, where $j = 1, \ldots, m$. Using these functions and the two solutions u and u^h we define the net function

$$\eta^h = u^h - u - \sum_{j=1}^{m} h^j v_j \quad \text{on } \overline{\Omega}_h. \tag{2.10}$$

solve (2.10) for u^h and substitute the resulting expression into equation (2.3). We get

$$L_h u + \sum_{j=1}^{m} h^j L_h v_j + L_h \eta^h = f \quad \text{on } \check{\Omega}_h,$$

$$l_h u + \sum_{j=1}^{m} h^j l_h v_j + l_h \eta^h = g \quad \text{on } D_h. \tag{2.11}$$

---

† Any sum with its upper limit less than its lower limit is taken to be zero; any such product is assumed to be 1.

Using Condition C we can write down the expansions

$$L_h u = f + \sum_{i=1}^{m} h^i a_{0,i} + \sigma_0^h \quad \text{on } \check{\Omega}_h,$$

$$l_h u = g + \sum_{i=1}^{m} h^i b_{0,i} + \rho_0^h \quad \text{on } D_h \tag{2.12}$$

and

$$L_h v_j = L v_j + \sum_{i=1}^{m-j} h^i a_{j,i} + \sigma_j^h \quad \text{on } \check{\Omega}_h,$$

$$l_h v_j = l v_j + \sum_{i=1}^{m-j} h^i b_{j,i} + \rho_j^h \quad \text{on } D_h. \tag{2.13}$$

Here

$$a_{j,i} \in M_{m-j-i}(\Omega), \qquad b_{j,i} \in N_{m-j-i}(D), \tag{2.14}$$

$a_{j,i}, b_{j,i}$ are independent of h and the remainders obey

$$\|\sigma_j^h\|_{\check{\Omega}_h} \le c_{j,1} h^{m-j+\beta}, \qquad \|\rho_j^h\|_{D_h} \le c_{j,2} h^{m-j+\beta}, \tag{2.15}$$

with constants $c_{j,1}$ and $c_{j,2}$ independent of h. Using expansions (2.12) and (2.13) we can rewrite (2.11) in the following form

$$f + \sum_{j=1}^{m} h^j L v_j + \sum_{j=0}^{m} h^j \sum_{i=1}^{m-j} h^i a_{j,i} + \sum_{j=0}^{m} h^j \sigma_j^h + L_h \eta^h = f \quad \text{on } \check{\Omega}_h,$$

$$g + \sum_{j=1}^{m} h^j l v_j + \sum_{j=0}^{m} h^j \sum_{i=1}^{m-j} h^i b_{j,i} + \sum_{j=0}^{m} h^j \rho_j^h + l_h \eta^h = g \quad \text{on } D_h. \tag{2.16}$$

Assuming

$$\xi^h = \sum_{j=0}^{m} h^j \sigma_j^h, \qquad \zeta^h = \sum_{j=0}^{m} h^j \rho_j^h$$

and using (2.15) we have

$$\|\xi^h\|_{\check{\Omega}_h} \le h^{m+\beta} c_4, \qquad \|\zeta^h\|_{D_h} \le h^{m+\beta} c_5, \tag{2.17}$$

where

$$c_4 = \sum_{j=0}^{m} c_{j,1}, \qquad c_5 = \sum_{j=0}^{m} c_{j,2}.$$

By some simple manipulations we can rewrite (2.16) in the following form

$$\sum_{j=1}^{m} h^j \left(L v_j + \sum_{i=1}^{j} a_{j-i,i} \right) + \xi^h + L_h \eta^h = 0 \quad \text{on } \check{\Omega}_h,$$

$$\sum_{j=1}^{m} h^j \left(l v_j + \sum_{i=1}^{j} b_{j-i,i} \right) + \zeta^h + l_h \eta^h = 0 \quad \text{on } D_h. \tag{2.18}$$

Thus (2.18), with remainders ξ^h and ζ^h, satisfying (2.17) for an arbitrary set of functions $v_j \in P_{m-j}(\overline{\Omega})$ and with η^h defined by equality (2.10).

Now we choose the functions $v_j, j = 1, 2, \ldots, m$, to be the solutions of the differential equations

$$
\begin{aligned}
Lv_j &= -\sum_{i=1}^{j} a_{j-i,i} \quad \text{in } \Omega, \\
lv_j &= -\sum_{i=1}^{j} b_{j-i,i} \quad \text{on } D.
\end{aligned}
\tag{2.19}
$$

For example, the function v_1 is found from

$$
\begin{aligned}
Lv_1 &= -a_{0,1} \quad \text{in } \Omega, \\
lv_1 &= -b_{0,1} \quad \text{on } D.
\end{aligned}
$$

Using Condition C for (2.12) it follows that $a_{0,1} \in M_{m-1}(\Omega)$ and $b_{0,1} \in N_{m-1}(D)$. Therefore the function v_1 is uniquely defined and $v_1 \in P_{m-1}(\overline{\Omega})$ (see Condition A). Suppose that the functions $v_j \in P_{m-j}(\overline{\Omega})$ have already been found for $j = 1, \ldots, k$, where $1 \le k \le m$. Then by virtue of Condition C (2.13) is valid for $j = 1, \ldots, k$ and satisfies conditions (2.14). Write down (2.19) for $j = k + 1$:

$$
\begin{aligned}
Lv_{k+1} &= -\sum_{i=1}^{k+1} a_{k-i+1,i} \quad \text{in } \Omega, \\
lv_{k+1} &= -\sum_{i=1}^{k+1} b_{k-i+1,i} \quad \text{on } D.
\end{aligned}
\tag{2.20}
$$

According to (2.14) the right-hand sides belong to $M_{m-k-1}(\Omega)$ and $N_{m-k-1}(D)$, respectively. Therefore from Condition A it follows that (2.20) has the unique solution $v_{k+1} \in P_{m-k-1}(\overline{\Omega})$. The independence of v_{k+1} from h is obvious.

Thus we have exhibited a method for constructing the functions $v_j \in P_{m-j}(\overline{\Omega})$ independent of h, for $j = 1, \ldots, m$. These v_j satisfy (2.18) with (2.17), and from (2.19) the relations (2.18) take the following form:

$$
\begin{aligned}
L_h \eta^h &= -\xi^h \quad \text{on } \check{\Omega}_h, \\
l_h \eta^h &= -\zeta^h \quad \text{on } D_h.
\end{aligned}
$$

From Condition B the inequality

$$
\|\eta^h\|_{\check{\Omega}_h} \le c(\|\xi^h\|_{\check{\Omega}_h} + \|\zeta^h\|_{D_h})
$$

follows. Using (2.17) we obtain (2.9), where $c_3 = c(c_4 + c_5)$. Substituting u^h from (2.10) we have (2.8), which has the required properties. The theorem is proved. $\square$

In the next section we will study a method for increasing the accuracy of difference solutions based on these expansions. Let us now return to Condition C. Very often the coefficients a_j and b_j in (2.6) will turn out to be zero for odd indices j. This occurred, for example, in the previous section. As a consequence the expansion (2.8) will have only even powers of h. This property allows us to substantially reduce the number of computations needed. Therefore we will modify Condition C when only even powers of h occur and prove a corresponding expansion theorem.

Condition D. For any function $\varphi \in P_{m-2k}(\overline{\Omega})$ where $k = 0, 1, \ldots, s$, $s = [m/2]$, the expansions

$$L_h \varphi = L\varphi + \sum_{j=1}^{s-k} h^{2j} a_j + \sigma^h \quad \text{on } \check{\Omega}_h,$$

$$l_h \varphi = l\varphi + \sum_{j=1}^{s-k} h^{2j} b_j + \rho^h \quad \text{on } D_h \tag{2.21}$$

hold, where $a_j \in M_{m-2k-2j}(\Omega)$, $b_j \in N_{m-2k-2j}(D)$, a_j, b_j are independent of h and the remainder obey the estimates

$$\|\sigma^h\|_{\check{\Omega}_h} \leq c_6 h^{m-2k+\beta}, \qquad \|\rho^h\|_{D_h} \leq c_7 h^{m-2k+\beta}, \tag{2.22}$$

where the constants c_6, c_7 are independent of h, and β is independent of h, k, and φ.

Theorem 2.2. *Assume Conditions A, B, and D hold for (2.2) and (2.3), and $f \in M_m(\Omega)$, $g \in N_m(D)$. Then the difference solution u^h has the expansion*

$$u^h = u + \sum_{j=1}^{s} h^{2j} v_j + \eta^h \quad \text{on } \overline{\Omega}_h. \tag{2.23}$$

Here the v_j are independent of h, $v_j \in P_{m-2j}(\overline{\Omega})$, and the remainder η^h obeys the estimate

$$\|\eta^h\|_{\overline{\Omega}_h} \leq c_8 h^{m+\beta}, \tag{2.24}$$

with constant c_8 independent of h.

The proof of (2.23), (2.24) is similar to the proof of the previous theorem. The only difference here is that only even powers of h occur in the expansions.

Remark. In a number of situations the difference equation $l_h u^h = g$ will coincide with the boundary condition $lu = g$ on D_h, so that in (2.6) and (2.21) the functions b_j and ρ^h vanish. In such a case Condition B can be simplified in the following way without invalidating the results of theorems 2.1 and 2.2:

Condition B′. Suppose the net function ψ^h is defined on $\overline{\Omega}_h$ and solves

$$\begin{aligned} L_h \psi^h &= f^h \quad \text{on } \check{\Omega}_h, \\ l_h \psi^h &= 0 \quad \text{on } D_h, \end{aligned} \tag{2.25}$$

where f^h is a net function with domain of definition $\check{\Omega}_h$. Then ψ^h obeys the estimate

$$\|\psi^h\|_{\bar{\Omega}_h} \leq c\|f^h\|_{\check{\Omega}_h}. \tag{2.26}$$

This condition also leads to the existence and uniqueness of the solution to (2.3).

1.3. Acceleration of Convergence

Now we use the expansions of Theorems 2.1 and 2.2 to correct the approximate solutions of (2.3). We introduce the uniform norm for net functions defined on $\bar{\Omega}_h$

$$\|v\|_{\bar{\Omega}_h} = \max_{x \in \bar{\Omega}_h} |v(x)|. \tag{3.1}$$

Other cases will be considered for other concrete difference schemes.

Assume the conditions of Theorem 2.1 are satisfied for the set of net regions $\bar{\Omega}_{h_k}$ with parameters $h_1 > h_2 > \cdots > h_{m+1} > 0$. We will assume that these net regions have nonempty intersection

$$\bar{\Omega}_H = \bigcap_{k=1}^{m+1} \bar{\Omega}_{h_k} \neq \varnothing.$$

According to Condition B the finite difference system

$$\begin{aligned}
L_h u^h &= f \quad \text{on } \check{\Omega}_h, \\
l_h u^h &= g \quad \text{on } D_h
\end{aligned} \tag{3.2}$$

has a unique solution for each value of the parameter $h = h_k$. Denote this solution by u^{h_k}. All the u^{h_k} are determined on $\bar{\Omega}_H$.

Consider the system

$$\sum_{k=1}^{m+1} \gamma_k = 1,$$

$$\sum_{k=1}^{m+1} \gamma_k h_k^j = 0, \qquad j = 1, \ldots, m \tag{3.3}$$

From §7.2 it follows that the determinant of this system is not zero and therefore this system has a unique solution. Form the linear combination with these weights:

$$V^H = \sum_{k=1}^{m+1} \gamma_k u^{h_k} \quad \text{on } \bar{\Omega}_H. \tag{3.4}$$

Let us prove that this solution V^H has a higher accuracy than each u^{h_k}.

Theorem 3.1. *Assume that the regions $\overline{\Omega}_{h_k}$ with parameters $h_1 > \cdots > h_m > h_{m+1} > 0$ satisfy the conditions of Theorem 2.1 with uniform norm (3.1) and that the inequality*

$$h_k/h_{k+1} \geq 1 + d_1, \qquad k = 1, \ldots, m, \tag{3.5}$$

with constant $d_1 > 0$ independent of h_k holds. Then on the nonempty intersection of these regions $\overline{\Omega}_H$ the estimate

$$\max_{\overline{\Omega}_H} |V^H - u| \leq d_2 h_1^{m+\beta} \tag{3.6}$$

holds, where V^H is the extrapolated solution (3.4) with weights γ_k from (3.3), u is solution of the differential equation (2.2), and d_2 is a constant independent of h_k.

PROOF. Fix an arbitrary point x of $\overline{\Omega}_H$. From Theorem 2.1 at this point we have the expansion

$$u^{h_k}(x) = u(x) + \sum_{j=1}^{m} h_k^j v_j(x) + \eta^{h_k}(x), \qquad k = 1, 2, \ldots, m + 1, \tag{3.7}$$

where the $v_j(x)$ do not depend on h_k and the remainders satisfy

$$|\eta^{h_k}(x)| \leq \|\eta^{h_k}\|_{\overline{\Omega}_{h_k}} \leq c_3 h_k^{m+\beta}. \tag{3.8}$$

Rewrite the right-hand side of (3.4) using (3.7). We have

$$V^H(x) = \sum_{k=1}^{m+1} \gamma_k u(x) + \sum_{k=1}^{m+1} \sum_{j=1}^{m} \gamma_k h_k^j v_j(x) + \sum_{k=1}^{m+1} \gamma_k \eta^{h_k}(x). \tag{3.9}$$

Since the $v_j(x)$ do not depend on k from (3.3) it follows that

$$\sum_{k=1}^{m+1} \sum_{j=1}^{m} \gamma_k h_k^j v_j(x) = \sum_{j=1}^{m} v_j(x) \sum_{k=1}^{m+1} \gamma_k h_k^j = 0.$$

In addition,

$$\sum_{k=1}^{m+1} \gamma_k u(x) = u(x).$$

Therefore we can rewrite (3.9) as

$$V^H(x) = u(x) + \sum_{k=1}^{m+1} \gamma_k \eta^{h_k}(x).$$

Hence

$$|V^H(x) - u(x)| \leq \sum_{k=1}^{m+1} |\gamma_k| |\eta^{h_k}(x)|. \tag{3.10}$$

To estimate $|\gamma_k|$ we use Lemma 2.3 of §7.2. From this lemma and (3.5) it follows that

$$|\gamma_k| \leq (1 + 1/d_1)^m, \qquad k = 1, \ldots, m + 1.$$

Using this estimate and (3.8) we can rewrite (3.10) as

$$|V^H(x) - u(x)| \le \sum_{k=1}^{m+1} (1 + 1/d_1)^m c_3 h_k^{m+\beta} \le c_3 m(1 + 1/d_1)^m h_1^{m+\beta}.$$

Put

$$d_2 = c_3 m(1 + 1/d_1)^m.$$

Then

$$|V^H(x) - u(x)| \le d_2 h_1^{m+\beta}, \qquad x \in \overline{\Omega}_H.$$

The estimate (3.6) follows from this relation. The theorem is proved. $\qquad\square$

Let us consider two methods for refining difference nets which are widely spaced. In the first method we take a sequence of nets $\overline{\Omega}_{h_k}$ with parameters $h_k = h/k$ where $h > 0, k = 1, 2, \ldots, m + 1$. In this case condition (3.5) is satisfied with constant $d_1 = 1/m$ for any $h > 0$. For these parameters (3.3) takes the form

$$\sum_{k=1}^{m+1} \gamma_k = 1,$$

$$\sum_{k=1}^{m+1} \gamma_k k^{-j} = 0, \qquad j = 1, \ldots, m.$$

From Lemma 2.1 of §7.2 we can write the solution of this system down explicitly

$$\gamma_k = \frac{(-1)^{m-k+1} k^{m+1}}{k! \, (m - k + 1)!}, \qquad k = 1, \ldots, m + 1. \tag{3.11}$$

These weights γ_k for several values of m are given in Table 1.2.

The data presented in Table 1.2 show that the weights γ_k increase rapidly in m. This leads to large round-off-errors and to other large irregular error terms as a result of inaccuracies in the solution of (3.2). For computers which use low precision this effect often predominates. Therefore we will present another method for refining the difference nets for which the growth of the weights γ_k is not noticeable.

Let us choose parameters $h_k = h/2^{k-1}$ where $h > 0, k = 1, \ldots, m + 1$. In this case (3.5) is satisfied for any $h > 0$ with constant $d_1 = 1$. The system (3.3) can be written as

$$\sum_{k=1}^{m+1} \gamma_k = 1,$$

$$\sum_{k=1}^{m+1} \gamma_k 2^{j(1-k)} = 0, \qquad j = 1, \ldots, m. \tag{3.12}$$

Table 1.2

m	γ_1	γ_2	γ_3	γ_4	γ_5	γ_6
1	-1	2				
2	$\dfrac{1}{2}$	-4	$\dfrac{9}{2}$			
3	$-\dfrac{1}{6}$	4	$-\dfrac{27}{2}$	$\dfrac{32}{3}$		
4	$\dfrac{1}{24}$	$-\dfrac{8}{3}$	$\dfrac{81}{4}$	$-\dfrac{128}{3}$	$\dfrac{625}{24}$	
5	$-\dfrac{1}{120}$	$\dfrac{4}{3}$	$-\dfrac{81}{4}$	$\dfrac{256}{3}$	$-\dfrac{3125}{24}$	$\dfrac{324}{5}$
Parameters:	h	$\dfrac{h}{2}$	$\dfrac{h}{3}$	$\dfrac{h}{4}$	$\dfrac{h}{5}$	$\dfrac{h}{6}$

The solutions for several m are given in Table 1.3. It is easy to see that in this case the γ_k grow more slowly.

Note, that one can calculate the sum

$$\sum_{i=1}^{m+1} \gamma_k u^{h_k}(x)$$

Table 1.3

m	γ_1	γ_2	γ_3	γ_4	γ_5	γ_6
1	-1	2				
2	$\dfrac{1}{3}$	-2	$\dfrac{8}{3}$			
3	$-\dfrac{1}{21}$	$\dfrac{2}{3}$	$-\dfrac{8}{3}$	$\dfrac{64}{21}$		
4	$\dfrac{1}{315}$	$-\dfrac{2}{21}$	$\dfrac{8}{9}$	$-\dfrac{64}{21}$	$\dfrac{1024}{315}$	
5	$-\dfrac{1}{9765}$	$\dfrac{2}{315}$	$-\dfrac{8}{63}$	$\dfrac{64}{63}$	$-\dfrac{1024}{315}$	$\dfrac{32768}{9765}$
Parameters:	h	$\dfrac{h}{2}$	$\dfrac{h}{4}$	$\dfrac{h}{8}$	$\dfrac{h}{16}$	$\dfrac{h}{32}$

using the Neville algorithm without first solving system (3.3). To do this consider the extrapolation diagram in Figure 1.2. First assume

$$T_j^{(0)} = u^{h_j}(x), \qquad j = 1, 2, \ldots, m + 1.$$

$$
\begin{array}{llll}
T_1^{(0)} \longrightarrow T_1^{(1)} & \cdots & T_1^{(m-1)} & T_1^{(m)} \\
T_2^{(0)} \nearrow T_2^{(1)} & \cdots & T_2^{(m-1)} & \\
\ \vdots \qquad\ \ \vdots & & & \\
T_m^{(0)} \qquad T_m^{(1)} & & & \\
T_{m+1}^{(0)} & & &
\end{array}
$$

Figure 1.2

We then compute the elements of columns with indices $i = 1, 2, \ldots, m$ recursively:

$$T_j^{(i)} = (h_{i+1} T_{j+1}^{(i-1)} - h_1 T_j^{(i-1)})/(h_{i+1} - h_1), \qquad j = 1, \ldots, m - i + 1.$$

As a result we have

$$T_1^{(m)} = \sum_{k=1}^{m+1} \gamma_k u^{h_k}(x).$$

The proof of this equality is given in §7.2.

The Neville algorithm is particularly useful for large m when the computation of the γ_k from (3.3) is difficult, or when the order m of the Richardson extrapolation is chosen during the computation. When the mesh-sizes of the nets are chosen according to

$$h_i = ha^{1-i}, \quad \text{where } h > 0, a > 1, i = 1, 2, \ldots, m + 1.$$

the Neville algorithm gives Romberg's rule (see [110]).

When the expansions (2.8) have many terms in the regular part the second method for choosing the parameters h_k results in a higher accuracy. But when we solve (3.2) for $h, h/2, h/4, \ldots$ the number of computations is much greater than in the first case. For ordinary differential equations halving the mesh-size doubles the amount of computation, as a rule. For multi-dimensional problems the effect is more pronounced.

Therefore, it is convenient to use the gap series (see [25, 124])

$$h, h/2, h/3, h/4, h/6, h/8, h/12, \ldots$$

This does not increase the amount of computations much while decreasing the mesh-sizes. At the same time one can take the constant d_1 in (3.5), which insures the boundness of $|\gamma_k|$, to be $\frac{1}{3}$, independent of the length of the gap series.

Let us now formulate our results when the regular part of the expansion contains only even powers of h.

Let m be a natural number and $s = [m/2]$. For parameters $h_1 > \cdots > h_{s+1} > 0$ construct net regions $\overline{\Omega}_{h_k}$. We assume that conditions of Theorem 2.2 are satisfied. Then the solutions u^{h_k} of (3.2) are defined on each net $\overline{\Omega}_{h_k}$. If the intersection of the net regions is nonempty, i.e.,

$$\overline{\Omega}_H = \bigcap_{k=1}^{s+1} \overline{\Omega}_{h_k} \neq \emptyset$$

then all solutions are defined on $\overline{\Omega}_H$.

Consider the system

$$\sum_{k=1}^{s+1} \gamma_k = 1,$$

$$\sum_{k=1}^{s+1} \gamma_k h_k^{2j} = 0, \qquad j = 1, \ldots, s. \tag{3.13}$$

Since $h_k \neq h_j$ for $k \neq j$, then from §7.2 it follows that the determinant of this system is not zero. Therefore, there is a unique solution $\gamma_1, \gamma_2, \ldots, \gamma_{s+1}$. Form the linear combination with weights γ_k:

$$V^H(x) = \sum_{k=1}^{s+1} \gamma_k u^{h_k}(x), \qquad x \in \overline{\Omega}_H. \tag{3.14}$$

We will prove that V^H approximates the solution u with greater accuracy than u^{h_k}.

Theorem 3.2. *Suppose the parameters* $h_1 > \cdots > h_{s+1} > 0$ *satisfy*

$$h_k/h_{k+1} \geq 1 + d_3, \qquad k = 1, \ldots, s, \tag{3.15}$$

with the constant $d_3 > 0$ *independent of* h_k. *Assume that* (3.2) *and the net regions* $\overline{\Omega}_{h_k}$ *satisfy the conditions of Theorem 2.2 with uniform norm* (3.1). *Then in the intersection* $\overline{\Omega}_H$ *of these net regions the estimate*

$$\max_{\overline{\Omega}_H} |V^H - u| \leq d_4 h_1^{m+\beta} \tag{3.16}$$

holds, where V^H *is the extrapolated solution* (3.14) *with weights* γ_k *from* (3.13), u *is the solution of* (2.2), *and* d_4 *is a constant independent of* h_k.

PROOF. Let us fix an arbitrary $x \in \overline{\Omega}_H$. From Theorem 2.2 the expansions

$$u^{h_k}(x) = u(x) + \sum_{j=1}^{s} h_k^{2j} v_j(x) + \eta^{h_k}(x), \qquad k = 1, \ldots, s+1$$

hold at this point. Substitute these expansions in the right-hand side of (3.14). We have

$$V^H(x) = \sum_{k=1}^{s+1} \gamma_k u(x) + \sum_{k=1}^{s+1} \sum_{j=1}^{s} h_k^{2j} \gamma_k v_j(x) + \sum_{k=1}^{s+1} \gamma_k \eta^{h_k}(x). \tag{3.17}$$

The $v_j(x)$ do not depend on k, and the γ_k satisfy (3.13), therefore

$$\sum_{k=1}^{s+1} \sum_{j=1}^{s} h_k^{2j} \gamma_k v_j(x) = \sum_{j=1}^{s} v_j(x) \sum_{k=1}^{s+1} \gamma_k h_k^{2j} = 0.$$

From the first equation of (3.13) it follows that

$$\sum_{k=1}^{s+1} \gamma_k u(x) = u(x),$$

and therefore

$$V^H(x) = u(x) + \sum_{k=1}^{s+1} \gamma_k \eta^{h_k}(x).$$

Hence

$$|V^H(x) - u(x)| \le \sum_{k=1}^{s+1} |\gamma_k| |\eta^{h_k}(x)|. \qquad (3.18)$$

From Theorem 2.2 the remainders η^{h_k} obey

$$|\eta^{h_k}(x)| \le c_8 h_k^{m+\beta}.$$

From (3.15) the conditions of Lemma 2.4 of §7.2 are satisfied for (3.13). This makes it possible to estimate the weights γ_k:

$$|\gamma_k| \le (1 + 1/(2d_3 + d_3^2))^s, \qquad k = 1, \dots, s + 1.$$

Using these inequalities to estimate the right-hand side of (3.18) we have

$$|V^H(x) - u(x)| \le \sum_{k=1}^{s+1} (1 + 1/(2d_3 + d_3^2))^s c_8 h_k^{m+\beta}$$

$$\le c_8 (1 + 1/(2d_3 + d_3^2))^s h_1^{m+\beta}.$$

Put

$$d_4 = c_8 (1 + 1/(2d_3 + d_3^2))^s.$$

The constant d_4 does not depend on h_k, therefore

$$|V^H(x) - u(x)| \le d_4 h_1^{m+\beta} \quad \forall x \in \overline{\Omega}_H.$$

The theorem is proved. $\qquad\qquad\qquad\qquad\qquad\qquad\qquad\qquad\qquad\qquad\qquad\square$

Let us now investigate the behavior of solutions to (3.13) when the two methods for decreasing the parameter h are used. The first method takes $h_k = h/k$ where $h > 0$, $k = 1, \dots, s + 1$. Condition (3.15) is easily varified: the constant d_3 is $1/s$ for each $h > 0$. For this sequence of parameters (3.13) has the form

$$\sum_{k=1}^{s+1} \gamma_k = 1,$$

$$\sum_{k=1}^{s+1} \gamma_k k^{-2j} = 0, \qquad j = 1, \dots, s.$$

Table 1.4

m	s	γ_1	γ_2	γ_3	γ_4	γ_5
$\left.\begin{matrix}2\\3\end{matrix}\right\}$	1	$-\dfrac{1}{3}$	$\dfrac{4}{3}$			
$\left.\begin{matrix}4\\5\end{matrix}\right\}$	2	$\dfrac{1}{24}$	$-\dfrac{16}{15}$	$\dfrac{81}{40}$		
$\left.\begin{matrix}6\\7\end{matrix}\right\}$	3	$-\dfrac{1}{360}$	$\dfrac{16}{45}$	$-\dfrac{729}{280}$	$\dfrac{1024}{314}$	
$\left.\begin{matrix}8\\9\end{matrix}\right\}$	4	$\dfrac{1}{8640}$	$-\dfrac{64}{945}$	$\dfrac{6561}{4480}$	$-\dfrac{16384}{2835}$	$\dfrac{390625}{72576}$
Parameters:		h	$\dfrac{h}{2}$	$\dfrac{h}{3}$	$\dfrac{h}{4}$	$\dfrac{h}{5}$

From Lemma 2.2 of §7.2 there follows

$$\gamma_k = \frac{2(-1)^{s-k+1}k^{2s+2}}{(s+k+1)!\,(s-k+1)!}, \qquad k = 1, \ldots, s+1.$$

These weights for several m are given in Table 1.4.

Note that the absolute values of weights γ_k grow more slowly in m than in general cases where the regular part of the expansion contains odd powers of h.

The second method takes $h_k = h \cdot 2^{1-k}$, $k = 1, 2, \ldots, s+1$, where $h > 0$, is a certain initial value. It is easy to see that the constant d_3 in (3.15) can be assumed to be 1. After a slight rearrangement (3.13) takes the form

$$\sum_{k=1}^{s+1} \gamma_k = 1,$$

$$\sum_{k=1}^{s+1} \gamma_k 2^{2j(1-k)} = 0, \qquad j = 1, \ldots, s. \tag{3.19}$$

The solutions for several m are shown in Table 1.5.

The increase in the absolute value of the coefficients γ_k with m is less marked than in the previous case. For large m a modified Neville algorithm should be used instead of calculating the sum

$$U^H(x) = \sum_{k=1}^{s+1} \gamma_k u^{h_k}(x)$$

by calculating the weights directly. To do this we return to the diagram in Figure 1.2. The zero-th column is given by

$$T_j^{(0)} = u^{h_j}(x), \qquad j = 1, 2, \ldots, s+1. \tag{3.20}$$

Table 1.5

m	s	γ_1	γ_2	γ_3	γ_4	γ_5
$\left.\begin{array}{l}2\\3\end{array}\right\}$	1	$-\dfrac{1}{3}$	$\dfrac{4}{3}$			
$\left.\begin{array}{l}4\\5\end{array}\right\}$	2	$\dfrac{1}{45}$	$-\dfrac{4}{9}$	$\dfrac{64}{45}$		
$\left.\begin{array}{l}6\\7\end{array}\right\}$	3	$-\dfrac{1}{2835}$	$\dfrac{4}{135}$	$-\dfrac{64}{135}$	$\dfrac{4096}{2835}$	
$\left.\begin{array}{l}8\\9\end{array}\right\}$	4	$\dfrac{1}{722925}$	$-\dfrac{4}{8505}$	$\dfrac{64}{2025}$	$-\dfrac{4096}{8505}$	$\dfrac{1048576}{722925}$
Parameters:		h	$\dfrac{h}{2}$	$\dfrac{h}{4}$	$\dfrac{h}{8}$	$\dfrac{h}{16}$

We calculate all the remaining columns by the recurrence formula

$$T_j^{(i)} = (h_{i+1}^2 T_{j+1}^{(i-1)} - h_1^2 T_j^{(i-1)})/(h_{i+1}^2 - h_1^2),$$
$$j = 1, \ldots, s - i + 1, \quad i = 1, 2, \ldots, s. \tag{3.21}$$

The result is

$$T_1^{(s)} = \sum_{k=1}^{s+1} \gamma_k u^{h_k}(x).$$

The proof of this equality is given in §7.2.

1.4. Correction by Higher-Order Differences

Using the notation of §1.2 we will now consider the problem of obtaining higher-order approximations

$$S_h u^h = f \quad \text{on } \check{\Omega}_h,$$
$$s_h u^h = g \quad \text{on } D_h. \tag{4.1}$$

Here S_h, s_h are linear operators approximating the differential operators L, l on the sets $\check{\Omega}_h$, D_h to a higher order of accuracy. As a rule, they have a more complicated form than L_h, and l_h. We assume the following for these operators.

Condition E. For any function $\varphi \in P_k(\overline{\Omega})$, where $0 \le k \le m$, the inequalities

$$\|S_h \varphi - L\varphi\|_{\check{\Omega}_h} \le c_9 h^{k+\beta},$$
$$\|s_h \varphi - l\varphi\|_{D_h} \le c_{10} h^{k+\beta}$$

hold.

If one succeeds in proving Condition E for (4.1) it will then have a unique solution u^h. In addition, the theorem on convergence (see, for example, [87]) tells us that the stability of the approximation implies the convergence of u^h to the exact solution u to a higher order. For example, when we have $u \in P_m(\overline{\Omega})$ we get the estimate

$$\|u^h - u\|_{\overline{\Omega}_h} \le c_{11} h^{m+\beta}. \tag{4.2}$$

But actually solving (4.1) may be difficult because of the complicated structure of matrix for this problem. In the case when the stability of problem (4.1) cannot be proved its solution may be practically impossible to find because the matrix is ill-conditioned or singular.

Usually (2.3) is easier to solve, although it will approximate the initial problem with a rather low order of accuracy. With this in mind, let us consider several iterations of the process [104, 105].

$$L_h u_{k+1}^h = f + L_h u_k^h - S_h u_k^h \quad \text{on } \check{\Omega}_h, \tag{4.3}$$

$$l_h u_{k+1}^h = g + l_h u_k^h - s_h u_k^h \quad \text{on } D_h, \qquad k = 0, 1, \ldots ; \tag{4.4}$$

$$u_0^h = 0 \quad \text{on } \overline{\Omega}_h. \tag{4.5}$$

At each step of this process a system like (2.3) is solved. Let us show that under certain conditions the solutions u_k^h obtained approximate the function u with great accuracy. To this end we formulate one more condition.

Condition F. The solution of

$$L_h w^h = L_h v^h - S_h v^h \quad \text{on } \check{\Omega}_h,$$
$$l_h w^h = l_h v^h - s_h v^h \quad \text{on } D_h,$$

with arbitrary net function v^h defined on $\overline{\Omega}_h$ obeys estimate

$$\|w^h\|_{\Omega_h} \le c_{12} \|v^h\|_{\overline{\Omega}_h}.$$

The above is sufficient to prove the increase in the accuracy of the u_k^h.

Theorem 4.1. *Assume that Conditions A, B, C, E, and F hold true for the operators* L, l, L_h, l_h, S_h, s_h *and that in (2.1) we have* $f \in M_m(\Omega), g \in N_m(D)$. *Then the solutions* u_k^h ($k = 1, \ldots, m+1$) *of (4.3)–(4.5) can be expanded*

$$u_k^h = u + \sum_{j=k}^{m} h^j v_{j,k} + \eta_k^h \quad \text{on } \overline{\Omega}_h. \tag{4.6}$$

Here the functions $v_{j,k}$ *are independent of* h, $v_{j,k} \in P_{m-j}(\overline{\Omega})$ *and the remainder* η_k^h *satisfies*

$$\|\eta_k^h\|_{\overline{\Omega}_h} \le d_k h^{m+\beta}. \tag{4.7}$$

PROOF. Note that u_1^h coincides with the solution u^h of (2.3). Therefore by Theorem 2.1 the expansion (4.6) holds for this function. We now suppose that (4.6) and (4.7) hold for some $k \ge 1$ and prove this for $k + 1$.

Consider an arbitrary set of the functions $v_{j,k+1} \in P_{m-j}(\overline{\Omega})$ independent of h, where $j = k+1, \ldots, m$. We construct the net function

$$\eta_{k+1}^h = u_{k+1}^h - u - \sum_{j=k+1}^{m} h^j v_{j,k+1} \quad \text{on } \overline{\Omega}_h. \tag{4.8}$$

Solve for u_{k+1}^h and substitute it into the left-hand side of equation (4.3). Substitute (4.6) into the right-hand side of this equation. We have

$$L_h u + \sum_{j=k+1}^{m} h^j L_h v_{j,k+1} + L_h \eta_{k+1}^h$$

$$= f + L_h u + \sum_{j=k}^{m} h^j L_h v_{j,k} + L_h \eta_k^h - S_h u - \sum_{j=k}^{m} h^j S_h v_{j,k} - S_h \eta_k^h.$$

Using Condition E and making some cancellations we have

$$\sum_{j=k+1}^{m} h^j L_h v_{j,k+1} + L_h \eta_{k+1}^h$$

$$= \sum_{j=k}^{m} h^j L_h v_{j,k} + L_h \eta_k^h - \sum_{j=k}^{m} h^j L_h v_{j,k} + \zeta_1^h - S_h \eta_k^h, \tag{4.9}$$

where

$$\|\zeta_1^h\|_{\tilde{\Omega}_h} \leq c_{14} h^{m+\beta}. \tag{4.10}$$

Keeping in mind the smoothness of the functions $v_{j,k}, v_{j,k+1}$, from Condition C we can write

$$L_h v_{j,k} = L v_{j,k} + \sum_{i=1}^{m-j} h^i A_{j,i} + \sigma_{j,k}^h \quad \text{on } \check{\Omega}_h, \tag{4.11}$$

$$L_h v_{j,k+1} = L v_{j,k+1} + \sum_{i=1}^{m-j} h^i B_{j,i} + \sigma_{j,k+1}^h \quad \text{on } \check{\Omega}_h. \tag{4.12}$$

Here $A_{j,i}, B_{j,i} \in M_{m-j-i}(\Omega)$ are independent of h and the remainders obey

$$\|\sigma_{j,k}^h\|_{\tilde{\Omega}_h} \leq c_{15} h^{m-j+\beta}, \qquad \|\sigma_{j,k+1}^h\|_{\tilde{\Omega}_h} \leq c_{16} h^{m-j+\beta}. \tag{4.13}$$

Using (4.11) and (4.12), we transform (4.9) as follows:

$$\sum_{j=k+1}^{m} h^j \left(L v_{j,k+1} + \sum_{i=1}^{j-k-1} B_{j-i,i} - \sum_{i=1}^{j-k} A_{j-i,i} \right) + L_h \eta_{k+1}^h$$

$$= L_h \eta_k^h - S_h \eta_k^h + \zeta_2^h \quad \text{on } \check{\Omega}_h. \tag{4.14}$$

From (4.10) and (4.13) there follows

$$\|\zeta_2^h\|_{\tilde{\Omega}_h} \leq c_{17} h^{m+\beta}. \tag{4.15}$$

Using a similar procedure one can obtain from

$$\sum_{j=k+1}^{m} h^j \left(l v_{j,k+1} + \sum_{i=1}^{j-k-1} b_{j-i,i} - \sum_{i=1}^{j-k} a_{j-i,i} \right) + l_h \eta_{k+1}^h$$

$$= l_h \eta_k^h - s_h \eta_k^h + \rho_2^h \quad \text{on } D_h. \quad (4.16)$$

Here the remainder ρ_2^h satisfies the estimate

$$\| \rho_2^h \|_{D_h} \le c_{18} h^{m+\beta}. \quad (4.17)$$

The functions $a_{j,i}, b_{j,i}$ are independent of h; $a_{j,i}, b_{j,i} \in N_{m-j-i}(D)$. They are taken from the expansions given by Condition C:

$$l_h v_{j,k} = l v_{j,k} + \sum_{i=1}^{m-j} h^i a_{j,i} + \varepsilon_{j,k}^h \quad \text{on } D_h, \quad (4.18)$$

$$l_h v_{j,k+1} = l v_{j,k+1} + \sum_{i=1}^{m-j} h^i b_{j,i} + \varepsilon_{j,k+1}^h \quad \text{on } D_h. \quad (4.19)$$

Thus for an arbitrary set of functions $v_{j,k+1} \in P_{m-j}(\overline{\Omega})$ and the function η^h defined by (4.8) we obtain (4.14), (4.16) with remainders ζ_2^h, ρ_2^h satisfying (4.15), (4.17).

Let us choose the functions $v_{j,k+1}$ $(j = k+1, \ldots, m)$ to be the solutions of the differential equations

$$Lv_{j,k+1} = \sum_{i=1}^{j-k} A_{j-i,i} - \sum_{i=1}^{j-k-1} B_{j-i,i} \quad \text{in } \Omega,$$

$$l v_{j,k+1} = \sum_{i=1}^{j-k} a_{j-i,i} - \sum_{i=1}^{j-k-1} b_{j-i,i} \quad \text{on } D. \quad (4.20)$$

In particular, the problem equation for $v_{k+1,k+1}$ can be written as

$$Lv_{k+1,k+1} = A_{k,1} \quad \text{in } \Omega,$$

$$l v_{k+1,k+1} = a_{k,1} \quad \text{on } D. \quad (4.21)$$

We have supposed that the functions $v_{j,k}$ satisfy Condition C and have derived expansions (4.11), (4.15). Therefore $A_{k,1}, a_{k,1}$ are fully defined by $v_{k,k}$ and are independent of h and $A_{k,1} \in M_{m-k-1}(\Omega)$, $a_{k,1} \in N_{m-k-1}(D)$. Equation (4.21) has the unique solution $v_{k+1,k+1} \in P_{m-k-1}(\overline{\Omega})$ from Condition A.

Let us suppose that for $n = k+1, \ldots, j-1$, where $k+2 \le j \le m+1$, the functions $v_{n,k+1} \in P_{m-n}(\overline{\Omega})$ are already known. Then from Condition C (4.12), (4.19) are valid for $n = k+1, \ldots, j-1$. Consider (4.20) for the function $v_{j,k+1}$. The terms $A_{j-i,i}, a_{j-i,i}$ on the right-hand side of this equation are defined in terms of the known functions $v_{j,k}$ from (4.11), and the terms $B_{j-i,i}, b_{j-i,i}$ for $i = 1, \ldots, j-k-1$ are fully and completely determined by the functions $v_{k+1,k+1}, \ldots, v_{j-1,k+1}$ from (4.12): $A_{j-i,i}, B_{j-i,i} \in M_{m-j}(\Omega)$, $a_{j-i,i}, b_{j-i,i} \in N_{m-j}(D)$. Therefore it follows from Condition A that (4.20) has a unique solution $v_{j,k+1} \in P_{m-j}(\overline{\Omega})$. It is obvious that it is independent of h.

Thus the functions $v_{j,k+1}$ $(j = k + 1, \ldots, m)$ having the required properties are well defined. The identities (4.14), (4.16) are thus valid. They can be simplified using (4.20):

$$L_h \eta_{k+1}^h = L_h \eta_k^h - S_k \eta_k^h + \xi_2^h \quad \text{on } \check{\Omega}_k,$$

$$l_h \eta_{k+1}^h = l_h \eta_k^h - s_h \eta_k^h + \rho_2^h \quad \text{on } D_h.$$

In order to estimate η_{k+1}^h, let us consider the two auxiliary systems

$$L_h \varepsilon_1^h = L_h \eta_k^h - S_h \eta_k^h \quad \text{on } \check{\Omega}_h,$$
$$l_h \varepsilon_1^h = l_h \eta_k^h - s_h \eta_k^h \quad \text{on } D_h; \tag{4.22}$$

$$L_h \varepsilon_2^h = \xi_2^h \quad \text{on } \check{\Omega}_h,$$
$$l_h \varepsilon_2^h = \rho_2^h \quad \text{on } D_h. \tag{4.23}$$

From Condition B the existence and uniqueness of the solution for both problems follows. This we can write η_{k+1}^h in the form $\varepsilon_1^h + \varepsilon_2^h$ from which there follows

$$\|\eta_{k+1}^h\|_{\bar{\Omega}_h} \le \|\varepsilon_1^h\|_{\bar{\Omega}_h} + \|\varepsilon_2^h\|_{\bar{\Omega}_h}.$$

For (4.22) Condition F gives us the estimate

$$\|\varepsilon_1^h\|_{\bar{\Omega}_h} \le c_{12} \|\eta_k^h\|_{\bar{\Omega}_h} \le c_{12} d_k h^{m+\beta}.$$

Using (4.15), (4.17) and Condition B for (4.23) we have

$$\|\varepsilon_2^h\|_{\bar{\Omega}_h} \le c(\|\xi_2^h\|_{\bar{\Omega}_h} + \|\rho_2^h\|_{D_h}) \le c(c_{17} + c_{18}) h^{m+\beta}.$$

The last three inequalities taken together give us an estimate of the form (4.7) for η_{k+1}^h. Solving for u_{k+1}^h in (4.8) for $k + 1$ we have (4.6). Theorem 4.1 is proved.   □

Thus at each step of (4.3)–(4.5) an increase in the order of accuracy takes place:

$$\|u_k^h - u\|_{\bar{\Omega}_h} = O(h^k), \qquad k = 1, 2, \ldots, m;$$

$$\|u_{m+1}^k - u\|_{\bar{\Omega}_h} = O(h^{m+\beta}).$$

Taking many iterations might appear to give a better result. This is not the case however. The iterative process (4.3)–(4.5) cannot give an order of accuracy greater than the order of approximation of the initial difference scheme. When one carries out many iterations two situations usually arise. When $c_{12} < 1$ in Condition F one can show that a unique solution of (4.1) will exist and find a stability estimate of form (2.5) for it. In this case as the number of iterations increases the process converges to a solution of (4.1) which is of the same order of accuracy as u_{m+1}^h. The number of operations increases proportionally to k and so the effectiveness of the method decreases.

If the constant c_{12} cannot be taken to be less than 1 the iterative process (4.3)–(4.5) diverges, i.e., it gives an unbounded sequence u_k^h even when the initial system was stable. When $c_{12} = 1$ a bounded sequence u_k^h can arise with order of accuracy not greater than the order of approximation of (4.1).

Thus, as the number of iterations increases beyond its optimal value, the order of accuracy fails to increase (it may even decrease) and the amount of computer time grows. This leads to a decrease in the effectivness of the method. It often happens that Condition C, rather than Condition D, holds so that the regular parts of the expansions contain only even powers of h. In this case we have the following result.

Theorem 4.2. *Assume that Conditions* A, B, D, E, *and* F *hold for the operators* L, l, L_h, l_h, S_h, s_h *and that in* (2.1) $f \in M_m(\Omega)$, $g \in N_m(D)$. *Then the solutions* u_k^h ($k = 1, \ldots, p + 1$; $p = [m/2]$) *of* (4.3)–(4.5) *can be expanded*

$$u_k^h = u + \sum_{j=k}^{p} h^{2j} v_{j,k} + \eta_k^h \quad on \ \overline{\Omega}_h. \tag{4.24}$$

Here $v_{j,k}$ *is independent of* h, $v_{j,k} \in P_{m-2j}(\overline{\Omega})$ *and the remainder satisfies*

$$\|\eta_k^h\|_{\overline{\Omega}_h} \leq d_k h^{m+\beta}. \tag{4.25}$$

The proof of this theorem is similar to that of Theorem 4.1. Here, however the regular part of the expansions will have only even powers of h. It is apparent that (4.3)–(4.5) will give a more rapid increase in the order of accuracy from iteration to iteration, in this case:

$$\|u_k^h - u\|_{\overline{\Omega}_h} = O(h^{2k}), \quad k = 1, \ldots, p;$$
$$\|u_{p+1}^h - u\|_{\overline{\Omega}_h} = O(h^{m+\beta}).$$

Thus to attain the same accuracy one should use half as many iterations as for Theorem 4.1.

Remark 1. In a number of problems the boundary condition $l_h u^h = g$ coincides on D_h with the boundary conditions $s_h u^h = g$ and $lu = g$. In this case the stability Condition B can be replaced by Condition B' from §1.2 without invalidating the results of Theorems 4.1 and 4.2. Conditions E and F are also simplified in this case.

Remark 2. In multidimensional problems Condition F is the most difficult to check. For the usual norms the constant c_{12} can depend on h. In this case the constants d_k in Theorems 4.1 and 4.2 also depend on h, where $d_k = O((1 + c_{12}(h))^{k-1})$. Hence according to Theorem 4.1, it follows that

$$\|u_k^h - u\|_{\overline{\Omega}_h} = O(h^k(1 + c_{12}(h))^{k-1}), \quad k = 1, \ldots, m;$$
$$\|u_{m+1}^h - u\|_{\overline{\Omega}_h} = O(h^{m+\beta}(1 + c_{12}(h))^m).$$

Therefore the effect of the sequential improvement in accuracy in (4.3)–(4.5) in this case will only occur when $c_{12}(h) = O(h^{-1})$. The estimates for (4.3)–(4.5) take the form

$$\|u_k^h - u\|_{\bar{\Omega}_h} = O(h^{2k}(1 + c_{12}(h))^{k-1}); \qquad k = 1, \ldots, p;$$

$$\|u_{k+1}^h - u\|_{\bar{\Omega}_h} = O(h^{m+\beta}(1 + c_{12}(h))^p).$$

Therefore the increase in accuracy will occur when

$$c_{12}(h) = O(h^{-2}).$$

1.5. Various Extrapolation Methods

Linear extrapolation is not the only possible way of accelerating the convergence of the approximate solutions u^h as $h \to 0$. We could replace the unknown function $u^h(x)$ (with argument h) by the interpolation polynomial

$$f(h) = \sum_{i=0}^{k-1} \gamma_i h^{pi}, \qquad p = 1 \quad \text{or} \quad p = 2. \tag{5.1}$$

The value $f(0)$ of this polynomial is then taken as the approximate value of the limit $\lim_{h \to 0} u^h(x)$. Taking various other classes of interpolating functions leads naturally to other extrapolation methods.

1.5.1. Rational Extrapolation

Let us use rational functions of the form

$$g(h) = \varphi(h^p)/\psi(h^p), \qquad p = 1 \quad \text{or} \quad p = 2, \tag{5.2}$$

as interpolating functions. Here the maximum exponents in the polynomials $\varphi(t)$, $\psi(t)$ do not exceed $[k/2]$ and $[(k + 1)/2]$, respectively (the sum of these numbers is k). The class of functions (5.2) will include the polynomials in h and consequently can be used with expansions of the forms (2.8) and (2.23).

It is obvious that the computation of the coefficients of the polynomials φ and ψ is rather complicated in the nonlinear case. There is a simple recurrent procedure for computing value $g(0)$ (due to Bulirsh–Stoer, see [24]).Consider the extrapolation diagram in Figure 1.3.

Assume

$$T_j^{(-1)} = 0 \quad \forall j = 2, \ldots, k; \qquad T_j^{(0)} = u^{hj} \quad \forall j = 1, 2, \ldots, k;$$

and calculate the recurrent sequence

$$T_j^{(i)} = T_{j+1}^{(i-1)} + (T_{j+1}^{(i-1)} - T_j^{(i-1)})/\{(h_j/h_{i+j})^p$$
$$\times [1 - (T_{j+1}^{(i-1)} - T_j^{(i-1)})/(T_{j+1}^{(i-1)} - T_{j+1}^{(i-2)})] - 1\},$$
$$j = 1, \ldots, k - i; \quad i = 1, \ldots, k - 1. \tag{5.3}$$

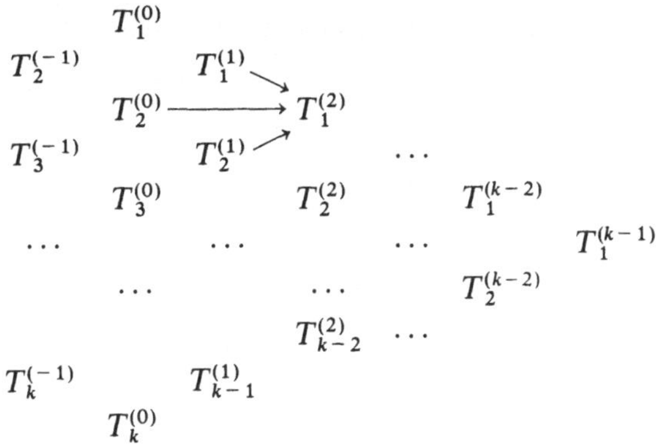

$$T_1^{(0)}$$

$$T_2^{(-1)} \qquad T_1^{(1)}$$

$$T_2^{(0)} \longrightarrow T_1^{(2)}$$

$$T_3^{(-1)} \qquad T_2^{(1)} \nearrow$$

$$T_3^{(0)} \qquad\qquad T_2^{(2)} \qquad\qquad T_1^{(k-2)}$$

$$\cdots \qquad\qquad \cdots \qquad\qquad \cdots \qquad\qquad T_1^{(k-1)}$$

$$\cdots \qquad\qquad \cdots \qquad\qquad T_2^{(k-2)}$$

$$T_{k-2}^{(2)} \quad \cdots$$

$$T_k^{(-1)} \qquad\qquad T_{k-1}^{(1)}$$

$$T_k^{(0)}$$

Figure 1.3.

The last value $T_1^{(k-1)}$ gives $g(0)$ for a rational function of the form (5.2), which, for the given set of h_i, is:

$$g(h_i) = u^{h_i}, \qquad i = 1, 2, \ldots, k. \tag{5.4}$$

It is shown in [24] that rational extrapolation gives the same accuracy as linear extrapolation, with approximately the same amount of computation, in general.

We will compare these methods using

$$u(h) = \alpha_0 + \alpha_1 h^2 + \alpha_2 h^4. \tag{5.5}$$

Using linear extrapolation for two values of the parameter, namely, h, $h/2$ results in:

$$u_L \equiv \tfrac{4}{3}u(h/2) - \tfrac{1}{3}u(h) = \alpha_0 + \frac{\alpha_2}{4} h^4. \tag{5.6}$$

Rational extrapolation with the same values yields

$$u_R \equiv u(h/2) + [u(h/2) - u(h)]/\{4(1 - [u(h/2) - u(h)]/u(h)) - 1\}$$
$$= 3u(h)u(h/2)/[4u(h) - u(h/2)]. \tag{5.7}$$

Using the explicit form of the function u we have

$$u_R = \alpha_0 + [(\alpha_1^2 - \alpha_0\alpha_2)h^4/4 + 5\alpha_1\alpha_2 h^6/16$$
$$+ \alpha_2^2 h^8/16]/[\alpha_0 + 5\alpha_1 h^2/4 + 21\alpha_2 h^4/16]. \tag{5.8}$$

We first assume that $\alpha_0 \neq 0$ and consider the main error terms in the extrapolation in (5.6) and (5.8). The last relation can be written in the form

$$u_R = \alpha_0 + (\alpha_1^2/(4\alpha_0) - \alpha_2/4)h^4 + O(h^6). \tag{5.9}$$

Therefore the main error term for rational extrapolation is $(\alpha_1^2/\alpha_0 - \alpha_2)h^4/4$. If α_0 and α_2 have the same sign then this is less than the error in (5.6). If they are of opposite sign it appears to be greater than the error in the linear extrapolation. Since the signs are not known *a priori* it is impossible to predict which method will have greater accuracy.

However, nonlinear extrapolation methods will have singular points appear in the neighborhood of which the accuracy of the extrapolation will decrease considerably. For rational extrapolation this occurs at the point where the desired solution changes sign. For the test function (5.5) this takes place at $\alpha_0 = 0$. In this case rational extrapolation yields:

$$u_R = \tfrac{1}{5}\alpha_1 h^2 + O(h^4).$$

The exact value is zero. This means that in this case rational extrapolation does not increase the accuracy compared to $u(h)$ and $u(h/2)$. The linear extrapolation u_L will have error $O(h^4)$ independent of the sign of α_0.

1.5.2. Exponential Extrapolation

Let us take the linear combinations of exponentials

$$g(s) = a_0 + \sum_{i=1}^{k-1} a_i q_i^s \tag{5.10}$$

as our interpolating functions. Here a_i, q_i are free parameters. To compute limits of sequences with principal part like (5.10) as $s \to \infty$, one uses an Aitken transformation (or a δ^2-transformation) or its generalization—the Shanks transformation (see [68]). These transformations are based on the fact that when knots are uniformly spaced the coefficient a_0 in $g(s)$ can be computed in a simple way.

Assume the function $g(s)$ of (5.10) takes a form such as

$$g(s) = u^{hs}, \qquad s = n - k + 1, n - k + 2, \ldots, n + k - 1; \quad n \geq k. \tag{5.11}$$

Assume $\Delta g_i = g(i + 1) - g(i)$ and compute the two $k \times k$ determinants

$$D = \det \begin{bmatrix} 1 & 1 & \cdots & 1 \\ \Delta g_{n-k+1} & \Delta g_{n-k+2} & \cdots & \Delta g_n \\ \Delta g_{n-k+2} & \Delta g_{n-k+3} & \cdots & \Delta g_{n+1} \\ \cdots & \cdots & \cdots & \cdots \\ \Delta g_{n-1} & \Delta g_n & \cdots & \Delta g_{n+k-2} \end{bmatrix},$$

$$D^* = \det \begin{bmatrix} g(n - k + 1) & g(n - k + 2) & \cdots & g(n) \\ \Delta h_{n-k+1} & \Delta g_{n-k+2} & \cdots & \Delta g_n \\ \cdots & \cdots & \cdots & \cdots \\ \Delta g_{n-1} & \Delta g_n & \cdots & \Delta g_{n+k-2} \end{bmatrix}$$

From these the value of a_0 can be determined in the following way:

$$a_0 = D^*/D. \tag{5.12}$$

If we assume that $n = k, k + 1, k + 2, \ldots$, then we obtain the new sequence $a_0^{(k)}, a_0^{(k+1)}, a_0^{(k+2)}, \ldots$. This is said to have been obtained from the sequence u^{h_s} by a Shanks transformation of the $(k - 1)$th order. If as $s \to \infty$ u^{h_s} looks like (5.10) from, then the sequence $a_0^{(n)}, n = k, k + 1, \ldots$ converges faster than u^{h_s}.

For $k = 1$ this method gives the well-known Aitken transformation

$$a_0 = \det\begin{bmatrix} g(n) & g(n + 1) \\ \Delta g_{n-1} & \Delta g_n \end{bmatrix} : \det\begin{bmatrix} 1 & 1 \\ \Delta g_{n-1} & \Delta g_n \end{bmatrix}$$

$$= \{g(n + 1)g(n - 1) - g^2(n)\}/\{g(n + 1) - 2g(n) + g(n - 1)\}. \quad (5.13)$$

To determine whether it is possible to use these transformations for our purposes we will show that (2.8) and (2.33) can be written in the form (5.8) for a certain choice of mesh-sizes h_i. Assume

$$h_s = h_0 b^{-s}, \qquad s = 1, 2, \ldots \qquad (5.14)$$

Here h_0 is a certain initial mesh-size, and $b > 1$ is the coefficient of refinement for the sequence of nets. In order that as many knots as possible coincide, the coefficient b should be taken to be an integer. The simplest case, when $b = 2$, has been discussed in §1.3.

Assuming condition (5.14) the expansions (2.8) and (2.33) result in

$$u^{h_s} = u + \sum_{j=1}^{k-1} h_0^{jp} v_j b^{-sjp} + \eta^{h_s}, \qquad (5.15)$$

where p is equal to 1 or 2 as above. Introducing the following notation $a_0 = u$, $a_j = h_0^{pj} v_j$ and $q_j = b^{-pj}$ we conclude that (5.10) approximates (2.8) and (2.33) with an accuracy which depends on the remainder η^{h_s}.

To compare these transformation with linear extrapolation we once more exploit the test function (5.5). To use the Aitken transformation three values of this function $u(h_1), u(h_2), u(h_3)$ are necessary. In accordance with (5.14) we take $h_1 = h, h_2 = h/2, h_3 = h/4$, i.e., $b = 2, h_0 = h$. Then we have

$$u(h) = \alpha_0 + \alpha_1 h^2 + \alpha_2 h^4,$$

$$u(h/2) = \alpha_0 + \alpha_1 h^2/4 + \alpha_2 h^4/16,$$

$$u(h/4) = \alpha_0 + \alpha_1 h^2/16 + \alpha_2 h^4/256.$$

From (5.12) we obtain the extrapolated value

$$u_E \equiv \alpha_0 + 9\alpha_1\alpha_2 h^4/(144\alpha_1 + 225\alpha_2 h^2)$$

$$= \alpha_0 + \alpha_2 h^4/16 + O(h^6). \qquad (5.16)$$

Thus the Aitken transformation does, in fact, give a value with accuracy $O(h^4)$. But linear extrapolation with two values $u(h/2), u(h/4)$ for this test function gives the following value:

$$u_L = \alpha_0 - \alpha_2 h^4/64.$$

The coefficient of h^4 has one fourth the magnitude than in (5.16).

Thus, to obtain the same accuracy the Aitken transformation requires more auxiliary solutions than linear extrapolation. This is because the values q_i in (5.10) are free and additional approximate solutions are needed to define them. This shows up more strongly in the Shanks transformation for $k > 1$. The denominator in (5.12) and (5.13) can near zero as well. This substantially increases the contribution of calculation errors.

1.5.3. The ε-Algorithm and Its Generalizations

We now consider the ε-algorithm (see [145]) which is often used to accelerate the convergence of sequences. It is based on the extrapolation diagram given in Figure 1.4. The first two columns are as in rational extrapolation

$$\varepsilon_{-1}^{(i)} = 0, \quad i = 2, 3, \ldots; \qquad \varepsilon_0^{(i)} = u^{h_i}, \quad i = 1, 2, \ldots \qquad (5.17)$$

The elements of the other columns are given in the following way:

$$\varepsilon_{s+1}^{(i)} = \varepsilon_{s-1}^{(i+1)} + 1/(\varepsilon_s^{(i+1)} - \varepsilon_s^{(i)}), \qquad i = 1, 2, \ldots, \quad s = 0, 1, \ldots \qquad (5.18)$$

A detailed investigation of this algorithm [145] proved that the $\varepsilon_{2k}^{(m)}$ give the Shanks transformations of order k, that is, that they coincide with the $a_0^{(m-k+1)}$ in (5.10)–(5.12). For odd subindices the values $\varepsilon_s^{(m)}$ have nothing to do with the approximation of u^{h_s}. Several generalizations of the ε-algorithm have appeared. For example, [21] proposes a "ρ-algorithm". This is also based on the extrapolation diagram given in Figure 1.4. The first two columns are given by (5.17) as before, while the other columns are given by the following formula

$$\varepsilon_{s+1}^{(i)} = \varepsilon_{s-1}^{(i+1)} + (x_{s+i+1} - x_i)/(\varepsilon_s^{(i+1)} - \varepsilon_s^{(i)}). \qquad i = 1, 2, \ldots; \quad s = 0, 1, \ldots;$$

$$(5.19)$$

where x_n is a sequence such that $\lim_{n \to \infty} x_n \to \infty$. We will assume that $x_n = h_n^{-p}$. Then from [21] we conclude that the $\varepsilon_{2k}^{(1)}$ coincide with the $T_1^{(2k)}$

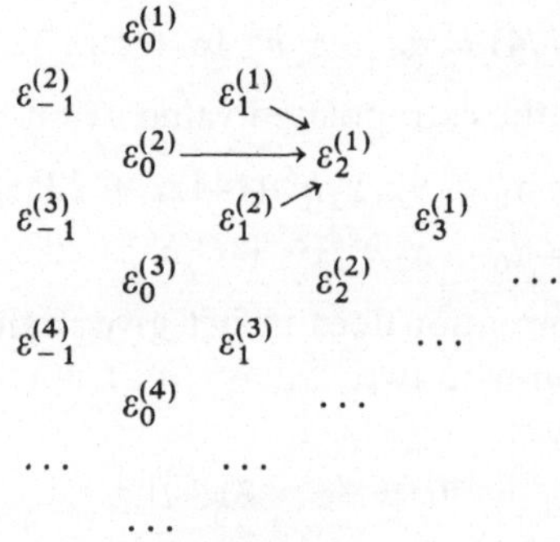

Figure 1.4.

from the Bulirsch–Stoer algorithm (5.3). Thus this "ρ-algorithm" is an economical form of rational extrapolation.

This algorithm, in somewhat different notation, is also discussed in [144]. This work also describes another simple algorithm for constructing a rational extrapolation.

Since these algorithms are only more economical forms of the exponential and rational extrapolations they have the same advantages and disadvantages in comparison with linear extrapolation.

1.6. The Effects of Computational Errors

When we solve the difference equations

$$L_h u^h = f^h \quad \text{on } \check{\Omega}_h,$$

$$l_h u^h = g^h \quad \text{on } D_h, \tag{6.1}$$

we usually do not get an exact solution u^h; rather, we get a net function $\tilde{u}^h$ with some computational error

$$\varepsilon^h = \tilde{u}^h - u^h \quad \text{on } \overline{\Omega}_h. \tag{6.2}$$

This error will consist of round-off errors, errors obtained by replacing non-arithmetic operations by arithmetic ones when computing the coefficients and the right-hand sides, as well as the errors which appear when (6.1) is solved iteratively. Iterative methods will be required, for example, for non-linear systems and for difference analogs of boundary-value problems for elliptic equations. One can compensate for the first two types of error by increasing the length of mantissa in the calculations. This can usually be accommodated within the algorithm so that it does not, substantially effect the amount of time required to solve the problem. To decrease third type of error one must also increase the number of iterations, which can have a negative effect on the computation time.

We will investigate the effect of error on the final result of a linear extrapolation. Let $\tilde{u}^{h_k}$, $k = 1, 2, \ldots, s + 1$, be a set of approximate solutions of (6.1) with net parameters $h_1, h_2, \ldots, h_{s+1}$. Denote their error by

$$\varepsilon^{h_k} = \tilde{u}^{h_k} - u^{h_k}.$$

Then instead of our usual corrected solution

$$u^H = \sum_{k=1}^{s+1} \gamma_k u^{h_k} \tag{6.3}$$

we have

$$\tilde{u}^H = \sum_{k=1}^{s+1} \gamma_k \tilde{u}^{h_k} = u^H + \sum_{k=1}^{s+1} \gamma_k \varepsilon^{h_k}. \tag{6.4}$$

Under the condition

$$h_k/h_{k+1} \geq 1 + d_1, \qquad k = 1, 2, \ldots, s,$$

from Lemma 2.3 of §7.2 it follows that

$$|\gamma_k| \leq d_2.$$

Therefore from (6.4) the estimate

$$|\tilde{u}^H - u^H| \leq d_2 \sum_{k=1}^{s+1} |\varepsilon^{h_k}| \tag{6.5}$$

follows.

Thus if none of the parameters h_k is very close to another the error of the linear extrapolation will be of the same order as the sum of errors of the solutions $\tilde{u}^{h_k}$. We have not taken into account the round-off errors which appear when calculating the sum (6.3). It is easy to show that their net effect is insignificant.

The total error when using the method of correction by higher-order differences arises in a different way. Let us consider the $(m + 1)$-step using the notation of §1.4:

$$\begin{aligned} L_h u_{k+1}^h &= f + L_h u_k^h - S_h u_k^h \quad \text{on } \check{\Omega}_h, \\ l_h u_{k+1}^h &= g + l_h u_k^h - s_h u_k^h \quad \text{on } D_h, \end{aligned} \tag{6.6}$$

where Condition F of §1.4 is satisfied. $u_0^h = 0$ on $\bar{\Omega}_h$ is taken as the initial approximation.

Assume that (6.6) for each $k = 0, 1, \ldots, m$ has been solved approximately so that we are using the approximate value $\tilde{u}_k^h$ with error $\eta_k^h = \tilde{u}_k^h - u_k^h$ instead of u_k^h. Therefore instead of (6.6) we have to solve

$$\begin{aligned} L_h v_{k+1} &= f + L_h \tilde{u}_k^h - S_h \tilde{u}_k^h \quad \text{on } \check{\Omega}_h, \\ l_h v_{k+1} &= g + l_k \tilde{u}_k^h - s_h \tilde{u}_k^h \quad \text{on } D_h. \end{aligned} \tag{6.7}$$

The solution of this problem differs from u_{k+1}^h by $\rho_{k+1} = v_{k+1} - u_{k+1}^h$. For this we have

$$\begin{aligned} L_h \rho_{k+1} &= L_h \eta_k^h - S_h \eta_k^h \quad \text{on } \check{\Omega}_h, \\ l_h \rho_{k+1} &= l_h \eta_k^h - s_h \eta_k^h \quad \text{on } D_h. \end{aligned}$$

From Condition F of §1.4 it follows that

$$\|\rho_{k+1}\|_{\bar{\Omega}_h} \leq c_{12} \|\eta_k^h\|_{\bar{\Omega}_h}.$$

But the solution of (6.7) is also only an approximation. As as result instead of v_{k+1} we have $\tilde{u}_{k+1}^h$ with error $\varepsilon_{k+1} = \tilde{u}_{k+1}^h - v_{k+1}$. Thus at the kth step we have $\tilde{u}_{k+1}^h$ with error

$$\eta_{k+1}^h = \tilde{u}_{k+1}^h - u_{k+1}^h$$

for which we have the estimate

$$\|\eta_{k+1}^h\|_{\bar{\Omega}_h} \leq c_{12} \|\eta_k^h\|_{\bar{\Omega}_h} + \|\varepsilon_{k+1}\|_{\bar{\Omega}_h}. \tag{6.8}$$

We will first consider the total error under the assumption that (6.7) for all $k = 0, 1, \ldots, m$ has been solved with an error which does not exceed δ: $\|\varepsilon_k\|_{\bar{\Omega}_h} \leq \delta, k = 1, 2, \ldots, m + 1$. For $k = 0$ we take $\tilde{u}_0^h = u_0^h = 0$ and consequently $\|\eta_0^h\|_{\bar{\Omega}_h} = 0$. Therefore from (6.8) it follows that

$$\|\eta_{m+1}^h\|_{\bar{\Omega}_h} \leq \delta(c_{12}^{m+1} - 1)/(c_{12} - 1) \quad \text{for } c_{12} \neq 1,$$

$$\|\eta_{m+1}^h\|_{\bar{\Omega}_h} \leq \delta(m + 1) \qquad\qquad \text{for } c_{12} = 1.$$

Usually the constant c_{12} is several times greater than 1. In this case (6.7) must to be solved with a certain "spare" accuracy for each $k = 0, 1, \ldots, m$. One convenient strategy is to solve (6.7) with a decrease in accuracy so that

$$\|\varepsilon_k\|_{\bar{\Omega}_h} \leq \delta/c_{12}^{m-k+1}, \qquad k = 1, 2, \ldots, m + 1.$$

Then the final error can be estimated independent of c_{12}:

$$\|\eta_{m+1}^h\|_{\bar{\Omega}_h} \leq \delta(m + 1).$$

Thus the method of increasing accuracy by higher-order differences is more sensitive to computational errors than is the Richardson extrapolation (which is, of course, linear).

CHAPTER 2

First-Order Ordinary Differential Equations

When one solves differential equations numerically one faces the problem of finding those solutions with as great an accuracy as possible in terms of the mesh-size of the net on which the reduction of the differential equation to a difference equation is carried out. At present many methods exist for finding approximate solutions with a given degree of accuracy when the solution of the differential equation is smooth. The Runge–Kutta method is one of the most widely used since it allows one to create an algorithm for the solution of the problem in a simple and straightforward manner. The theoretical foundations for the construction of the algorithm are well understood as well, making this a most attractive method of numerical mathematics.

In recent years semiexplicit and implicit algorithms for solving systems of ordinary differential equations have gained attention, especially when the higher derivatives depend on a small parameter. Such systems are referred to as "stiff"; their solutions, as a rule, exhibit rapid exponential increase (or decrease) near the origin while behaving comparatively smoothly elsewhere. A series of excellent algorithms have been developed through the use of algorithms which adjust themselves to the optimal regimes. Rosenbrock's methods, which one can interpret as one particular regularization of the Runge–Kutta algorithms, should be distinguished from the others (see [31, 111]).

In this chapter we will use the simplest algorithms and use Richardson extrapolation to realize them on different nets. A linear combination of the solutions obtained will thus enable us to obtain a solution with the maximal accuracy allowed by smoothness of data. Such accuracy from our conventional approach follows, as usual, from an analysis of the expansion of the solution in powers of a small parameter, which turns out to be the mesh-size.

Therefore, the smoothness of the solution is the main factor in constructing a numerical approximation with the given accuracy. The extrapolation of first-order ordinary differential equations has been thoroughly investigated. The survey [53] presents the theoretical results obtained up to 1971. In [122, 96, 123, 33, 118] a detailed comparison of many numerical methods is presented. Various modifications of the extrapolation method turns out to be more effective for nonstiff systems when either very high accuracy is required or when it is possible to readily estimate the size of the integration step needed for a given level of accuracy. If such an estimate cannot be made then most of the computing time will be spent in finding the initial step size. Nevertheless, the extrapolation method itself is very economical.

2.1. The Crank–Nicholson Scheme

In the previous chapter we considered the simple linear equations. We will consider general first-order differential equations. We suppose that the input data satisfy certain requirements which provide for sufficient smoothness of the solution, and consequently allow approximate solutions with a high order of accuracy.

Finally, we will study two special questions which arise when Richardson extrapolation is used. The first question concerns the construction of accurate solutions at points not on the net using Lagrange interpolation polynomials. The second question is connected with the choice of the relations between the mesh-sizes in the sequence of nets.

2.1.1. Extrapolation for Nonstiff Problems

Consider the system

$$\frac{du}{dt} = f(t, u), \qquad t \in (0, 1), \tag{1.1}$$

$$u(0) = u_0. \tag{1.2}$$

Introduce the class $C^m(\overline{\Omega})$, which denotes the set of functions which are m times continuously differentiable on the set $\overline{\Omega}$. On $C^m[0, 1]$ we define the norm

$$\|\varphi\|_{C^m[0, 1]} = \max_{0 \leq k \leq m} \max_{[0, 1]} |\varphi^{(k)}|.$$

Concerning the right-hand side we assume that

$$f \in C^r([0, 1] \times (-\infty, \infty)) \tag{1.3}$$

with the integer constant $r \geq 2$ and

$$\sup_{\substack{t \in [0, 1] \\ y \in (-\infty, \infty)}} \left| \frac{\partial^i f}{\partial y^i} (t, y) \right| \leq c_1, \qquad i = 0, 1, 2. \tag{1.4}$$

Condition (1.4) is rather restrictive since it is not realized for very simple functions, for example, for $f(t, y) = y^2$. In the case when there is a bounded solution of (1.1), (1.2) then condition (1.4) can be realized from (1.3). Indeed assume the estimate

$$\|u\|_{C[0, 1]} \leq c_2 \tag{1.5}$$

holds for a solution u of (1.1)–(1.3). Then the function $f(t, y)$ can be re-defined outside the rectangle $[0, 1] \times [-c_2, c_2]$ so that the extended function is still in $C'([0, 1] \times (-\infty, \infty))$ and in such a way that condition (1.4) holds, assuming, for example, that $f(t, y) = 0 \; \forall t \in [0, 1], |y| \geq 2c_2$. It is obvious that (1.1)–(1.3) with the new f will have the function u as its unique solution.

The converse result is also valid (see, for example, [26]): from (1.3) and (1.4) the existence of a unique solution of (1.1), (1.2) follows directly; from the continuity of the solution u on the interval $[0, 1]$ it follows that

$$u \in C^{r+1}[0, 1]. \tag{1.6}$$

Let us construct the uniform net

$$\bar{\omega}_\tau = \{ t_j = j\tau, j = 0, 1, \ldots, M \} \tag{1.7}$$

with the mesh-size $\tau = 1/M$ and introduce the midpoints of the intervals

$$\breve{\omega}_\tau = \{ t_{j+1/2} = (j + 1/2)\tau, j = 0, 1, \ldots, M - 1 \}. \tag{1.8}$$

To solve (1.1), (1.2) numerically we use the implicit central difference method which is often referred to as the Crank–Nicholson method:

$$u_i^\tau = f(t, u_i^\tau) \quad \text{on } \breve{\omega}_\tau, \tag{1.9}$$

$$u^\tau(0) = u_0. \tag{1.10}$$

Let us prove that for sufficiently small τ condition (1.4) guarantees that the system of nonlinear equations obtained has a solution. To do this we will give a technique for calculating the sequence of values $u^\tau(t)$ for $t = \tau, 2\tau, 3\tau, \ldots$. For example, suppose $u^\tau(t)$ is already known for some $t \in \bar{\omega}_\tau$. Let us write down the difference equation (1.9) corresponding to the argument $t + \tau/2$:

$$\{ u^\tau(t + \tau) - u^\tau(t) \}/\tau = f(t + \tau/2, \{ u^\tau(t + \tau) + u^\tau(t) \}/2). \tag{1.11}$$

In order to solve this for $u^\tau(t + \tau)$ consider the Newton method [54] non-linear equations

$$P(x) = 0, \tag{1.12}$$

where P is a smooth function with real argument. Let x_0 be the initial approximation; the subsequent values are calculated from the formula

$$x_{n+1} = x_n - \{P'(x_n)\}^{-1} P(x_n), \qquad n = 0, 1, \ldots . \qquad (1.13)$$

Under certain conditions the sequence x_n converges to the solution of (1.12). We present sufficient conditions from monograph [68].

Theorem 1.1. *Assume that the following conditions hold:*

(1) *the function $P(x)$ defined in the closed circle*

$$|x - x_0| \le \delta, \qquad (1.14)$$

is twice differentiable in this circle, and its second derivative is bounded:

$$|P''(x)| \le K, \qquad |x - x_0| \le \delta; \qquad (1.15)$$

(2) $$|P'(x_0)| \le B; \qquad (1.16)$$

(3) $$|P(x_0)/P'(x_0)| \le \eta; \qquad (1.17)$$

(4) *the values K, B, η satisfy the condition*

$$h = K\eta/B \le 1/2; \qquad (1.18)$$

(5) *for some number δ the inequality*

$$(1 - \sqrt{1 - 2h})\eta/h \le \delta \qquad (1.19)$$

holds. Then:

(1) *in circle (1.14) the equation $P(x) = 0$ has a unique solution x^*;*
(2) *in (1.13) the approximation x_n can be constructed for any n, all x_n lie in the circle (1.14), and $x_n \to x^*$ when $n \to \infty$.*

Let us check that the conditions of this theorem are fulfilled for equation (1.11), which we write in the following form:

$$P(x) \equiv (x - x_0)/\tau - f(t + \tau/2, (x + x_0)/2) = 0, \qquad x_0 = u^\tau(t). \quad (1.20)$$

The function $P(x)$ is defined on the whole real axis and for the second derivative we have the estimate

$$|P''(x)| \le \left| \frac{\partial^2 f}{\partial u^2}(t + \tau/2, (x + x_0)/2) \right| / 4 \le c_1/4 \quad \forall x \in (-\infty, \infty).$$

Thus one can take the constant δ as large as necessary, and $k = c_1/4$. For the second condition it is sufficient to take $\tau < 2/c_1$. Assume, for example, $\tau \le 1/c_1$. Then

$$|P'(x_0)| = \left| 1/\tau - \frac{1}{2} \frac{\partial f}{\partial u}(t + \tau/2, x_0) \right| \le c_1/2.$$

Thus, B can be taken equal to $c_1/2$. It remains to find η. Consider the inequality

$$|P(x_0)/P'(x_0)| = \tau|f(t + \tau/2, x_0)| \Big/ \left\{1 - \frac{\tau}{2}\frac{\partial f}{\partial u}(t + \tau/2, x_0)\right\}$$

$$\leq 2\tau|f(t + \tau/2, x_0)| \leq 2\tau c_1.$$

Therefore, we assume $\eta = 2\tau c_1$. For inequality (1.18) it is necessary that

$$h = (c_1/4) \cdot 2\tau c_1 \cdot 2/c_1 = \tau c_1 \leq \tfrac{1}{2}.$$

Hence $\tau \leq 1/(2c_1)$. Inequality (1.19) is automatically satisfied because δ is arbitrary. Therefore in order that all conditions of Theorem 1.1 are satisfied it is sufficient to have

$$\tau \leq 1/(2c_1). \tag{1.21}$$

Then from this theorem it follows that equation (1.19), and this also (1.11), has a unique solution x^*, which we denote by $u^\tau(t + \tau)$.

The Newton method (1.13), (1.20) converges quadraticly [68]. At the expense of our initial approximation $x_0 = u^\tau(t)$ we have the estimate $|x^* - x_0| = O(\tau)$. On the basis of quadratic convergence

$$|x^* - x_1| = O(\tau^2)$$

$$|x^* - x_2| = O(\tau^4),$$

$$|x^* - x_3| = O(\tau^8), \ldots.$$

Thus for smooth solutions, if condition (1.21) is true, then there is no need for many iterations. Nevertheless it is necessary to choose some criterion for terminating this process. Consider, for example, the following condition: the iterations continue until, for given sufficiently small $\varepsilon > 0$, the condition

$$|P(x_n)| \leq \varepsilon \tag{1.22}$$

holds this value x_n is then taken as the approximate value of $u^\tau(t + \tau)$. In this case instead of solving (1.9), (1.10) we actually solve the nearby system

$$v_t^\tau = f(t, v_t^\tau) + \rho \quad \text{on } \check{\omega}_\tau, \tag{1.23}$$

$$v^\tau(0) = u_0, \tag{1.24}$$

where

$$|\rho(t)| \leq 2\varepsilon \quad \text{on } \check{\omega}_\tau; \tag{1.25}$$

ρ contains the round-off errors, the remainder from Newton's method, and the errors in the calculation of f. Let us prove that the solution v^τ of this system differs from the solution of the initial difference problem by a small quantity.

Theorem 1.2. *Assume condition* (1.4) *holds for* f, *and* τ *obeys* (1.21). *Then the estimate*

$$\|u^\tau - v^\tau\|_{C,\tau} \le 2\varepsilon(e^{2c_1} - 1)/c_1 \qquad (1.26)$$

holds.

PROOF. Let us show that the inequality

$$|u^\tau(t) - v^\tau(t)| \le \{\tau\varepsilon/(1 - \tau c_1/2)\} \sum_{j=0}^{t/\tau - 1} \{(1 + \tau c_1/2)/(1 - \tau c_1/2)\}^j \quad (1.27)$$

holds. For $t = 0$ it is obvious because $v^\tau(0) - u^\tau(0) = 0$ and the sum on the right-hand side with these limits is zero. Suppose that this inequality holds for some $t \in \bar\omega_\tau$; we will prove it for $t + \tau$. Subtract (1.9) from (1.23) and take the modulus of both sides. We have

$$|v_{\hat{t}}^\tau - u_{\hat{t}}^\tau| \le |f(t, v_{\hat{t}}^\tau) - f(t, u_{\hat{t}}^\tau)| + |\rho|. \qquad (1.28)$$

Now we use the following simple inequality, which follows from (1.4):

$$|f(t, v_{\hat{t}}^\tau) - f(t, u_{\hat{t}}^\tau)| = |v_{\hat{t}}^\tau - u_{\hat{t}}^\tau| \left|\frac{\partial f}{\partial u}(t, \xi)\right| \le c_1|v_{\hat{t}}^\tau - u_{\hat{t}}^\tau|.$$

Applying this inequality and the estimate (1.25) to (1.28) we have

$$|v_{\hat{t}}^\tau - u_{\hat{t}}^\tau| \le c_1|v_{\hat{t}}^\tau - u_{\hat{t}}^\tau| + 2\varepsilon.$$

Consider the above inequality at the point $t + \tau/2$. After some simple transformations we have

$$(1 - \tau c_1/2)|v^\tau(t + \tau) - u^\tau(t + \tau)| \le (1 + \tau c_1/2)|v^\tau(t) - u^\tau(t)| + 2\tau\varepsilon.$$

Due to (1.21) $1 - \tau c_1/2$ is positive. Divide both sides of the inequality by this factor and use (1.27):

$$|v^\tau(t + \tau) - u^\tau(t + \tau)| \le \{(1 + \tau c_1/2)/(1 - \tau c_1/2)\}$$
$$\times |v^\tau(t) - u^\tau(t)| + 2\tau\varepsilon/(1 - \tau c_1/2)$$
$$\le \{2\tau\varepsilon/(1 - \tau c_1/2)\} \sum_{j=0}^{t/\tau} \{(1 + \tau c_1/2)/(1 - \tau c_1/2)\}^j.$$

Thus (1.27) is proved. We can simplify the right-hand side using the inequality

$$(1 + x)/(1 - x) \le e^{4x} \quad \forall\, x \in [0, 1/4]$$

and $\tau c_1/2 \le 1/4$, which follows from (1.21):

$$|v^\tau(t) - u^\tau(t)| \le \{2\tau\varepsilon/(1 - \tau c_1/2)\}$$
$$\times [\{(1 + \tau c_1/2)/(1 - \tau c_1/2)\}^{t/\tau} - 1]/\{(1 + \tau c_1/2)/(1 - \tau c_1/2) - 1\}$$
$$\le 2\varepsilon(e^{2tc_1} - 1)/c_1.$$

Taking the maximum over $\bar\omega_\tau$ we get (1.26). The theorem is proved. $\square$

A more complicated proof would allow us to reduce the exponent in (1.26) by two.

Let us return to the difference system (1.9), (1.10). We will prove that its solution can be expanded in even powers of τ (see also [80]).

Theorem 1.3. *Assume that conditions* (1.3), (1.4) *hold for* (1.1), (1.2). *Then the solution of* (1.9), (1.10) *can be expanded under condition* (1.21):

$$u^\tau = u + \sum_{j=1}^{m} \tau^{2j} v_j + \eta^\tau \quad \text{on } \bar\omega_\tau, \tag{1.29}$$

where $m = [(r-1)/2]$, v_j is independent of τ, $v_j \in C^{r-2j+1}[0, 1]$ and the remainder obeys

$$\|\eta^\tau\|_{C,\tau} \le c_3 \tau^r. \tag{1.30}$$

PROOF. Assume $v_0 \equiv u$ and find m functions which solve the differential equations

$$v_j' - \frac{\partial f}{\partial u}(t, u)v_j = \sum_{k=1}^{j} \left(-\frac{v_{j-k}^{(2k+1)}}{4^k(2k+1)!} + \frac{v_{j-k}^{(2k)}}{4^k(2k)!} \cdot \frac{\partial f}{\partial u}(t, u) \right)$$

$$+ \sum_{l=2}^{j} \frac{1}{l!} \frac{\partial^l f}{\partial y^l}(t, u) \sum_{t_1 + \cdots + t_l = j} \prod_{i=1}^{l} \left(\sum_{k=0}^{t_i} \frac{v_{i-k}^{(2k)}}{4^k(2k)!} \right) \quad \text{on } (0, 1), \tag{1.31}$$

$$v_j(0) = 0, \quad j = 1, 2, \ldots, m. \tag{1.32}$$

For example, the function v_1 is defined by

$$v_1' - \frac{\partial f}{\partial u}(t, u)v_1 = -u'''/24 + \frac{\partial f}{\partial u}(t, u)u''/8 \quad \text{on } (0, 1), \tag{1.33}$$

$$v_1(0) = 0. \tag{1.34}$$

The right-hand side of this equation contains only the function u and is $r - 2$ times continuously differentiable. Due to the linearity of the problem there is a unique solution $v_1 \in C^{r-1}[0, 1]$. Now assume that the functions $v_1, \ldots, v_{j-1}$ have already been found and $v_k \in C^{r-2k+1}[0, 1]$. The right-hand side of (1.31) giving v_j does not contain the functions v_k with index greater than $j - 1$ and is $r - 2j$ times continuously differentiable with respect to $t \in [0, 1]$. Therefore the linear system (1.31), (1.32) for v_j has a unique solution in $C^{r-2j+1}[0, 1]$. Thus all the required functions can be determined.

Using these functions we define the function

$$w = \sum_{j=0}^{m} \tau^{2j} v_j \quad \text{on } \bar\omega_\tau.$$

Substitute this in the difference operator of equation (1.9):

$$w_{\bar{t}} - f(t, w_{\bar{t}}) = \sum_{j=0}^{m} \tau^{2j}(v_j)_{\bar{t}} - f\left(t, \sum_{j=0}^{m} \tau^{2j}(v_j)_{\bar{t}}\right). \tag{1.35}$$

Use the expansions from Lemma 1.1 of §7.1 in order to transform the terms on the right-hand side:

$$\sum_{j=0}^{m} \tau^{2j}(v_j)_{\bar{t}} = \sum_{j=0}^{m} \tau^{2j} \sum_{k=0}^{m-j} \tau^{2k} \frac{v_j^{(2k+1)}}{4^k(2k+1)!} + \tau^r \theta_1$$

$$= \sum_{j=0}^{m} \tau^{2j} \sum_{k=0}^{j} \frac{v_{j-k}^{(2k+1)}}{4^k(2k+1)!} + \tau^r \theta_1,$$

$$\sum_{j=0}^{m} \tau^{2j}(v_j)_{\bar{t}} = \sum_{j=0}^{m} \tau^{2j} \sum_{k=0}^{j} \frac{v_{j-k}^{(2k)}}{4^k(2k)!} + \tau^r \theta_2, \tag{1.36}$$

where $|\theta_i| \le c_4$ for $i = 1, 2$ on $\check{\omega}_\tau$. Use the Taylor series expansion for the function $f(t, w_{\bar{t}})$:

$$f(t, w_{\bar{t}}) = f\left(t, \sum_{j=0}^{m} \tau^{2j} \sum_{k=0}^{j} \frac{v_{j-k}^{(2k)}}{4^k(2k)!}\right) + \tau^r \theta_2 \frac{\partial f}{\partial u}(t, \xi)$$

$$= f(t, u) + \sum_{l=1}^{m} \left\{ \sum_{j=1}^{m} \tau^{2j} \sum_{k=0}^{j} \frac{v_{j-k}^{(2k)}}{4^k(2k)!} \right\}^l \frac{1}{l!} \frac{\partial^l f}{\partial u^l}(t, u) + \tau^r \theta_3.$$

From the boundedness of the functions v_j and f and of their derivatives it follows that $|\theta_3| \le c_5$ on $\check{\omega}_\tau$. Expand the expression in brackets and collect terms with like powers of τ:

$$f(t, w_{\bar{t}}) = f(t, u) + \sum_{j=1}^{m^2} \tau^{2j} \left\{ \sum_{l=1}^{m} \frac{1}{l!} \frac{\partial^l f}{\partial u^l}(t, u) \right.$$

$$\left. \times \sum_{t_1 + \cdots + t_l = j} \prod_{i=1}^{l} \left(\sum_{k=0}^{t_i} \frac{v_{t_i-k}^{(2k)}}{4^k(2k)!} \right) \right\} + \tau^r \theta_3.$$

Leave all terms of order τ^{2j}, $j \le m$ unchanged; write the others in the form $\tau^r \theta_4$, where $|\theta_4| \le c_6$ on $\check{\omega}_\tau$. Using the equality thus obtained, together with (1.36), in (1.35) we have

$$w_{\bar{t}} - f(t, w_{\bar{t}}) = u' - f(t, u) + \sum_{j=1}^{m} \tau^{2j} \left\{ \sum_{k=0}^{j} \frac{v_{j-k}^{(2k+1)}}{4^k(2k+1)!} \right.$$

$$\left. - \sum_{l=1}^{j} \frac{1}{l!} \frac{\partial^l f}{\partial u^l}(t, u) \sum_{t_1 + \cdots + t_l = j} \prod_{i=1}^{l} \left(\sum_{k=0}^{t_i} \frac{v_{t_i-k}^{(2k)}}{4^k(2k)!} \right) \right\}$$

$$+ \tau^r \theta_1 - \tau^r \theta_3 - \tau^r \theta_4.$$

According to the definition of v_j and u all terms inside the brackets exactly cancel, as does $u' - f(t, u)$. As a result we have

$$w_{\bar{t}} = f(t, w_{\bar{t}}) + \tau^r \theta_1 - \tau^r \theta_3 - \tau^r \theta_4 \quad \text{on } \check{\omega}_\tau. \tag{1.37}$$

From the initial condition (1.2), (1.10), (1.32) it follows that $w(0) = u^h(0) = u(0) = u_0$. Therefore Theorem 1.2 can be used for ω. This theorem implies the estimate $\|\eta^\tau\|_{C,\tau} \leq c_3 \tau^r$ for $\eta^\tau \equiv u^h - w$. The theorem is proved. □

We will use the expansion obtained to justify the increased accuracy claimed for the Richardson extrapolation method of §1.3. Assume conditions (1.3) and (1.4) hold. Assume $m = [(r - 1)/2]$. Fix integers $0 < N_1 < \cdots < N_{m+1}$ and construct the nets $\bar{\omega}_{\tau_k}$ with mesh-sizes $\tau_k = 1/(N_k M)$, where the natural number M can in principle be as large as required, and

$$M \geq 2c_1/N_1. \tag{1.38}$$

This condition guarantees that the difference system (1.9), (1.10) has a solution on each net $\bar{\omega}_{\tau_k}$. All solutions $u^{\tau k}$ are defined on the net with mesh-size $\tau = 1/M$.

Consider the system

$$\sum_{k=1}^{m+1} \gamma_k = 1.$$

$$\sum_{k=1}^{m+1} \gamma_k N_k^{-2j} = 0, \qquad j = 1, \ldots, m. \tag{1.39}$$

Since the N_k are distinct this system is regular and has a unique solution. Form the linear combination

$$U^H = \sum_{k=1}^{m+1} \gamma_k u^{\tau k} \quad \text{on } \bar{\omega}_\tau. \tag{1.40}$$

We will show that when $\tau \to 0$ the solution U^H is of a higher order of accuracy than any $u^{\tau k}$.

Theorem 1.4. *Assume that conditions* (1.3) *and* (1.4) *hold for the system* (1.1), (1.2). *Then the corrected solution* (1.40) *whose weights are given by the system* (1.39), *assuming* (1.38), *obeys the following estimate*

$$\|U^H - u\|_{C,\tau} \leq c_7 \tau^r. \tag{1.41}$$

PROOF. Since the N_i are fixed we have:

$$\tau_k/\tau_{k+1} \geq 1 + d_3, \qquad k = 1, \ldots, m,$$

where

$$d_3 = \min_{1 \leq k \leq m} (N_{k+1}/N_k) - 1.$$

The rest of the proof is the same as the proof of Theorem 3.2 of §1.3. □

Let us note that the system (1.23), (1.24) is the one we are actually solving with the remainder of (1.25) rather than (1.9), (1.10). The fact that we are using Newton's method to approximate the system (1.9), (1.10) is the main source of the remainder we will find a criterion for optimizing the number of iterations in this method. Let the expected error of the corrected solution which appears, because of the discretization (assuming the conditions of Theorems 1.3 and 1.4) be δ. It is reasonable to use Newton's method (1.13), (1.20) so long as the error thus incurred does not exceed this value. To do this it suffices to terminate iterations when (1.22) is attained with

$$\varepsilon \sim c_1 \delta \Big/ \Big\{ (e^{2c_1} - 1) \sum_{k=1}^{m+1} |\gamma_k| \Big\}.$$

From Theorem 1.2 the contribution of the error in computing each solution u^{τ_k} by the Newton method does not exceed

$$\delta \Big/ \sum_{k=1}^{m+1} |\gamma_k|.$$

From §1.6 it follows that the computational error in the corrected solution will be of order δ.

We have focused on the role of the error caused which arises from the Newton method for a good reason. When we use iterative methods we must not assume that the iteration error is less than round-off error, or the errors in calculating nonarithmetical functions. These errors are random in nature, so that many iterations may be needed to achieve a desired accuracy.

Remark. When

$$\frac{\partial f}{\partial u} (t, u) \le 0$$

the restriction (1.21) on τ can be removed without altering the stability of (1.9), (1.10) or the convergence of the Newton method (1.13), (1.20). In inequality (1.26) of Theorem 1.2 the exponential factor can be deleted, as well. This usually happens for "stiff" systems, which will be discussed in greater detail later.

2.1.2. Computing the Extrapolated Solution at an Arbitrary Point

We have seen that at the knots of $\bar{\omega}_\tau$ the required accuracy can be attained. A problem arises if we use this method for three or more nets. This is caused by the fact that the corrected solution is calculated at points which are common knots for all the nets. The number of such points may seem small. It might also be necessary to find an approximate solution at points which are not knots. At such points it is necessary to interpolate by splines, by trigonometrical polynomials, and so forth. We will consider a simple situation using Lagrange interpolation polynomials.

Let us extend difference functions u^{τ_k} from the difference net $\bar{\omega}_{\tau_k}$ to the whole interval $[0, 1]$ in the following way. Take an arbitrary elementary interval $[t_j, t_{j+1}]$ of $\bar{\omega}_{\tau_k}$. The solution is defined only at the endpoints t_j, t_{j+1}. Choose $r - 2$ more adjacent knots of $\bar{\omega}_{\tau_k}$ and define the value of $u^{\tau_k}(t)$ on $[t_j, t_{j+1}]$ to be the value of the Lagrange interpolation polynomial on these r knots. If we interpolate over all elementary intervals we get a continuous functions which coincides with u^{τ_k} on the knots of $\bar{\omega}_{\tau_k}$. We will also denote this by u^{τ_k}.

We will use these interpolants to calculate an approximate solution of the form

$$U^H = \sum_{k=1}^{m+1} \gamma_k u^{\tau_k} \quad \text{on } [0, 1], \tag{1.42}$$

where the weights γ_k are found from (1.39).

Theorem 1.5. *Assume* (1.1) *and* (1.2) *satisfy* (1.3). *Then* (1.42) (*whose weights are found from* (1.39)) *obeys the estimate*

$$\|U^H - u\|_{C[0, 1]} \leq c_8 \tau^r. \tag{1.43}$$

PROOF. From Theorem 1.3 at the knots of $\bar{\omega}_{\tau_k}$ we have the expansion

$$u^{\tau_k}(t) = u(t) + \sum_{j=1}^{m} \tau_k^{2j} v_j(t) + \eta^{\tau_k}(t), \tag{1.44}$$

where v_j does not depend on τ, $v_j \in C^{r-2j+1}[0, 1]$, and the remainder η^{τ_k} obeys the estimate

$$\|\eta^{\tau_k}\|_{C, \tau_k} \leq c_9 \tau_k^r. \tag{1.45}$$

We will prove that this expansion is valid for the functions u^{τ_k} on the whole interval $[0, 1]$. Take an arbitrary point $t \in [0, 1]$. It will belong to some elementary interval $[t_j, t_{j+1}]$. On this interval we have interpolated at $t_i, t_{i+1}, \ldots, t_{i+r-1}$; therefore

$$u^{\tau_k}(t) = \sum_{l=0}^{r-1} \alpha_l(t) u^{\tau_k}(t_{i+l}),$$

where the α_l are polynomials of degree $r - 1$. Using (1.44) at the knots we can write this formula in the form

$$u^{\tau_k}(t) = \sum_{l=0}^{r-1} \alpha_l(t) u(t_{i+l}) + \sum_{j=1}^{m} \tau_k^{2j} \sum_{l=0}^{r-1} \alpha_l(t) v_j(t_{i+l}) + \sum_{l=0}^{r-1} \alpha_l(t) \eta^{\tau_k}(t_{i+l}). \tag{1.46}$$

Since $u \in C^{r+1}[0, 1]$ its interpolation formula has accuracy of order τ_k^r. Therefore

$$\sum_{l=0}^{r-1} \alpha_l(t) u(t_{i+l}) = u(t) + \rho_0(t), \tag{1.47}$$

where

$$|\rho_0| \leq c_{10}\,\tau_k^r.$$

The functions v_j are somewhat less smooth; therefore, interpolation by r knots results in a lower order of accuracy (Lemma 3.1 of §7.3):

$$\sum_{l=0}^{r-1} \alpha_l(t)v_j(t_{i+l}) = v_j(t) + \rho_j(t), \qquad (1.48)$$

where

$$|\rho_j| \leq c_{11}\tau_k^{r-2j}.$$

For functions η^{τ_k} obey the simple estimate

$$\left| \sum_{l=0}^{r-1} \alpha_l(t)\eta^{\tau_k}(t_{i+l}) \right| \leq c_{12}\,\tau_k^r.$$

This follows from (1.45) and the boundedness of the coefficients α_l (Lemma 3.2 of §7.3).

Replacing the terms in (1.46) by their expansions in (1.47) and (1.48) and assuming

$$\eta^{\tau_k}(t) = \rho_0(t) + \sum_{j=1}^{m} \tau_k^{2j}\rho_j(t) + \sum_{l=0}^{r-1} \alpha_l(t)\eta^{\tau_k}(t_{i+l}),$$

we obtain an expansion of the form (1.44) on the whole interval $[0, 1]$ for $k = 1, \ldots, m + 1$, where

$$|\eta^{\tau_k}(t)| \leq (c_{10} + mc_{11} + c_{12})\tau_k^r. \qquad (1.49)$$

Summing up these expansions with weights γ_k we get

$$U^H(t) = \sum_{k=1}^{m+1} \gamma_k u(t) + \sum_{j=1}^{m} \left(\sum_{k=1}^{m+1} \gamma_k \tau_k^{2j} \right) v_j(t) + \sum_{k=1}^{m+1} \gamma_k \eta^{\tau_k}(t).$$

Because the γ_k satisfy equations (1.39) we obtain

$$U^H(t) = u(t) + \sum_{k=1}^{m+1} \gamma_k \eta^{\tau_k}(t),$$

from which there follows:

$$|U^H(t) - u(t)| \leq \sum_{k=1}^{m+1} |\gamma_k| |\eta^{\tau_k}(t)|. \qquad (1.50)$$

The mesh-sizes τ_k obey

$$\tau_k/\tau_{k+1} = N_{k+1}/N_k.$$

At $M \to \infty$ the numbers N_k are fixed, and therefore

$$\tau_k/\tau_{k+1} \geq 1 + c_{13},$$

where

$$c_{13} = \min_{1 \le k \le m} (N_{k+1}/N_k) - 1 > 0.$$

This allows us to apply Lemma 2.4 of §7.2 from which the boundedness of the γ_k follows:

$$|\gamma_k| \le \{1 + 1/(2c_{13} + c_{13}^2)\}^{m+1}.$$

The η^{τ_k} satisfy (1.49) as well. Therefore, from (1.50) it follows that

$$|U^H(t) - u(t)| \le \sum_{k=1}^{m+1} \tau_k^r \{1 + 1/(2c_{13} + c_{13}^2)\}^{m+1}(c_{10} + mc_{11} + c_{12}).$$

Because of $\tau_k \ge \tau_1 = \tau/N_1$ we have

$$|U^H(t) - u(t)| \le c_8 \tau^r \quad \forall\, t \in [0, 1],$$

where

$$c_8 = (m + 1)N_1^{-r}\{1 + 1/(2c_{13} + c_{13}^2)\}^{m+1}(c_{10} + mc_{11} + c_{12}).$$

From this (1.43) follows. Theorem 1.5 is proved.   □

2.1.3. On the Choice of the Ratio Between Mesh-Sizes

We will now attempt to determine the ratio of mesh-sizes of ω_{τ_k}, as determined by the integers $N_1, \ldots, N_{m+1}$, at which the highest accuracy is attained. To do this we must require that the right-hand side of (1.1), (1.2) has somewhat more smoothness:

$$f \in C^{2m+3}([0, 1] \times (-\infty, \infty)).$$

Let us show that this requirement allows to obtain additional information on the behavior of the main term of the error $U^H - u$. From Theorem 1.3 applied to the approximate solutions u^{τ_k} it follows that

$$u^{\tau_k}(t) = u(t) + \sum_{j=1}^{m+1} \tau_k^{2j} v_j(t) + \xi^{\tau_k}(t), \qquad t \in \bar{\omega}_{\tau_k}, \tag{1.51}$$

where

$$\|\xi^{\tau_k}\|_{C, \tau_k} = O(\tau_k^{2m+3}).$$

Summing up (1.51) with weights γ_k, which satisfy (1.39), we have

$$U^H(t) = u(t) + \sum_{k=1}^{m+1} \gamma_k \tau_k^{2m+2} v_{m+1}(t) + \rho^H(t), \tag{1.52}$$

where

$$\|\rho^H\|_{C, \tau} = O(\tau^{2m+3}).$$

Since v_{m+1} does not depend on τ_k or γ_k, the modulus of the main term of the error has coefficient

$$\sum_{k=1}^{m+1} \gamma_k \tau_k^{2m+2}.$$

This coefficient can be simplified if we use Lemma 2.6 of §7.2. From this lemma there follows:

$$\sum_{k=1}^{m+1} \gamma_k \tau_k^{2m+2} = (-1)^{m+2} \prod_{k=1}^{m+1} \tau_k^2.$$

We have already mentioned that the amount of computing time necessary to solve (1.9), (1.10) on one net $\bar{\omega}_{\tau_k}$ depends on the number of arithmetic operations (mainly in the computation of the right-hand side and is proportional to $1/\tau_k$. With a high degree of accuracy the amount of computing time necessary to solve the $m + 1$ approximate systems (1.9), (1.10) on nets with mesh-sizes $\tau_1, \ldots, \tau_{m+1}$ is equal to

$$\alpha \sum_{k=1}^{m+1} 1/\tau_k,$$

where α is a certain time constant (e.g., in seconds) which depends on the type of computer used, the efficiency of the code, and so on, but is independent of τ_k.

Consider the following problem: find a set of natural numbers $N_1, \ldots, N_{m+1}$ such that for fixed

$$\mu = \alpha \sum_{k=1}^{m+1} 1/\tau_k = \frac{\alpha}{\tau} \sum_{k=1}^{m+1} N_k$$

(which characterizes the time needed to compute the approximate solution (1.40)) we are required to find the minimum of the quantity

$$v = \prod_{k=1}^{m+1} \tau_k^2 = \tau^{2m+2} \prod_{k=1}^{m+1} N_k^{-2}$$

(which characterizes the main term in the error for the approximate solution (1.40)).

To solve this problem we use Lagrange multipliers with N_k as continuous variables. Construct the function

$$\phi(N_1, \ldots, N_{m+1}) = \tau^{2m+2} \prod_{k=1}^{m+1} N_k^{-2} + \lambda \left(\frac{\alpha}{\tau} \sum_{k=1}^{m+1} - \mu \right);$$

a necessary condition for an extremum is:

$$\frac{\partial \phi}{\partial N_l} = -2\tau^{2m+2} N_l^{-1} \prod_{k=1}^{m+1} N_k^{-2} + \lambda\alpha/\tau.$$

Since α and τ are positive, then λ is also positive. Hence

$$N_1 = \cdots = N_{m+1} = 2\tau^{2m+3} \frac{1}{\lambda\alpha} \prod_{k=1}^{m+1} N_k^{-2}.$$

Thus, the nearer the N_i are to each other the lower the value of v is at a fixed time μ. For real processes the limiting case $N_1 = \cdots = N_{m+1}$ is impossible since the system (1.39) would become inconsistent, and (1.40) meaningless. If the values of the N_k are too close it will lead to a sharp increase in the γ_k and therefore the round-off errors not considered in our model could become the main term in the error in (1.40).

To illustrate these statements we will present the results of some numerical calculations for several values of the N_k.

The system

$$u' + tu = (t^2 + t + 1)e^t, \qquad t \in (0, 2), \tag{1.53}$$

$$u(0) = 0.$$

is solved by

$$u(t) = te^t.$$

It is easy to verify that the assumptions, which allow us to extrapolate with two or more mesh-sizes τ_k, are in fact fulfilled for this problem. First, we solve the difference problems with mesh-sizes τ_k and, using the known errors at $t = 2$, we construct the graph of

$$\zeta(M_k) = |u^{\tau_k}(2) - u(2)|$$

versus M_k, the number of points of the net $\bar\omega_{\tau_k}$. This graph (in the logarithmic coordinates) is given in Figure 2.1.

We now construct the corrected solutions, extrapolating the two approximate solutions, using different ratios between the mesh-sizes of two nets. In Figure 2.1 we give three graphs of the dependence of the error of the corrected solution at the point $t = 2$ on the total number of knots in the two nets. The graphs correspond to the ratios $N_2 : N_1$ between the mesh-sizes of the first and second nets being, respectively, $16 : 15$, $2 : 1$, $10 : 1$. To take into account the influence of the round-off errors computations were carried on a computer in floating point made with a six digit mantissa.

By comparing the graphs we see that the extrapolation method is the most economical with ratio $16 : 15$ until round-off errors first appear. Note, however that the profit versus a ratio of $2 : 1$ is negligible, although the round-off errors show up more strongly at smaller mesh-sizes. In the second case the first number of common points among the nets is much higher than in the first case, and therefore the extrapolation errors (not taken into consideration here) usually exceed the savings obtained from the choice of the ratio $N_2 : N_1$. We should also note that extrapolating at $N_2 : N_1 = 10 : 1$ can not compete with the first two choices.

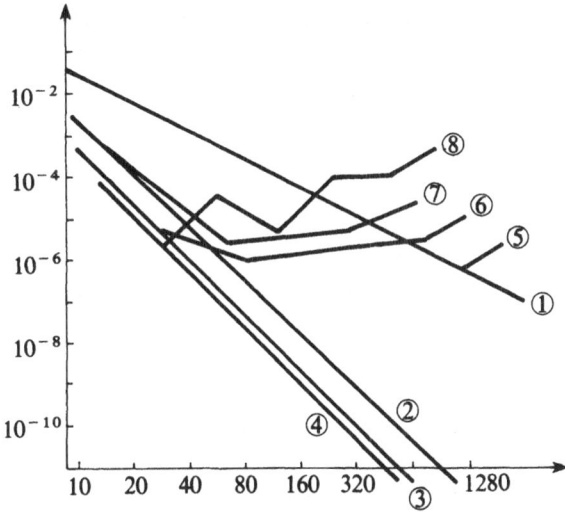

Figure 2.1. The error in the approximate solutions to (1.53) at the knot $t = 2$ with and without round-off errors.

(1) Graph of the error in the solution using the Crank–Nicholson scheme (1.9), (1.10) without round-off errors; (2) graph of the error in the extrapolated solution (1.40) with the ratio of mesh-sizes $N_2 : N_1 = 10 : 1$ without round-off errors; (3) same as (2) but with ratio $N_2 : N_1 = 2 : 1$; (4) the same as (2) but with ratio $N_2 : N_1 = 16 : 15$ (5) graph of the error in the solution using the Crank–Nicholson scheme (1.9), (1.10) with round-off errors; (6) graph of the error in the extrapolated solution (1.40) with the ratio of mesh-sizes $N_2 : N_1 = 10 : 1$ with round-off errors; (7) same as (6), but $N_2 : N_1 = 2 : 1$; (8) same as (6), but $N_2 : N_1 = 16 : 15$.

2.1.4. Some Results for Stiff Problems

The results of this section involve the systems of ordinary differential equations

$$\frac{du}{dt} = f(t, u), \qquad t \in (0, 1). \tag{1.54}$$

$$u(0) = u_0,$$

where u_0, $u(t)$ are vectors with n components, f is a vector-function of $n + 1$ arguments: $f(t, u) = f(t, u_1, \ldots, u_n)$. It is necessary to assume conditions (1.3), (1.4) hold for each component f_i of the vector-function f. Then Theorem 1.3 can be reformulated quite simply in the case of vectors.

Let us consider the question of applying extrapolation methods to stiff systems. In such problems the differential equation has particular solutions with rapid exponential decrease while the exact solution changes somewhat more moderately on a larger portion of the integration interval. The most common systems (1.54) are those with Jacobian $J(t)$ with elements $J_{ij} = \partial f_i/\partial u_j$, $i, j = 1, \ldots, n$, with distinct eigenvalues $\lambda_i(t)$ so that

$$\text{Re}(\lambda_i) < 0 \quad \text{and} \quad \max_{i=1,\ldots,n} \text{Re}(-\lambda_i) \gg 1 \quad \forall\, t \in (0, 1). \tag{1.55}$$

Stiff systems usually satisfy conditions (1.3), (1.4) with very large c_1, so that the restriction on the step size $\tau \leq 1/2c_1$ becomes very troublesome. Nevertheless for stiff systems with (1.55) it is possible to justify the stability of (1.9), (1.10) with an arbitrary mesh-size $\tau > 0$. In order to prove this as well as the convergence of the Newton method somewhat different methods are needed, which we will not consider here (see [96] and its bibliography).

Let us point out two reasons [80] why the implicit method of rectangles is preferable to the method of trapezoids for stiff problems:

$$u_{\bar{t}}^\tau(t) = f_{\bar{t}}(t, u^\tau(t)) \quad \text{on } \check{\omega}_\tau,$$
$$u^\tau(0) = u_0, \tag{1.56}$$

where

$$f_{\bar{t}}(t, u(t)) = \{f(t + \tau/2, u(t + \tau/2)) + f(t - \tau/2, u(t - \tau/2))\}/2.$$

For the latter method the results of §2.1.1 can also be justified. But using the trapezoid rule after we have defined $u^\tau(t + \tau/2)$ using the Newton method it is necessary to compute $f(t + \tau/2, u^\tau(t + \tau/2))$ in order to use it in the next step. Therefore the trapezoid rule requires a greater number of calculations for the right-hand side. Another reason arises as follows. Consider the model equation

$$u' = \lambda(t)u, \tag{1.57}$$

where $\lambda(t) < 0$ so that $u(t)$ decreases with increasing t. From the method of trapezoids it follows that

$$u^\tau(t + \tau/2) = (\{1 + \tau\lambda(t - \tau/2)/2\}/\{1 - \tau\lambda(t + \tau/2)/2\})u^\tau(t - \tau/2).$$

Therefore in order that the inequality $|u^\tau(t + \tau/2)| \leq |u^\tau(t - \tau/2)|$ hold it is necessary that

$$\tau\{\lambda(t + \tau/2) - \lambda(t - \tau/2)\} \leq 4.$$

If the function $\lambda(t)$ increases on a certain part of the interval of integration then this condition imposes a constraint on the possible step length. There is no such limitation when we use the implicit method of rectangles, though the problem of approximating the rapidly decreasing solutions remains.

We will assume in (1.57) that λ is some complex constant with $\text{Re}(\lambda) < 0$. Then it follows that

$$u^\tau(t + \tau/2) = \{(1 + \tau\lambda/2)/(1 - \tau\lambda/2)\}u^\tau(t - \tau/2). \tag{1.58}$$

If

$$Re(\lambda\tau) \ll 0 \qquad (1.59)$$

then the difference solution $u^\tau(t)$ is approximately equal to $(-1)^{t/j}u(0)$. Instead of the rapidly vanishing solution $|u(t)| = e^{Re\,(\lambda)t}$ we obtain an oscillating approximate solution. To avoid this, Lindberg [79] suggested that one smooth out the approximate solution by

$$\tilde{u}^\tau(t) = u_{\bar{t}\bar{t}}^\tau(t), \qquad (1.60)$$

where $u_{\bar{t}\bar{t}}(t) = \{u(t-\tau) + 2u(t) + u(t+\tau)\}/4$. In this case from (1.58) we have

$$\tilde{u}^\tau(t) = u^\tau(t)/\{1 - (\tau\lambda/2)^2\},$$

and from (1.59) the rapid decrease of the solution follows. If this smoothing out is used repeatedly (M times) and we use the result as the initial approximation in calculating the difference solution then we have

$$\tilde{u}^\tau(t) = u^\tau(t)/\{1 - (\tau\lambda/2)^2\}^M.$$

Thus smoothing enough times allows us to suppress the corresponding terms in the difference solution. For the stiff system of (1.54) the solution contains both rapidly decreasing and slowly varying terms. In this case the application (1.60) to the slowly varying terms preserves the form of the expansion (1.29). Therefore, the results on the accuracy of the extrapolation for the slowly varying terms of the solution remain true. Some modifications of the implicit method of rectangles and a comparison of it with other methods are described in [96].

2.2. Explicit Difference Schemes

The present section is concerned with the initial-value problem for an ordinary nonlinear differential equation of the first order. Obtaining approximate solutions of nonlinear equations is a very important problem. It is usually solved using the Runge–Kutta algorithm, or by linear multistep methods. It is also possible to solve nonlinear equations to a higher degree of accuracy by extrapolating simple difference solutions with respect to a mesh-size. We will concentrate on this latter method.

2.2.1. The Euler Method

In the Euler method explicit approximations of equations are constructed. This method occupies a special place among the many numerical methods for solving nonlinear problems. In spite of only first-order accuracy this method is attractive because of its extremely simple form.

Consider again the equation

$$\frac{du}{dt} = f(t, u), \qquad t \in (0, 1). \tag{2.1}$$

with initial condition

$$u(0) = u_0, \tag{2.2}$$

where $f(t, u)$ is a real function, and

$$f \in C^r([0, 1] \times (-\infty, \infty)), \qquad r \geq 2. \tag{2.3}$$

We assume that a solution exists and is unique, and that

$$u \in C^{r+1}[0, 1]. \tag{2.4}$$

To solve (2.1), (2.2) numerically, we use the Euler method on a regular net of mesh-size $\tau = 1/M$:

$$u_t^\tau = f(t, u^\tau), \qquad t \in \mathring{\omega}_\tau, \tag{2.5}$$

$$u^\tau(0) = u_0. \tag{2.6}$$

Since the resulting difference problem is nonlinear, a direct application of the general theorems of Chapter 1 is impossible. However, the steps we used to justify the greater accuracy of the extrapolated solution remain the same.

Assuming that the solution u of (2.1), (2.2) is known, we first form the system of equations†

$$\sum_{s=1}^{l+1} \frac{1}{s!} \frac{d^s v_{l-s+1}}{dt^s} = \sum_{s=1}^{l} \frac{1}{s!} \frac{\partial^s f}{\partial u^s}(t, u) \sum_{i_1 + \cdots + i_s = l} v_{i_1} \cdots v_{i_s}, \qquad t \in (0, 1), \tag{2.7}$$

with the initial conditions for the unknown functions $v_1, \ldots, v_{r-1}$

$$v_l(0) = 0, \qquad l = 1, \ldots, r - 1. \tag{2.8}$$

If we put $v_0 = u$, we can find the functions v_l iteratively in the index l.

Indeed, write out equation (2.7) for $l = 1$. We have

$$\frac{dv_1}{dt} - v_1 \frac{\partial f(t, u)}{\partial u} = -\frac{1}{2} \frac{d^2 v_0}{dt^2}, \qquad t \in (0, 1). \tag{2.9}$$

It is obvious that if $v_0 = u$ (2.9) is a linear equation with respect to the function v_1. Since the coefficients of (2.1) belong to $C^{r-1}[0, 1]$ by virtue of the linearity of the equation there is a unique solution $v_1 \in C^r[0, 1]$ with initial condition $v_1(0) = 0$. The subsequent functions v_l are determined in a similar way. Let the $(l - 1)$th function $v_{l-1} \in C^{r-l+2}[0, 1]$ be known. We rewrite (2.7) as

$$\frac{dv_l}{dt} - v_l \frac{\partial f}{\partial u}(t, u) = \sum_{s=2}^{l} \frac{1}{s!} \frac{\partial^s f}{\partial u^s}(t, u) \sum_{i_1 + \cdots + i_s = l} v_{i_1} \cdots v_{i_s}$$

$$- \sum_{s=2}^{l+1} \frac{1}{s!} \frac{d^s v_{l-s+1}}{dt^s}, \qquad t \in (0, 1). \tag{2.10}$$

† Here the indices $i_1, \ldots, i_s$ are positive integers and $\sum_{i_1 + \cdots + i_s = l}$ means summation over all combinations of indices whose sum is equal to l: if $s > l$, then the sum is taken to be zero.

It is easy to verify that the maximal index of the functions v_k on the right-hand side does not exceed $l - 1$ and each term is continuously differentiable on the interval $[0, 1]$, no less than $r - l$ times. Therefore the solution of the linear equation (2.10) under the initial condition $v_l(0) = 0$ exists, is unique and belongs to $C^{r-l+1}[0, 1]$. We now construct the net function

$$\eta^\tau = \tau^{-r}\left(u^\tau - \sum_{l=0}^{r-1} \tau^l v_l\right) \quad \text{on } \bar{\omega}_\tau. \tag{2.11}$$

Theorem 2.1. *If the solution of the difference problem* (2.5), (2.6) *satisfies* (2.3) *and* (2.4) *then the expansion*

$$u^\tau = u + \sum_{l=1}^{r-1} \tau^l v_l + \tau^r \eta^\tau \quad \text{on } \bar{\omega}_\tau \tag{2.12}$$

holds, where the functions v_l are found from (2.7), (2.8) *and are independent of τ. If the approximate solutions are uniformly bounded in τ†*

$$\|u^\tau\|_{C,\tau} \le c_2, \tag{2.13}$$

then the remainder η^τ is also bounded

$$\|\eta^\tau\|_{C,\tau} \le c_3. \tag{2.14}$$

PROOF. Let us replace the u^τ of equation (2.5) with (2.12). With $v_0 = u$ we have

$$\sum_{l=0}^{r-1} \tau^l (v_l)_t + \tau^r \eta_t^\tau = f\left(t, \sum_{l=0}^{r-1} \tau^l v_l + \tau^r \eta^\tau\right).$$

We use Lemma 1.1 of §7.1 to rewrite the left-hand side and use the Taylor formula for the second argument of f to rewrite the right-hand side:

$$\sum_{l=0}^{r-1} \tau^l \left(\sum_{s=0}^{r-l-1} \tau^s \frac{1}{(s+1)!} \frac{d^{s+1} v_l}{dt^{s+1}} + \tau^{r-l} \rho_{r-l}^\tau\right) + \tau^r \eta_t^\tau = f\left(t, \sum_{l=0}^{r-1} \tau^l v_l\right) + \tau^r \eta^\tau \sigma^\tau. \tag{2.15}$$

Here

$$\sigma^\tau = \frac{\partial f}{\partial u}(t, \xi^\tau),$$

where ξ^τ is a certain point in the interval with endpoints

$$\sum_{l=0}^{r-1} \tau^l v_l, \quad u^\tau.$$

† One condition which will give (2.13) (from [11]) is:

$$\max_{t \in [0, 1]} \sup_{u \in (-\infty, \infty)} \left| \frac{\partial f}{\partial u}(t, u) \right| \le c_1.$$

Note that due to the continuity of

$$\sum_{l=0}^{r-1} \tau^l v_l$$

on $[0, 1]$ and the uniform boundedness of the net function u^τ on $\bar{\omega}_\tau$ all the values ξ^τ belong to some finite interval $[-c_4, c_4]$ for any $\tau > 0$ and $t \in \mathring{\omega}_\tau$. Since the derivative $(\partial f / \partial u)(t, u)$ is continuous on the rectangle $[0, 1] \times [-c_4, c_4]$ it is bounded and

$$|\sigma^\tau| \leq c_5 \quad \text{on } \mathring{\omega}_\tau. \tag{2.16}$$

Due to the smoothness of the function v_l, the coefficients ρ_{r-l}^τ in the remainder terms are also bounded:

$$\sum_{l=0}^{r-1} |\rho_{r-l}^\tau| \leq c_6 \quad \text{on } \mathring{\omega}_\tau. \tag{2.17}$$

Now we change the order of summation on the left-hand side of (2.15) and again apply the Taylor formula to the right-hand side

$$\sum_{l=0}^{r-1} \tau^l \sum_{s=1}^{l+1} \frac{1}{s!} \frac{d^s v_{l-s+1}}{dt^s} + \tau^r \sum_{l=0}^{r-1} \rho_{r-l}^\tau + \tau^r \eta_t^\tau$$

$$= f(t, u) + \sum_{l=1}^{r-1} \left\{ \sum_{s=1}^{r-1} \tau^s v_s \right\}^l \frac{1}{l!} \frac{\partial^l f}{\partial u^l} (t, u) + \tau^r g^\tau + \tau^r \eta_t^\tau \sigma^\tau. \tag{2.18}$$

Here

$$g^\tau = \frac{1}{r!} \left\{ \sum_{s=1}^{r-1} \tau^{s-1} v_s \right\}^r \frac{\partial^r f}{\partial u^r} (t, \zeta^\tau),$$

where ζ^τ belongs to the interval with endpoints

$$u, \quad \sum_{l=0}^{r-1} \tau^l v_l.$$

From the continuity of the functions v_l on $[0, 1]$ it follows that they are bounded. Therefore ζ^τ is also uniformly bounded

$$|\zeta^\tau| \leq c_7 \quad \text{on } \mathring{\omega}_\tau. \tag{2.19}$$

The continuity of the function

$$\frac{\partial^r f}{\partial u^r} (t, u)$$

on the rectangle $[0, 1] \times [-c_7, c_7]$ assures the boundedness of the function g^τ:

$$|g^\tau| \leq c_8 \quad \text{on } \mathring{\omega}_\tau. \tag{2.20}$$

We transform the double sum on the right-hand side of (2.18) as follows: we expand the expression in braces and group like terms in powers of τ. This yields

$$\sum_{l=1}^{(r-1)^2} \tau^l \sum_{s=1}^{r-1} \frac{1}{s!} \frac{\partial^s f}{\partial u^s}(t, u) \sum_{i_1 + \cdots + i_s = l} v_{i_1} \cdots v_{i_s}.$$

We leave the terms of order τ^l, $l < r$, unchanged and write the other terms as a single term:

$$\tau^r b^\tau = \sum_{l=r}^{(r-1)^2} \tau^l \sum_{s=1}^{r-1} \frac{1}{s!} \frac{\partial^s f}{\partial u^s} \sum_{i_1 + \cdots + i_s = l} v_{i_1} \cdots v_{i_s}.$$

Divide both sides of the equality by τ^r and take the absolute values, keeping in mind that $\tau \le 1$. This yields

$$|b^\tau| \le \sum_{l=r}^{(r-1)^2} \sum_{s=1}^{r-1} \frac{1}{s!} \left| \frac{\partial^s f}{\partial u^s} \right| \sum_{i_1 + \cdots + i_s = l} v_{i_1} \cdots v_{i_s}.$$

Taking the functions $u(t)$, $v_i(t)$, $(\partial^s f/\partial u^s)(t, u(t))$ to be independent of τ and uniformly bounded on $[0, 1]$, we rewrite the inequality in the form

$$|b^\tau| \le c_9 \quad \text{on } \mathring{\omega}_\tau. \tag{2.21}$$

Thus the right-hand side of (2.18) reduces to

$$f(t, u) + \sum_{l=1}^{r-1} \tau^l \sum_{s=1}^{l} \frac{1}{s!} \frac{\partial^s f}{\partial u^s}(t, u) \sum_{i_1 + \cdots + i_s = l} v_{i_1} \cdots v_{i_s} + \tau^r b^\tau + \tau^r g^\tau + \tau^r \eta^\tau \sigma^\tau.$$

Using (2.7) and the results of rewriting both sides of (2.18) we arrive at

$$\tau^r \sum_{l=0}^{r-1} \rho_{r-l}^\tau + \tau^r \eta_t^\tau = \tau^r b^\tau + \tau^r g^\tau + \tau^r \eta^\tau \sigma^\tau \quad \text{on } \mathring{\omega}_\tau$$

or

$$\eta_t^\tau - \sigma^\tau \eta^\tau = b^\tau + g^\tau - \sum_{l=0}^{r-1} \rho_{r-l}^\tau \quad \text{on } \mathring{\omega}_\tau. \tag{2.22}$$

From this relation, with the help of (2.16), (2.17), (2.20), and (2.21) we obtain

$$|\eta^\tau(t + \tau)| \le |\eta^\tau(t)|(1 + \tau c_5) + \tau(c_6 + c_8 + c_9). \tag{2.23}$$

Now we exploit the following result, which can be easily established by induction (see, for example, [119]). $\qquad \square$

Lemma 2.2. *Suppose the net function ξ is defined on $\bar{\omega}_\tau$ and satisfies the inequality*

$$|\xi(t + \tau)| \le |\xi(t)|(1 + \tau \delta) + \tau B, \qquad t \in \mathring{\omega}_\tau,$$

where $B \ge 0$ and $\delta > 0$. Then

$$|\xi(t)| \le e^{\delta t} |\xi(0)| + (e^{\delta t} - 1) B/\delta, \qquad t \in \bar{\omega}_\tau.$$

The initial value $\eta^\tau(0)$ can be obtained from the definition of η^τ and (2.2), (2.6), (2.8):

$$\eta^\tau(0) = \tau^{-r}\left(u^\tau(0) - \sum_{l=0}^{r-1} \tau^l v_l(0)\right) = 0.$$

Therefore from (2.23) and Lemma 2.2 we get the estimate

$$|\eta^\tau(t)| \le (e^{c_5 t} - 1)(c_6 + c_8 + c_9)/c_5, \qquad t \in \bar{\omega}_\tau,$$

with $t \le 1$ and $c_3 = (e^{c_5} - 1)(c_6 + c_8 + c_9)/c_5$ we have

$$\|\eta^\tau\|_{c,\tau} \le c_3.$$

The theorem is proved.   □

By evaluating the terms in (2.22) one can prove the boundedness of the separated difference.

Corollary. *If the conditions of Theorem 2.1 are satisfied, then the following estimate*

$$|\eta^\tau_t(t)| \le e^{c_5}(c_6 + c_8 + c_9), \qquad t \in \mathring{\omega}_\tau, \tag{2.24}$$

is valid.

We use (2.12) to make our approximate solutions more accurate.

Assume conditions (2.3) and (2.4) are satisfied. For fixed integers $N_r > \cdots > N_1 > 0$ we construct nets $\bar{\omega}_{\tau_k}$ with mesh-sizes $\tau_k = 1/(N_k M)$ where the natural number M can increase indefinitely. We solve the difference problem (2.5), (2.6) on each net $\bar{\omega}_{\tau_k}$. All solutions u^{τ_k} are defined on the net $\bar{\omega}_\tau$ with mesh-size $\tau = 1/M$.

Consider the system

$$\sum_{k=1}^{r} \gamma_k = 1,$$

$$\sum_{k=1}^{r} \gamma_k \tau_k^j = 0, \qquad j = 1, \ldots, r - 1. \tag{2.25}$$

Since the τ_k are disjoint the determinant of this system is not zero, and so there is a unique solution $\gamma_1, \ldots, \gamma_k$. We form the linear combination with these weights

$$U^H = \sum_{k=1}^{r} \gamma_k u^{\tau_k} \quad \text{on } \bar{\omega}_\tau. \tag{2.26}$$

In spite of the fact that the accuracy of each solution u^{τ_k} is of order τ, the solution U^H has accuracy of order τ^r.

Theorem 2.3. *Assume conditions* (2.3) *and* (2.4) *are satisfied with integer* $r \geq 1$ *for* (2.1), (2.2). *Then the corrected solution* (2.26), *with the weights from* (2.25) *obeys the estimate*

$$\| U^H - u \|_{C, \tau} \leq c_{10} \tau^r. \tag{2.27}$$

PROOF. From Theorem 2.1 we can expand

$$u^{\tau k} = u + \sum_{l=1}^{r-1} \tau_k^l v_l + \tau_k^r \eta^{\tau k} \tag{2.28}$$

at each knot of $\bar{\omega}_\tau$. Adding these up with weights γ_k we have

$$U^H = \sum_{k=1}^{r} \gamma_k u^{\tau k} = \sum_{k=1}^{r} \gamma_k u + \sum_{l=1}^{r-1} \sum_{k=1}^{r} \gamma_k \tau_k^l v_l + \sum_{k=1}^{r} \gamma_k \tau_k^r \eta^{\tau k}.$$

The right-hand side can be simplified because the functions u and v_l are independent of k and the weights γ_k satisfy (2.25). As a result we have

$$U^H = u + \sum_{k=1}^{r} \gamma_k \tau_k^r \eta^{\tau k} \quad \text{on } \bar{\omega}_\tau. \tag{2.29}$$

The estimate for $|\eta^{\tau k}|$ follows from (2.14) and the estimate for $|\gamma_k|$ of the form

$$|\gamma_k| \leq (1 + 1/c_{11})^r, \qquad c_{11} = \min_{1 \leq k \leq r-1} (N_{k+1}/N_k) - 1$$

follows from Lemma 2.3 of §7.2. Therefore from (2.29) we have

$$|U^H - u| \leq c_3 (1 + 1/c_{11})^r \sum_{k=1}^{r} \tau^r N_k^{-r}$$

$$\leq r c_3 (1 + c_{11})^r N_1^{-r} c_{11}^{-r} \tau^r \quad \text{on } \bar{\omega}_\tau.$$

Writing

$$c_{10} = r c_3 (1 + 1/c_{11})^r N_1^{-r}$$

we get (2.27). The theorem is proved. $\qquad\qquad\qquad\qquad\qquad\qquad\qquad\square$

By applying Lagrange interpolation one can construct an algorithm for increasing the accuracy at points which are not common to all nets. This method is discussed at length in the previous section.

2.2.2. An Explicit Scheme Using Central Differences

Note that in the Euler scheme the expansion of the approximate solution in powers of τ contains odd powers. To eliminate each term in the regular part of this expansion we have to solve an auxiliary difference problem of the form (2.5), (2.6). Therefore, we should look for an explicit method for expanding the approximate solution in powers of τ which would contain only even powers in the regular part.

We now consider a scheme using central differences, often referred to as the method of central rectangles:

$$u^{\tau}_{\hat{t}} = f(t, u^{\tau}), \qquad t \in \omega_{\tau}, \tag{2.30}$$

where

$$u^{\tau}_{\hat{t}}(t) = \{u^{\tau}(t + \tau) - u^{\tau}(t - \tau)\}/(2\tau).$$

For each t, equation (2.30) contains values of the function u^{τ} at three knots; therefore, the explicit computation of successive values of u^{τ} requires two initial conditions. One of them follows from (2.2):

$$u^{\tau}(0) = u_0. \tag{2.31}$$

The other initial condition must be such that the expansion of the approximate solution u^{τ} will contain only even powers of τ. Such a condition has been formulated by Gragg (see [47]):

$$u^{\tau}(\tau) = u_0 + \tau f(0, u_0). \tag{2.32}$$

The initial conditions (2.31), (2.32) and the equation (2.30) for $t = \tau$, $2\tau, \ldots, 1$ allow us to find the values of $u^{\tau}(t)$ for $t = 0, \tau, \ldots, 1 + \tau$. Note that the two initial values could be taken at $-\tau, 0$.

Moreover, if we assume

$$u^{\tau}(-\tau) = u_0 - \tau f(0, u_0), \qquad u^{\tau}(0) = u_0, \tag{2.33}$$

then from (2.30) we obtain the same difference solution as previously, because from (2.33) and (2.30) for $t = 0$ we have equation (2.32):

$$u^{\tau}(\tau) = u^{\tau}(-\tau) + 2\tau f(0, u^{\tau}(0)) = u_0 + \tau f(0, u_0).$$

Assume that for (2.1), (2.2) the conditions (2.3), (2.4) with integer $r \geq 0$ are satisfied. Suppose $s = [(r - 1)/2]$. Assuming that the solution u is known we construct the linear system of equations†

$$
\begin{aligned}
v'_l - \frac{\partial f}{\partial u}(t, u)w_l &= -\frac{u^{(2l+1)}}{(2l+1)!} - \sum_{k=1}^{l-1} \frac{v^{(2k+1)}_{l-k}}{(2k+1)!} \\
&\quad + \sum_{k=2}^{l} \frac{1}{k!} \frac{\partial^k f}{\partial u^k}(t, u) \sum_{i_1 + \cdots + i_k = l} w_{i_1} \cdots w_{i_k}, \\
w'_l - \frac{\partial f}{\partial u}(t, u)v_l &= -\frac{u^{(2l+1)}}{(2l+1)!} - \sum_{k=1}^{l-1} \frac{w^{(2k+1)}_{l-k}}{(2k+1)!} \\
&\quad + \sum_{k=2}^{l} \frac{1}{k!} \frac{\partial^k f}{\partial u^k}(t, u) \sum_{i_1 + \cdots + i_k = l} v_{i_1} \cdots v_{i_k}, \qquad t \in (0, 1), \quad l = 1, 2, \ldots, s.
\end{aligned}
\tag{2.34}
$$

† We recall that $\sum_{i_1 + \cdots + i_k = l}$ means summation over all possible sets $(i_1, \ldots, i_k)$ of natural numbers for which $i_1 + \cdots + i_k = l$.

The initial conditions are

$$v_l(0) = -\frac{u^{(2l)}(0)}{(2l)!} - \sum_{k=1}^{l-1} \frac{v_{l-k}^{(2k)}(0)}{(2k)!}, \qquad l = 1, 2, \ldots, s. \qquad (2.35)$$

$$w_l(0) = 0. \qquad (2.36)$$

From this system of equations we can find v_l, w_l, iteratively. For example, for $l = 1$ we have

$$v_1' - \frac{\partial f}{\partial u}(t, u)w_1 = -u'''/6,$$

$$w_1' - \frac{\partial f}{\partial u}(t, u)v_1 = -u'''/6, \qquad t \in (0, 1), \qquad (2.37)$$

with initial conditions

$$v_1(0) = -u''(0)/2, \qquad w_1(0) = 0. \qquad (2.38)$$

Since both the coefficients and the right-hand side of (2.37) belong to class $C^{r-2}[0, 1]$, then the linearity of the equations shows that there is a unique pair of functions $v_1, w_1 \in C^{r-1}[0, 1]$, which satisfy (2.37), (2.38). The subsequent functions are defined in the same way. Assume the functions v_k, $w_k \in C^{r+1-2k}[0, 1]$ for $k = 1, \ldots, l-1$ have been found. Then it is easy to see that on the right-hand side of (2.34) and (2.36) there are no functions v_k, w_k with indices greater than $l-1$. All terms on the right-hand side of (2.34) are continuously differentiable not less than $(r - 2l)$ times as well. Therefore, there is a unique solution

$$v_l, w_l \in C^{r-2l+1}[0, 1].$$

We now continue the functions u, v_l, w_l a distance τ beyond $[0, 1]$ preserving the smoothness. This can be done using a Taylor-series of appropriate length. For example,

$$u(t) = \sum_{k=0}^{r+1} \frac{t^k}{k!} u^{(k)}(0) \qquad \forall\, t \in [-\tau, 0],$$

$$u(t) = \sum_{k=0}^{r+1} \frac{(t-1)^k}{k!} u^{(k)}(1) \qquad \forall\, t \in [1, 1+\tau].$$

Theorem 2.4. *If the requirements* (2.3), (2.4) *for difference system* (2.33), (2.30) *are satisfied then the expansions*

$$u^\tau = u + \sum_{l=1}^{s} \tau^{2l} v_l + \eta^\tau, \qquad t = -\tau, \tau, 3\tau, \ldots, \qquad (2.39)$$

$$u^\tau = u + \sum_{l=1}^{s} \tau^{2l} w_l + \eta^\tau, \qquad t = 0, 2\tau, 4\tau, \ldots \qquad (2.40)$$

are valid. Here $s = [(r - 1)/2]$; *the functions* v_l, w_l *are defined by* (2.34)–(2.36) *and are independent of* τ. *If the approximate solutions are bounded uniformly in* τ:

$$\|u^\tau\|_{C,\tau} \le c_{12} \tag{2.41}$$

then the remainder η^τ obeys

$$\|\eta^\tau\|_{C,\tau} \le c_{13}\tau^r. \tag{2.42}$$

PROOF. Since the functions u^τ, u, v_l, w_l are given the expansions (2.39), (2.40) unambiguously define a net function η^τ at $-\tau, 0, \tau, \ldots, 1 + \tau$. We thus have to prove estimate (2.42). With this in mind we substitute the expansions (2.39), (2.40) in (2.30). We first consider the case where t/τ is even, where $t \in \overline{\omega}_\tau$. We have

$$u_{\bar{t}\bar{t}} + \sum_{l=1}^{s} \tau^{2l}(v_l)_{\bar{t}\bar{t}} + \eta^\tau_{\bar{t}\bar{t}} = f\left(t, u + \sum_{l=1}^{s} \tau^{2l}w_l + \eta^\tau\right).$$

We rewrite the left-hand side of this equality according to Lemma 1.1 of §7.1. The right-hand side we expand by the Taylor formula

$$\sum_{k=0}^{s} \frac{\tau^{2k}}{(2k+1)!} u^{(2k+1)} + \tau^r\rho_0 + \sum_{l=1}^{s} \tau^{2l}\left\{\sum_{k=0}^{s-l} \frac{\tau^{2k}}{(2k+1)!} v_l^{(2k+1)}\right.$$

$$\left. + \tau^{r-2l}\rho_l\right\} + \eta^\tau_{\bar{t}\bar{t}} = f\left(t, u + \sum_{l=1}^{s} \tau^{2l}w_l\right) + \eta^\tau\sigma^\tau. \tag{2.43}$$

Here $\sigma^\tau = (\partial f/\partial u)(t, \xi)$, where ξ belongs to the interval with endpoints

$$u^\tau, \quad u + \sum_{l=1}^{s} \tau^{2l}w_l.$$

Keeping in mind the smoothness of the functions u, w_l, v_l, f, we obtain

$$|\sigma^\tau| \le c_{14}, \qquad \sum_{l=0}^{s} |\rho_l| \le c_{15} \quad \forall\, t = 0, 2\tau, 4\tau, \ldots \tag{2.44}$$

Now we change the order of summation on the left-hand side of (2.43); on the right-hand side we again use the Taylor formula:

$$u' + \sum_{l=1}^{s} \tau^{2l}\left\{\frac{u^{(2l+1)}}{(2l+1)!} + \sum_{k=0}^{l-1} \frac{v_{l-k}^{(2k+1)}}{(2k+1)!}\right\} + \tau^r\sum_{l=0}^{s} \rho_l + \eta^\tau_{\bar{t}\bar{t}}$$

$$= f(t, u) + \sum_{k=1}^{s} \frac{1}{k!}\left(\sum_{l=1}^{s} \tau^{2l}w_l\right)^k \frac{\partial^k f}{\partial u^k}(t, u) + \xi_1^\tau + \eta^\tau\sigma^\tau. \tag{2.45}$$

For ξ_1^τ the estimate

$$|\xi_1^\tau| \le c_{16}\tau^r \tag{2.46}$$

holds. This estimate follows from the boundedness of the functions w_l and the continuity of $\partial^{s+1}f/\partial u^{s+1}$. In the double sum on the right-hand side of

(2.45). We expand the expression in brackets and cancel like terms in τ. As a result we get

$$\sum_{l=1}^{s^2} \tau^{2l} \sum_{k=1}^{s} \frac{1}{k!} \frac{\partial^k f}{\partial u^k}(t, u) \sum_{i_1 + \cdots + i_k = l} w_{i_1} \cdots w_{i_k}.$$

We leave the terms of orders $\tau^2, \ldots, \tau^{2s}$ unchanged, and rewrite the rest in the form

$$\xi_2^\tau = \sum_{l=s+1}^{s^2} \tau^{2l} \sum_{k=1}^{s} \frac{1}{k!} \frac{\partial^k f}{\partial u^k}(t, u) \sum_{i_1 + \cdots + i_k = l} w_{i_1} \cdots w_{i_k}.$$

It is clear that

$$|\xi_2^\tau| \leq \tau^r c_{17}. \tag{2.47}$$

Thus (2.45) is brought to the form

$$u' + \sum_{l=1}^{s} \tau^{2l} \left\{ \frac{u^{(2l+1)}}{(2l+1)!} + \sum_{k=0}^{l-1} \frac{v_{l-k}^{(2k+1)}}{(2k+1)!} \right\} + \tau^r \sum_{l=0}^{s} \rho_l + \eta_{\bar{t}\bar{t}}^\tau$$

$$= f(t, u) + \sum_{l=1}^{s} \tau^{2l} \sum_{k=1}^{l} \frac{1}{k!} \frac{\partial^k f}{\partial u^k}(t, u) \sum_{i_1 + \cdots + i_k = l} w_{i_1} \cdots w_{i_k}$$

$$+ \xi_1^\tau + \xi_2^\tau + \sigma^\tau \eta^\tau.$$

Using equations (2.1) and (2.34) we cancel terms and get

$$\eta_{\bar{t}\bar{t}}^\tau - \sigma^\tau \eta^\tau = \xi_3^\tau \quad \forall\, t = 0, 2\tau, 4\tau, \ldots. \tag{2.48}$$

where the right-hand side, from (2.44), (2.46), and (2.47) satisfies

$$|\xi_3^\tau| \leq c_{18} \tau^r, \qquad c_{18} = c_{15} + c_{16} + c_{17}, \quad \forall\, t = 0, 2\tau, 4\tau, \ldots \tag{2.49}$$

In the same way we obtain (2.48) and (2.49) for $t = \tau, 3\tau, 5\tau, \ldots$. Therefore (2.48) and (2.49) are fulfilled for all $t \in \bar{\omega}_\tau$.

Now let us consider the initial conditions for (2.48). For $t = 0$

$$\eta^\tau(0) = u^\tau(0) - u(0) - \sum_{l=1}^{s} \tau^{2l} w_l(0).$$

Using the initial conditions (2.2), (2.31), (2.36) we come to

$$\eta^\tau(0) = 0. \tag{2.50}$$

For $t = \pm\tau$ we have the expansions

$$\eta^\tau = u^\tau - u - \sum_{l=1}^{s} \tau^{2l} v_l.$$

Let us take their average; using (2.32) and (2.33) we have

$$\{\eta^\tau(\tau) + \eta^\tau(-\tau)\}/2 = u_0 - \{u(\tau) + u(-\tau)\}/2 - \sum_{l=1}^{s} \tau^{2l} \{v_l(\tau) + v_l(-\tau)\}/2.$$

Utilizing Lemma 1.1 of §7.1, we rewrite the right-hand side:

$$\{\eta^\tau(\tau) + \eta^\tau(-\tau)\}/2 = u_0 - \sum_{l=0}^{s} \tau^{2l} \frac{u^{(2l)}(0)}{(2l)!}$$

$$+ \sum_{l=1}^{s} \tau^{2l} \sum_{k=0}^{s-l} \frac{\tau^{2k}}{(2k)!} v_l^{(2k)}(0) + \tau^r \zeta_4^\tau$$

$$= - \sum_{l=1}^{s} \tau^{2l} \left\{ \frac{u^{(2l)}(0)}{(2l)!} + \sum_{k=0}^{l-1} \frac{v_{l-k}^{(2k)}(0)}{(2k)!} \right\} + \tau^r \zeta_4^\tau.$$

Here the constant ζ_4^τ satisfies

$$|\zeta_4^\tau| \leq c_{19}. \tag{2.51}$$

Taking the initial conditions (2.35) into account we have

$$\{\eta^\tau(\tau) + \eta^\tau(-\tau)\}/2 = \tau^r \zeta_4^\tau. \tag{2.52}$$

From (2.48) for $t = 0$ it follows that

$$\{\eta^\tau(\tau) - \eta^\tau(-\tau)\}/(2\tau) = \sigma^\tau(0)\eta^\tau(0) + \tau^r \zeta_3^\tau(0).$$

Multiply this by τ or $-\tau$ and add this to (2.52). Using (2.50) we get two estimates

$$\eta^\tau(\pm\tau) = \tau^r \zeta_5^\tau, \quad \text{where } |\zeta_5^\tau| \leq c_{20}. \tag{2.53}$$

To estimate other values of η^τ we will use the following result.

Lemma 2.5. *Suppose the net function ξ satisfies the inequality*

$$|\xi(t + 2\tau)| \leq |\xi(t)| + |\xi(t + \tau)|\tau\delta + \tau B \quad \forall\, t = 0, \tau, 2\tau, \ldots,$$

where $B \geq 0, \delta \geq 0$. Then

$$|\xi(t)| \leq e^{\delta t}(|\xi(0)| + |\xi(\tau)| + tB). \tag{2.54}$$

PROOF. Estimate (2.54) is obvious for $t = 0, \tau$. We assume that it holds for some $t, t + \tau$ and check it for $t + 2\tau$. We have

$$|\xi(t + 2\tau)| \leq e^{\delta t}(|\xi(0)| + |\xi(\tau)| + tB)$$
$$+ e^{\delta(t+\tau)}(|\xi(0)| + |\xi(\tau)| + (t + \tau)B)\tau\delta + \tau B$$
$$\leq e^{\delta(t+2\tau)}(|\xi(0)| + |\xi(\tau)|) + (t + \tau)Be^{\delta(t+\tau)}(1 + \tau\delta)$$
$$\leq e^{\delta(t+2\tau)}(|\xi(0)| + |\xi(\tau)| + (t + 2\tau)B).$$

Since t is arbitrary the lemma is proved.

From the difference equation (2.48) we have

$$\eta^\tau(t + 2\tau) = \eta^\tau(t) + 2\tau\sigma^\tau(t + \tau)\eta^\tau(t + \tau) + 2\tau\zeta_3^\tau(t + \tau).$$

Using (2.44) and (2.49) we obtain

$$|\eta^\tau(t + 2\tau)| \leq |\eta^\tau(t)| + |\eta^\tau(t + \tau)|2\tau c_{14} + 2\tau^{r+1}c_{18}.$$

From Lemma 2.5 and estimate (2.53) we come to

$$|\eta^{\tau}(t)| \leq \tau^r e^{2c_{14}t}(c_{20} + 2tc_{18}) \quad \forall t \in \bar{\omega}_{\tau}.$$

Assuming $t = 1$ on the right-hand side we get (2.42). The proof of Theorem 2.4 is complete. □

Estimate (2.54) of Lemma 2.5 is rather simple but is not optimal. A more complicated proof halves the exponent (see [11]).

Let us return to the system of equations (2.34). For each l it can be transformed to two disjoint scalar equations.

To do this we consider the sum and the difference of equations (2.34). We have

$$z'_l - \frac{\partial f}{\partial u}(t, u)z_l = a_l, \qquad t \in [0, 1], \tag{2.55}$$

$$y'_l + \frac{\partial f}{\partial u}(t, u)y_l = b_l, \qquad t \in [0, 1], \tag{2.56}$$

where $z_l = v_l + w_l$, $y_l = v_l - w_l$ are new unknown functions and a_l, b_l are known functions which can be expressed in terms of $u, v_1, \ldots, v_{l-1}, w_1, \ldots, w_{l-1}$. We write the solutions of these equations in the following form

$$z_l(t) = e^{g(t)}\left\{z_l(0) + \int_0^t a_l(x)e^{-g(x)}\,dx\right\},$$
$$\tag{2.57}$$
$$y_l(t) = e^{-g(t)}\left\{y_l(0) + \int_0^t b_l(x)e^{g(x)}\,dx\right\},$$

where

$$g(x) = \int_0^x \frac{\partial f}{\partial u}(t, u(t))\,dt.$$

Now consider the case when $\partial f/\partial u$ has a negative value in the interval of integration. This case is the one which is most favorable for the stability of the initial-value problem (2.1), (2.2) since the effect of perturbations on the initial data and on the right-hand side decrease exponentially in t. At the same time from (2.57) we see that z_l grows slower than a_l and that y_l grows faster than b_l. Moreover, y_l tends to grow exponentially even if b_l is a bounded function. Since $v_l = (z_l + y_l)/2$, $w_l = (z_l - y_l)/2$ both functions grow rapidly in t. These functions are not important since they can be eliminated from (2.39), (2.40). But the remainder η^{τ} depends strongly on the value of the derivatives of these functions, which grow rapidly in t.

Generally speaking, the sum $z_l = v_l + w_l$ will grow more slowly than its term v_l, w_l, Therefore, instead of (2.39), (2.40) we consider an expansion in their sum. Since they are defined at different knots it is necessary to interpolate.

As a result the following smoothing is obtained [47]:

$$\tilde{u}^\tau(1) = \tfrac{1}{4}u^\tau(1 - \tau) + \tfrac{1}{2}u^\tau(1) + \tfrac{1}{4}u^\tau(1 + \tau) = u_{if}^\tau(1). \qquad (2.58)$$

Here $\tilde{u}^\tau(1)$ is a new "smoothed" value at the endpoint of the interval of integration. The form (2.39), (2.40) remains valid for this new value.

Theorem 2.6. *Assume the conditions of Theorem 2.4 hold. Then the " smoothed" value $\tilde{u}^\tau(1)$ has the expansion*

$$\tilde{u}^\tau(1) = u(1) + \sum_{l=1}^{s} \tau^{2l} g_l + O(\tau^r), \qquad (2.59)$$

where $s = [(r - 1)/2]$, and the constants g_l are independent of τ.

PROOF. Assume $1/\tau$ is an even number. Then for according to Theorem 2.4. $t = 1 \pm h$ can be expanded

$$u^\tau(t) = u(t) + \sum_{l=1}^{s} \tau^{2l} v_l(t) + \eta^\tau(t).$$

Sum these with weights $\tfrac{1}{2}$ and use Lemma 1.1 of §7.1; we have

$$\{u^\tau(1 + \tau) + u^\tau(1 - \tau)\}/2 = \sum_{l=0}^{s} \frac{\tau^{2l}}{(2l)!} u^{(2l)}(1)$$

$$+ \sum_{l=1}^{s} \tau^{2l} \sum_{j=0}^{s-l} \frac{\tau^{2j}}{(2j)!} v_l^{(2j)}(1) + O(\tau^r).$$

Add (2.40) to this and halve the sum:

$$u_{if}^\tau(1) = u(1) + \sum_{l=1}^{s} \tau^{2l} \left\{ \frac{u^{(2l)}(1)}{2(2l)!} + w_l(1)/2 + \sum_{j=0}^{l-1} \frac{v_{l-j}^{(2j)}(1)}{2(2j)!} \right\} + O(\tau^r).$$

Denoting the corresponding expression by g_l we get (2.59). The theorem is proved.   □

It is worthwhile to note that the problem of growth of the coefficients is only partially solved by this smoothing. Therefore, the scheme (2.30), (2.31), (2.58) taken together with one of the extrapolation methods in the limit is the most successful combination for a small interval of integration. Suppose it is necessary to solve (2.1), (2.2) on a large interval $[0, T]$. Then it is natural to divide the interval into several parts

$$[0, T] = [0, x_1] \cup [x_1, x_2] \cup \cdots \cup [x_{p-1}, x_p],$$

where $0 < x_1 < \cdots < x_p = T$. Instead of a single problem (2.1), (2.2) we consider the sequence of the problems

$$\begin{cases} u_1' = f(t, u_1) & \text{on } [0, x_1], \\ u_1(0) = u_0, \end{cases} \qquad (2.60)$$

$$\begin{cases} u_i' = f(t, u_i) & \text{on } [x_{i-1}, x_i], \\ u_i(x_{i-1}) = u_{i-1}(x_{i-1}), & i = 2, 3, \ldots, p. \end{cases} \qquad (2.61)$$

It is apparent that the exact solutions of these problems coincide with the function u on the appropriate intervals. Each of these problems can be solved with the help of (2.30), (2.31), (2.58), together with one of the extrapolation methods.

Instead of an exact initial value in (2.61) one should take an approximate value $u_{i-1}^\tau(x_{i-1})$ obtained from a numerical solution of the previous problem.

Bulirsh and Stoer have suggested that one choose the endpoints of the intervals of integration in the course of calculation, taking into account the required accuracy and the order of extrapolation. In the algorithms they suggest a rational extrapolation is used. These algorithms give good results for problems with smooth data when great accuracy is required. We refer the reader to [24, 25, 96].

2.2.3. Expansions for the Simplest Quadratures

In the present section we will construct expansions in even powers of the net parameter for two simple quadrature rules, namely, for the central rectangle rule and the trapezoidal rule. Usually these expansions involve the Bernoulli numbers [67, 102]. We will derive them using general results from §1.2. To do this we reduce the study of quadrature rules to the study of a difference scheme for an ordinary differential equation of the first order.

Indeed, let us consider the problem

$$y' = f \quad \text{on } [0, 1], \qquad y(0) = 0. \tag{2.62}$$

It is clear that

$$y(t) = \int_0^t f(t)\, dt.$$

To obtain a numerical solution to (2.62) we consider the Crank–Nicholson scheme

$$y_{\hat{t}}^\tau = f \quad \text{on } \breve{\omega}_\tau, \tag{2.63}$$

$$y^\tau(0) = 0. \tag{2.64}$$

From this the values of the difference function y^τ can be found explicitly

$$y^\tau(t) = \sum_{\substack{x \in \breve{\omega}_\tau \\ x \le t}} f(x)\tau, \qquad t \in \bar{\omega}_\tau, \tag{2.65}$$

or equivalently

$$y^\tau(t_j) = \sum_{i=0}^{j-1} f(t_{i+1/2})\tau, \qquad j = 1, 2, \dots, M. \tag{2.66}$$

Hence we can see that we have obtained the central-rectangles rule. We will prove that the corresponding expansion relative in even powers of the mesh-size τ is valid.

Theorem 2.7. *Let $f \in C^m[0, 1]$. Then the central-rectangles rule* (2.65), *or* (2.66), *has the expansion*

$$y^\tau(t) = \int_0^t f(x)\, dx + \sum_{j=1}^s h^{2j} v_j(t) + \eta^\tau(t) \quad \forall\, t \in \bar{\omega}_\tau, \qquad (2.67)$$

where $s = [(m - 1)/2]$, $v_j \in C^{m-2j+1}[0, 1]$ *and are independent of* τ. *The remainder* η^τ *obeys the estimates*

$$\|\eta^\tau\|_{C,\tau} \leq c_{21}\tau^m, \qquad (2.68)$$

$$\max_{\tilde{\omega}_\tau} |\eta_i^\tau| \leq c_{22}\tau^m. \qquad (2.69)$$

PROOF. Let us verify that the conditions of theorem 2.2 of Chapter 1 are satisfied. Condition A is evidently satisfied, with classes

$$M_k(\Omega) = C^k[0, 1],\ P_k(\bar{\Omega}) = C^{k+1}[0, 1],\ N_k(D) = R.$$

Considering the remark at the end of §1.2 we will verify Condition B′ rather than B. In (2.66) we take the absolute value of both sides and increase the range of summation on the right-hand side:

$$|y^\tau(t_j)| \leq \sum_{i=0}^{M-1} |f(t_{i+1/2})\tau| \leq \max_{\tilde{\omega}_\tau} |f|.$$

Hence

$$\|y^\tau\|_{C,\tau} \leq \max_{\tilde{\omega}_\tau} |f|. \qquad (2.70)$$

We now verify Condition D. The boundary-value operator l_h has an expansion of the form (2.21) of §1.2 with b_j, ρ^h taken to be zero. The difference operator of equation (2.63) has an expansion of the required form from Lemma 1.1 of Chapter 7. Therefore from Theorem 2.2 of §1.2 we see that the expansion (2.67) with the remainder bounded by (2.68) is valid. Now we verify Condition B′ for the norm $\max_{\tilde{\omega}_\tau} |y_i^\tau|$ in function space, with $y^\tau(0) = 0$. Functions satisfying this condition obey

$$\max_{\tilde{\omega}_\tau} |y_i^\tau| \leq \max_{\tilde{\omega}_\tau} |f|, \qquad (2.71)$$

from (2.63). Using this estimate we again obtain, by Theorem 2.2 of §1.2, the estimate (2.69). The theorem is proved. ☐

Now we obtain a similar expansion for the trapezoidal rule. We again reduce to the study of the Crank–Nickolson scheme for (2.62). To do this we replace the right-hand side of the difference equation by:

$$y_i^\tau = f_i \quad \text{on } \tilde{\omega}_\tau, \qquad (2.72)$$

$$y^\tau(0) = 0. \qquad (2.73)$$

From this y^τ can be found

$$y^\tau(x) = \sum_{\substack{t \in \bar\omega_\tau \\ t \le x}} f_i(t)\tau, \qquad x \in \bar\omega_\tau, \tag{2.74}$$

or

$$y^\tau(t_j) = \sum_{i=0}^{j-1} \{f(t_i) + f(t_{i+1})\}\tau/2, \qquad j = 1, 2, \ldots, M. \tag{2.75}$$

Theorem 2.8. *Let* $f \in C^m[0, 1]$. *Then the trapezoidal rule* (2.74), *or* (2.75), *has the expansion*

$$y^\tau(t) = \int_0^t f(x)\, dx + \sum_{j=1}^s h^{2j} w_j(t) + \mu^\tau(t) \quad \forall\, t \in \bar\omega_\tau, \tag{2.76}$$

where $s = [(m - 1)/2]$, w_j *are independent of* τ, $w_j \in C^{m-2j+1}[0, 1]$ *and the remainder satisfies*

$$\|\mu^\tau\|_{C,\tau} \le c_{23}\tau^m, \tag{2.77}$$

$$\max_{\bar\omega_\tau} |\mu_i^\tau| \le c_{24}\tau^m. \tag{2.78}$$

PROOF. We apply (1.2) of §7.1 to the right-hand side of (2.72):

$$f_i(t) = \sum_{i=0}^{[(m-1)/2]} \tau^{2i} \frac{1}{(2i)!\, 4^i} f^{(2i)}(t) + O(\tau^m).$$

Thus the solution y^τ of (2.72), (2.73) is the weighted average of the solutions of (2.63), (2.64) with right-hand sides $f^{(2i)}(t)$ and $O(\tau^m)$. If the right-hand side is equal to $f^{(2i)}$ we apply Theorem 2.7, while if it is $O(\tau^m)$ we use (2.70) and (2.71). As a result we obtain expansions of the form (2.76), with thé number of terms depending on the smoothness of $f^{(2i)}$. The sum of these expansions with the corresponding weights gives (2.76) with bounds (2.77), (2.78). The theorem is proved. □

The results allow us to apply method for accelerating convergence described in §1.3. As a result simple quadrature rule on two, three, and more nets augmented by Richardson extrapolation yield quadrature rules of higher order. For example, the trapezoidal rule on nets with mesh-sizes τ and $\tau/2$ results in Simpson's rule:

$$\int_t^{t+\tau} f(x)\, dx = \tfrac{4}{3}\{(f(t) + f(t + \tau/2))\tau/4$$

$$+ (f(t + \tau/2) + f(t + \tau))\tau/4\} - \tfrac{1}{3}\{f(t) + f(t + \tau)\}\tau/2$$

$$= \{f(t) + 4f(t + \tau/2) + f(t + \tau)\}\tau/6.$$

A review of the papers on constructing such new quadrature rules is presented in [53]. Formulas of the Newton–Cotes form are widely used to integrate functions given in tabular form on a net with a uniform mesh-size. But they are less efficient than Gaussian quadrature rules using special knots for functions defined analytically, whose values can be calculated at any point.

We will use the results of the present section to study integral equations.

2.3. The Splitting-Up Method for Systems of Equations

When one attempts to solve an initial-value problem for a system of linear differential equations it is sometimes useful to reduce the problem to a set of simpler initial-value problems. This can be done when the coefficient matrix of the system can be decomposed into a sum of simpler matrices with certain properties. In this case the reduction is carried out via the "splitting-up method" (see [87, 114, 146]). This method is effective when using a sequence of nets. As a result the splitting-up method will allow us to obtain solutions with greater accuracy.

Consider the system of linear differential equations

$$\frac{du}{dt} + Au = f, \qquad t \in (0, 1), \tag{3.1}$$

with initial condition

$$u(0) = u_0. \tag{3.2}$$

Here $A(t)$ is a square $m \times m$ matrix and $f(t)$, $u(t)$, u_0 are vectors with m components. Suppose that the elements of the matrix A and the vector f are functions in $C^r[0, 1]$. Then (see [26]) there is a unique solution of (3.1), (3.2), and the components of u belong to $C^{r+1}[0, 1]$.

Suppose the matrix A can be represented as a sum of $m \times m$ matrices with elements having the same smoothness:

$$A = \sum_{i=1}^{n} A_i, \tag{3.3}$$

where the $A_i(t)$ are nonnegative definite at every $t \in [0, 1]$:

$$(A_i(t)v, v) \geq 0 \quad \forall v \in E^m. \tag{3.4}$$

Here E^m is the m-dimensional vector space with the scalar product

$$(v, w) = \sum_{i=1}^{m} v_i w_i,$$

with

$$v = \begin{bmatrix} v_1 \\ \vdots \\ v_m \end{bmatrix}, \qquad w = \begin{bmatrix} w_1 \\ \vdots \\ w_m \end{bmatrix}$$

and with norm

$$\|v\| = (v, v)^{1/2}.$$

To solve (3.1), (3.2) numerically we will use the implicit splitting-up scheme (see [87]):

$$(I + \tau A_1(t))u^\tau(t - \tau(n - 1)/n) = u^\tau(t - \tau) + \tau f(t),$$
$$(I + \tau A_2(t))u^\tau(t - \tau(n - 2)/n) = u^\tau(t - \tau(n - 1)/n),$$
$$\dots\dots\dots\dots\dots\dots\dots\dots\dots\dots\dots\dots\dots\dots\dots\dots\dots \qquad (3.5)$$
$$(I + \tau A_n(t))u^\tau(t) = u^\tau(t - \tau/n), \qquad t \in \omega_\tau,$$

$$u^\tau(0) = u_0. \qquad (3.6)$$

Here I is the $m \times m$ identity matrix.

Note that at each step one has to solve a system of linear algebraic equations with matrix $I + \tau A_i(t)$. Therefore the decomposition of the matrix A into its summands must be made in such a way that the solution of these systems does not present any difficulties.

Let us prove the stability of (3.5), (3.6).

Theorem 3.1. *The problem* (3.5), (3.6) *obeys the a priori estimate*

$$\max_{t \in \bar{\omega}_\tau} \|u^\tau(t)\| \leq \|u_0\| + \max_{t \in \omega_\tau} \|f(t)\|. \qquad (3.7)$$

PROOF. Take the scalar product of each of (3.5) with $u^\tau(t - \tau(n - 1)/n), \dots,$ $u^\tau(t)$, respectively. From the first equation we have

$$\|u^\tau(t - \tau(n - 1)/n)\|^2 + \tau(A_1(t)u^\tau(t - \tau(n - 1)/n),$$

$$u^\tau(t - \tau(n - 1)/n)) = (u^\tau(t - \tau), u^\tau(t - \tau(n - 1)/n))$$
$$+ \tau(f(t), u^\tau(t - \tau(n - 1)/n)).$$

Using the nonnegative definitions of the matrix A_1, and Cauchy–Schwartz–Bunyakovsky inequality we have

$$\|u^\tau(t - \tau(n - 1)/n)\|^2 \leq \|u^\tau(t - \tau)\| \|u^\tau(t - \tau(n - 1)/n)\|$$
$$+ \tau\|f(t)\| \|u^\tau(t - \tau(n - 1)/n)\|.$$

From this inequality it is easy to see

$$\|u^\tau(t - \tau(n - 1)/n)\| \leq \|u^\tau(t - \tau)\| + \tau\|f(t)\|. \qquad (3.8)$$

Other equations similarly lead to a chain of inequalities:

$$\|u^{\tau}(t - \tau(n - 2)/n)\| \leq \|u^{\tau}(t - \tau(n - 1)/n)\|,$$

$$\cdots\cdots\cdots\cdots\cdots\cdots\cdots\cdots\cdots\cdots\cdots\cdots\cdots\cdots\cdots\cdots$$

$$\|u^{\tau}(t)\| \leq \|u^{\tau}(t - \tau/n)\|.$$

Combining these with (3.8) we have

$$\|u^{\tau}(t)\| \leq \|u^{\tau}(t - \tau)\| + \tau\|f(t)\|.$$

This inequality together with $\|u^{\tau}(0)\| = \|u_0\|$, from the initial condition (3.6), leads to the estimate

$$\|u^{\tau}(t)\| \leq \|u_0\| + t \max_{t \in \omega_{\tau}} \|f(t)\|.$$

Setting t to 1 we come to (3.7). The theorem is proved.   □

Everthing that follows in this section is a generalization of the results of §1 to the vector case. The main steps of the proof remain unchanged. We will focus our attention on the differences between the vector and scalar cases.

Construct the set of the differential equations†

$$\frac{dv_l}{dt} + Av_l = -\sum_{s=2}^{\min(l+1,n)} \left(\sum_{i_1 < i_2 < \cdots < i_s} A_{i_1} \cdots A_{i_s} \right) v_{l-s+1}$$

$$+ \sum_{i=0}^{l-1} \frac{(-1)^{l-i+1}}{(l-i+1)!} v_i^{(l-i+1)} \quad \text{on } (0, 1) \tag{3.9}$$

and

$$v_l(0) = 0,‡ \qquad l = 1, \ldots, r. \tag{3.10}$$

From this set the vector-functions v_l can be determined iteratively if we assume $v_0 = u$.

Indeed, assume the k vector-functions v_i, $i = 0, \ldots, k - 1$, with components from $C^{r+1-i}[0, 1]$ have already been defined. Then the maximal index of the functions on the right-hand side of (3.9) is $k - 1$; therefore the vector-function v_k can really be found from (3.9) with uniform initial condition. One can readily see that the components of the right-hand side are $r - k$ times continuously differentiable on the interval $[0, 1]$. Therefore the components of the solution v_k belong to $C^{r-k+1}[0, 1]$. Thus all v_l can be found from (3.9), (3.10); their components will belong to the classes $C^{r+1-l}[0, 1]$.

---

† Sums with lower limits greater than their upper limits are assumed to be zero. $\sum_{i_1 < i_2 < \cdots < i_s}$ denotes summation over all possible (different) products consisting of s factors $A_{i_1} A_{i_2} \cdots A_{i_s}$ with indices satisfying the condition $1 \leq i_1 < i_2 < \cdots < i_s \leq n$.

‡ Here 0 is zero element of the space E^m.

Let us now show that the following representation

$$u^{\tau} = \sum_{j=0}^{r-1} \tau^j v_j + \tau^r \eta^{\tau} \quad \text{on } \bar{\omega}_{\tau} \tag{3.11}$$

is valid for (3.5), with the vector-function η^{τ} defined at the knots $\bar{\omega}_{\tau}$ such that

$$\max_{t \in \bar{\omega}_{\tau}} \|\eta^{\tau}(t)\| \le c_1. \tag{3.12}$$

Since the functions u^{τ} and v_j have already been defined, the definition of η^{τ} follows from (3.11):

$$\eta^{\tau} = u^{\tau} - \sum_{j=0}^{r-1} \tau^j v_j \quad \text{on } \bar{\omega}_{\tau}. \tag{3.13}$$

The only thing which is left to prove is (3.12). We rewrite system (3.5) omitting the intermediate values of the vector-function u^{τ}. Then we have

$$(I + \tau A_1(t)) \cdots (I + \tau A_n(t))u^{\tau}(t) = u^{\tau}(t - \tau) + \tau f(t), \qquad t \in \bar{\omega}_{\tau}. \tag{3.14}$$

Substituting the expansion for u^{τ} from (3.11) in the above we have

$$(I + \tau A_1(t)) \cdots (I + \tau A_n(t))\left(\sum_{j=0}^{r-1} \tau^j v_j(t) + \tau^r \eta^{\tau}(t)\right)$$

$$= \sum_{j=0}^{r-1} \tau^j v_j(t - \tau) + \tau^r \eta^{\tau}(t - \tau) + \tau f(t). \tag{3.15}$$

Rewrite the right-hand side of this relation. For each of the functions v_j we apply the Taylor formula. We have

$$v_j(t - \tau) = \sum_{i=0}^{r-j} \frac{(-\tau)^i}{i!} v_j^{(i)}(t) + \tau^{r-j+1} \sigma_j^{\tau}(t),$$

where

$$\|\sigma_j^{\tau}(t)\| \le \frac{\sqrt{m}}{(r - j + 1)!} \max_{1 \le k \le m} \max_{[0, 1]} |v_{j,i}^{(r-j+1)}| \quad \forall t \in \omega_{\tau}. \tag{3.16}$$

Here $v_{j,i}$ is the ith component of the vector-function v_j. Taking into account the formula for $v_j(t - \tau)$ we rewrite expression (3.15) omitting the argument t when there is no misunderstanding. We have

$$(I + \tau A_1) \cdots (I + \tau A_n)\left(\sum_{j=0}^{r-1} \tau^j v_j + \tau^r \eta^{\tau}\right)$$

$$= \sum_{j=0}^{r-1} \tau^j \sum_{i=0}^{r-j} \frac{(-\tau)^i}{i!} v_j^{(i)} + \tau^r \eta^{\tau}(t - \tau) + \tau f + \tau^{r+1} \sum_{j=0}^{r-1} \sigma_j^{\tau}.$$

Multiply out the expressions on the left-hand side of this equality and collect like terms in the same power of τ. We have

$$\sum_{s=1}^{n} \tau^s \sum_{i_1 < i_2 < \cdots < i_s} A_{i_1} A_{i_2} \cdots A_{i_s} \sum_{j=0}^{r-1} \tau^j v_j + \tau^r (I + \tau A_1) \cdots (I + \tau A_n) \eta^\tau$$

$$= \sum_{j=0}^{r} \tau^j \sum_{i=0}^{j-1} \frac{(-1)^{j-i}}{(j-i)!} v_i^{(j-i)} + \tau^r \eta^\tau(t - \tau) + \tau f + \tau^{r+1} \sum_{j=0}^{r-1} \sigma_j^\tau.$$

This relation can be easily transformed to

$$\sum_{j=1}^{r+n-1} \tau^j \sum_{s=1}^{\min(j,n)} \sum_{i_1 < i_2 < \cdots < i_s} A_{i_1} A_{i_2} \cdots A_{i_s} v_{j-s} + \tau^r (I + \tau A_1) \cdots (I + \tau A_n) \eta^\tau$$

$$= \sum_{j=0}^{r} \tau^j \sum_{i=0}^{j-1} \frac{(-1)^{j-i}}{(j-i)!} v_i^{(j-i)} + \tau^r \eta^\tau(t - \tau) + \tau f + \tau^{r+1} \sum_{j=0}^{r-1} \sigma_j^\tau. \quad (3.17)$$

From the definition of the function v_l (3.9) it is easy to convince oneself of the validity of

$$\frac{dv_0}{dt} + A v_0 = f \quad \text{on } (0, 1),$$

since $v_0 = u$, and

$$\sum_{s=1}^{\min(l+1,n)} \sum_{i_1 < i_2 < \cdots < i_s} A_{i_1} A_{i_2} \cdots A_{i_s} v_{l-s+1}$$

$$= \sum_{i=0}^{l} \frac{(-1)^{l-i+1}}{(l-i+1)!} v_i^{(l-i+1)} \quad \text{on } (0, 1), \quad l = 1, \ldots, r-1.$$

Note that these last relations are actually another form of (3.9). We use these relations to simplify both sides of (3.17):

$$\sum_{j=r+1}^{r+n-1} \tau^j \sum_{s=1}^{\min(j,n)} \sum_{i_1 < i_2 < \cdots < i_s} A_{i_1} A_{i_2} \cdots A_{i_s} v_{j-s}$$

$$+ \tau^r (I + \tau A_1) \cdots (I + \tau A_n) \eta^\tau = \tau^r \eta^\tau(t - \tau) + \tau^{r+1} \sum_{j=0}^{r-1} \sigma_j^\tau.$$

Transferring the first group of terms from the right-hand side to the left-hand side we have

$$(I + \tau A_1) \cdots (I + \tau A_n) \eta^\tau = \eta^\tau(t - \tau) + \tau \rho^\tau, \qquad t \in \omega_\tau, \qquad (3.18)$$

where the net vector-function ρ^τ is uniformly bounded over ω_τ:

$$\max_{t \in \omega_\tau} \| \rho^\tau(t) \| \le c_2. \qquad (3.19)$$

The last inequality follows from the fact that ρ^τ is the sum of two kinds of terms

$$\rho^\tau = \sum_{j=0}^{r-1} \sigma_j^\tau - \sum_{j=r+1}^{r+n-1} \tau^{j-r-1} \sum_{s=1}^{\min(j,n)} \sum_{i_1<i_2<\cdots<i_s} A_{i_1} A_{i_2} \cdots A_{i_s} v_{j-s}. \qquad (3.20)$$

The first sum is uniformly bounded over ω_τ due to (3.16) and the smoothness of the components of v_j. Each term in the second sum is bounded, due to the uniform boundedness of the elements of A_i and v_{j-s}, since these elements are at least continuous so that they attain their maximum and minimum on the interval [0, 1]. The number of terms in the second group does not exceed $(n-1)2^n$. Let us estimate each term separately. Consider the expression $\|\tau^{j-r-1} A_{i_1} \cdots A_{i_s} v_{j-s}\|$. Since $j \geq r+1$ and $\tau < 1$ this expression does not exceed $\|A_{i_1} \cdots A_{i_s} v_{j-s}\|$. It has been shown that the components of v_{j-s} are uniformly bounded. Choose the maximum of these

$$c_3 = \max_{1 \leq k \leq m} \max_{[0,1]} |v_{j-s,k}|,$$

where $v_{j-s,k}$ is the kth component of the vector v_{j-s}. It is obvious that $\|v_{j-s}\| \leq c_3\sqrt{m}$. Now choose the maximum of the absolute values of the elements of A_i:

$$c_4 = \max_{1 \leq i \leq n} \max_{[0,1]} \max_{\substack{1 \leq k \leq m \\ 1 \leq l \leq m}} |A_{i,k,l}|,$$

where $A_{i,k,l}$ is the element of A_i in the kth row and lth column. Using the simple inequality (see, for example, [103]) for the Euclidean norm of the matrix A_i

$$\|A_i\| \leq mc_4,$$

we finally have

$$\|A_{i_1} \cdots A_{i_s} v_{j-s}\| \leq \|A_{i_1}\| \cdots \|A_{i_s}\| \|v_{j-s}\| \leq m^s c_4^s c_3.$$

Thus the required estimate of the arbitrary term in (3.20) has been found; therefore inequality (3.19) is valid. Using (3.13) one can obtain an initial condition for (3.16) at the point $t = 0$. From (3.2), (3.6), and (3.10) we obtain

$$\eta^\tau(0) = \tau^{-r}\left(u^\tau(0) - \sum_{j=0}^{r-1} \tau^j v_j(0)\right) = \tau^{-r}(u(0) - u(0)) = 0.$$

So for η^τ we obtain a difference equation for which the *a priori* estimate (3.7) is valid. It follows from this estimate that

$$\max_{t \in \bar{\omega}_\tau} \|\eta^\tau(t)\| \leq \max_{t \in \omega_\tau} \|\rho^\tau(t)\|.$$

From (3.19) we come to (3.12) where $c_1 = c_2$.

Let us formulate this result in the form of a theorem.

Theorem 3.2. *Suppose the matrix A of* (3.1), (3.2) *can be represented as a sum* (3.3) *of n summand satisfying condition* (3.4) *and so that the elements of the matrices A_i and of the vector f belong to $C^r[0, 1]$. Then the solution u^τ of* (3.5), (3.6) *has the expansion*

$$u^\tau = u + \sum_{j=1}^{r-1} \tau^j v_j + \tau^r \eta^\tau \quad \text{on } \bar{\omega}_\tau,$$

where the v_j are vector-functions with elements in $C^{r+1-j}[0, 1]$ which are independent of τ, and the net vector-function η^τ is bounded:

$$\max_{t \in \bar{\omega}_\tau} \|\eta^\tau(t)\| \le c_1. \tag{3.21}$$

Using this expansion we can justify the extrapolation method presented in §2.2, the only difference being that scalars are replaced substituted by vectors. Assume the conditions of Theorem 3.2 hold. Construct nets $\bar{\omega}_{\tau_k}$ with mesh-sizes $\tau_k = 1/(N_k M)$ for the fixed integers $0 < N_1 < \cdots < N_r$, where the natural number M can grow indefinitely. Construct the approximate problems (3.5), (3.6) and solve them on each net ω_{τ_k}. All such solutions $u^{\tau k}$ are defined on the net $\bar{\omega}_\tau$ with mesh-size $\tau = 1/M$. The determinant of the system of equations

$$\sum_{k=1}^{r} \gamma_k = 1,$$

$$\sum_{k=1}^{r} \gamma_k \tau_k^j = 0, \qquad j = 1, \ldots, r - 1, \tag{3.22}$$

is not zero, and so there is a unique solution $\gamma_1, \ldots, \gamma_r$. Form the linear combination

$$U^H(t) = \sum_{k=1}^{r} \gamma_k u^{\tau k}(t), \qquad t \in \bar{\omega}_\tau; \tag{3.23}$$

we will prove that it possesses improved accuracy when $\tau \to 0$.

Theorem 3.3. *Assume that conditions* (3.3), (3.4) *hold for* (3.1), (3.2) *and that the elements of the matrices A_i and of the vector f belong to $C^r[0, 1]$ with r a natural number. Then the corrected solution* (3.23) *with weights γ_k satisfying* (3.22) *obeys the estimate*

$$\max_{t \in \bar{\omega}_\tau} \|U^H(t) - u(t)\| \le c_5 \tau^m. \tag{3.24}$$

PROOF. Since the conditions of Theorem 3.2 hold, then at each knot of $\bar{\omega}_\tau$ we can expand

$$u^{\tau k} = u + \sum_{j=1}^{r-1} \tau^j v_j + \tau_k^r \eta^{\tau k}, \qquad k = 1, \ldots, r.$$

Sum these with weights γ_k and use (3.22). We have

$$U^H = u + \sum_{k=1}^{r} \gamma_k \tau_k^r \eta^{\tau k} \quad \text{on } \bar{\omega}_\tau. \tag{3.25}$$

In order to estimate $|\gamma_k|$ we use Lemma 2.3 of §7.2. Assume

$$c_6 = \min_{1 \le k \le r-1} N_{k+1}/N_k - 1.$$

Then

$$|\gamma_k| \le (1 + 1/c_6)^r, \quad k = 1, \ldots, r.$$

This inequality, together with (3.21), allows us to obtain the following inequality from (3.25):

$$\|U^H(t) - u(t)\| \le c_1(1 + 1/c_6)^r \sum_{k=1}^{r} \tau_k^r, \quad t \in \bar{\omega}_\tau.$$

But $\tau_k \le \tau \; \forall \, k = 1, \ldots, r$. Therefore introducing the notation

$$c_5 = rc_1(1 + 1/c_6)^r$$

we obtain (3.24). The theorem is proved. $\square$

We illustrate these results with a numerical example. Consider (3.1), (3.2) with symmetric A:

$$\begin{bmatrix} 9.6045364 & -0.2154636 & -0.1974636 & -0.7826544 & -1.7593524 \\ & 0.9645364 & -0.0174636 & -0.0626544 & -0.1393524 \\ & & 0.1005364 & 0.0093456 & 0.0226476 \\ & \text{(reflected elements)} & & 0.0761824 & 0.1553904 \\ & & & & 0.3652084 \end{bmatrix}$$

Its eigenvalues are $10, 1, 0.1, 0.01, 0.001$. The symmetry of A and the positivity of its eigenvalues imply that this matrix is nonnegative definite (see [134]). We separate A into two triangular matrices in the following way. Take the subdiagonal elements and half the diagonal elements as a lower triangular matrix A_1, and the superdiagonal elements and half the diagonal elements as an upper triangular matrix A_2. It is obvious that $A_1 = A_2^T$ and $A = A_1 + A_2$. Let us prove that A_1 and A_2 are nonnegative definite:

$$(A_i u, u) = \tfrac{1}{2}(A_i u, u) + \tfrac{1}{2}(u, A_i^T u) = \tfrac{1}{2}((A_i + A_i^T)u, u)$$
$$= \tfrac{1}{2}(Au, u) \ge 0 \quad \forall \, u \in E^5, \quad i = 1, 2.$$

The vector of initial values was taken equal to

$$(0.68, 0.68, 0.68, -0.28, -1.88)^T.$$

The exact solution of problem (3.1), (3.2) with the right-hand side equal to zero is the vector-function

$$u(t) = \begin{bmatrix} 0.98 & -0.02 & -0.02 & -0.08 & -0.18 \\ -0.02 & 0.98 & -0.02 & -0.08 & -0.18 \\ -0.02 & -0.02 & 0.98 & -0.08 & -0.18 \\ -0.08 & -0.08 & -0.08 & 0.68 & -0.72 \\ -0.18 & -0.18 & -0.18 & -0.72 & -0.62 \end{bmatrix} \begin{bmatrix} e^{-10t} \\ e^{-t} \\ e^{-0.1t} \\ e^{-0.01t} \\ e^{-0.001t} \end{bmatrix}$$

We give the error in the difference solutions of (3.5), (3.6) and the approximate solutions (3.23) extrapolated from them below.

Table 2.1

Component	Maximal error of the difference solution of (3.5), (3.6)			Maximal error of the extrapolated solution (3.23)	
	$\tau = \frac{1}{80}$	$\tau = \frac{1}{160}$	$\tau = \frac{1}{320}$	$\tau = \frac{1}{80}$ $\tau = \frac{1}{160}$	$\tau = \frac{1}{80}$ $\tau = \frac{1}{160}$ $\tau = \frac{1}{160}$ $\tau = \frac{1}{320}$
1	1.55×10^{-2}	7.87×10^{-3}	3.97×10^{-3}	2.35×10^{-4}	1.83×10^{-6}
2	2.85×10^{-3}	1.44×10^{-3}	7.21×10^{-4}	2.21×10^{-5}	1.86×10^{-7}
3	8.96×10^{-4}	4.54×10^{-4}	2.29×10^{-4}	1.28×10^{-5}	1.38×10^{-7}
4	3.42×10^{-3}	1.73×10^{-3}	8.72×10^{-4}	4.97×10^{-5}	5.40×10^{-7}
5	7.74×10^{-3}	3.93×10^{-3}	1.98×10^{-3}	1.18×10^{-4}	1.29×10^{-6}

2.4. Equations of Singularities

So far we have considered differential equations that have smooth solutions. In the present section we will touch upon differential equations with singular solutions. We prove that in this case we can also perform Richardson extrapolation using approximate solutions of relatively low accuracy defined on a sequence of nets. Our application of the Richardson method uses an expansion of the approximate solution in fractional powers of the mesh-size. We will give a method for obtaining the first few terms in this expansion [41, 92] by means of a concrete example. The extrapolation method also appears to be successful in some cases when the initial differential equations have singularities of other (e.g., logarithmic) types.

Let us consider the problem

$$u' = \sqrt{t}, \qquad t \in (0, 1), \tag{4.1}$$

$$u(0) = 0, \tag{4.2}$$

whose solution is

$$u = \tfrac{2}{3} t^{3/2}.$$ (4.3)

Despite the fact that the second derivative fails to be smooth on $[0, 1]$, we will try to find a solution of (4.1), (4.2) using the Crank–Nicholson scheme

$$u_{\bar{t}}^{\tau} = \sqrt{t}, \qquad t \in \breve{\omega}_{\tau},$$ (4.4)

$$u^{\tau}(0) = 0.$$ (4.5)

Consider the behaviour of the error $\psi^{\tau} = u^{\tau} - u$ on $\bar{\omega}_{\tau}$. Substituting $u^{\tau} = u + \psi^{\tau}$ into (4.4) we have

$$u_{\bar{t}} + \psi_{\bar{t}}^{\tau} = t^{1/2}, \qquad t \in \bar{\omega}_{\tau}.$$ (4.6)

The function u is continuously differentiable on any interval not containing the point $t = 0$. Therefore for all points of $\breve{\omega}_{\tau}$ except the first one, the expansion

$$u_{\bar{t}}(t) = u'(t) + \frac{\tau^2}{24} u'''(t) + \frac{\tau^4}{16 \cdot 120} u^{V}(\xi)$$

$$= t^{1/2} - \frac{\tau^2}{96} t^{-3/2} - \frac{\tau^4}{2048} \xi^{-7/2}$$

is valid, where $\xi \in [t - \tau/2, t + \tau/2]$. Keeping this in mind equation (4.6) can be written as

$$\psi_{\bar{t}}^{\tau} = \frac{\tau^2}{96} t^{-3/2} + \frac{\tau^4}{2048} \xi^{-7/2}.$$ (4.7)

Apparently, at all points of $\bar{\omega}_{\tau}$, except the first, the term of order τ^2 dominates. Let us try to extract the principal contribution to ψ^{τ}, as before, by solving the differential equation

$$v' = \frac{\tau^2}{96} t^{-3/2}, \qquad t \in (\tau, 1).$$ (4.8)

The function

$$v = -\frac{\tau^2}{48} t^{-1/2} + d$$ (4.9)

is the general solution of (4.8), where d is a constant independent of t. It is clear that whatever d may be, the initial condition $v(0) = 0$ cannot be satisfied. Therefore we find d from the first difference equation

$$u_{\bar{t}}^{\tau}(\tau/2) = \sqrt{\tau/2}.$$

Since $u^{\tau}(0) = 0$, then $u^{\tau}(\tau) = \tau^{3/2}/\sqrt{2}$. Hence

$$u(\tau) + \psi^{\tau}(\tau) = \tau^{3/2}/\sqrt{2}.$$

Thus, if we want v to describe the behaviour of the main term of the error it is necessary that

$$v(\tau) = \psi^\tau(\tau) = (3\sqrt{2} - 4)\tau^{3/2}/6,$$

whence

$$d = (24\sqrt{2} - 31)\tau^{3/2}/48.$$

Consequently, the main term of the error is described on ω_τ by

$$v(t) = (24\sqrt{2} - 31)\tau^{3/2}/48 - \tau^2 t^{-1/2}/48.$$

We present the results of a numerical experiment in the form of the graphs of

$$\eta_1(M) = |\psi^\tau(1)|, \tag{4.10}$$

$$\eta_2(M) = |\psi^\tau(1) - (24\sqrt{2} - 31)\tau^{3/2}/48| \tag{4.11}$$

versus the integer $M = 1/\tau$. These graphs are given in Figure 2.2 in logarithmic coordinates. The fact that the theoretical and experimental slopes coincide illustrates the behaviour of η_1 and η_2 as given previously.

Note that Richardson extrapolation can also be realized when the solution is expanded in fractional powers so that a corrected version can be constructed. Assume that l natural numbers $N_1 < \cdots < N_l$ are given and that for any natural number M at the knot t there are l expansions with net parameters $\tau_k = \tau/N_k$ where $\tau = 1/M$

$$u^{\tau_k}(t) = u(t) + \sum_{i=1}^{l-1} \tau_k^{\alpha_i} v_i(t) + \tau_k^{\alpha_l} \eta^{\tau_k}(t), \qquad k = 1, \ldots, l. \tag{4.12}$$

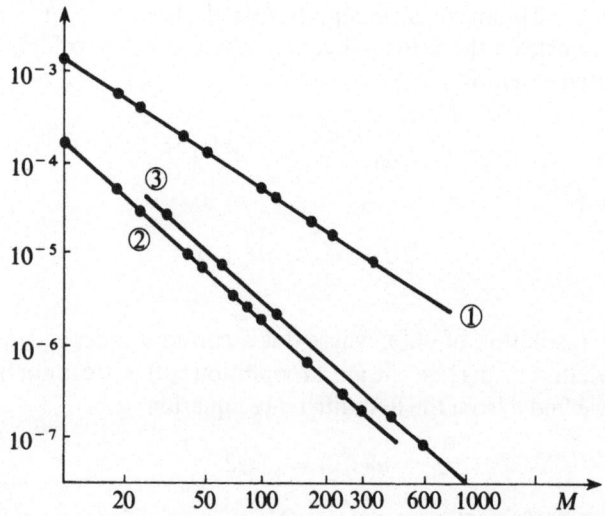

Figure 2.2. Error in the approximate solutions of (4.1), (4.2) at $t = 1$.
(1) Graph of η_1 in (4.10); (2) graph of η_2 in (4.11); (3) graph of η_3 in (4.17).

Here the functions u, v_i and the indices α_i, α_l are independent of τ and $0 < \alpha_1 < \alpha_2 < \cdots < \alpha_l$. The estimate

$$|\eta^{\tau_k}(t)| \leq c_1 \tag{4.13}$$

holds for the remainder.

Let us form the linear combination

$$U^H(t) = \sum_{k=1}^{l} \gamma_k u^{\tau_k}(t), \tag{4.14}$$

with weights satisfying

$$\sum_{k=1}^{l} \gamma_k = 1,$$

$$\sum_{k=1}^{l} \frac{\gamma_k}{N_k^{\alpha_i}} = 0, \qquad i = 1, \ldots, l-1. \tag{4.15}$$

We do not need many additional conditions to prove that the system (4.5) is solvable. We assume that (4.5) has a solution $\gamma_1, \ldots, \gamma_l$ for the set of N_k. One must note that the system and its solution do not depend on the parameter M. From (4.15) it follows that

$$\sum_{k=1}^{l} \gamma_k = 1,$$

$$\sum_{k=1}^{l} \gamma_k \tau_k^{\alpha_i} = 0, \qquad i = 1, \ldots, l-1. \tag{4.16}$$

According to (4.12) we have

$$U^H(t) = \sum_{k=1}^{l} \gamma_k u(t) + \sum_{k=1}^{l} \sum_{i=1}^{l-1} \gamma_k \tau_k^{\alpha_i} v_i(t) + \sum_{k=1}^{l} \gamma_k \tau_k^{\alpha_l} \eta^{\tau_k}(t).$$

Using (4.16) and the fact that the function u and v_i are independent of τ_k we have

$$U^H(t) = u(t) + \sum_{k=1}^{l} \gamma_k \tau_k^{\alpha_l} \eta^{\tau_k}(t).$$

Hence

$$|U^H(t) - u(t)| \leq \sum_{k=1}^{l} |\gamma_k| \tau_k^{\alpha_l} |\eta^{\tau_k}(t)|.$$

Apply (4.13) and use the definition τ_k to estimate the right-hand side of the inequality obtained. This yields

$$|U^H(t) - u(t)| \leq c_1 \tau^{\alpha_l} \sum_{k=1}^{l} |\gamma_k| N_k^{-\alpha_l}.$$

Since c_1, γ_k, and N_k do not depend on τ or M, the given estimate shows that the solution U^H has accuracy of order τ^{α_l} as $M \to \infty$.

To illustrate these results we return to (4.4), (4.5). Let us check the efficiency of the extrapolation in fractional powers for this problem. Assume that the solution of (4.4), (4.5) has an expansion (4.12) with parameters $l = 2$, $\alpha_1 = \frac{3}{2}$, $\alpha_2 = 2$. We assume $N_1 = 1$, $N_2 = 2$ and calculate the solutions u^{τ_k} of (4.4), (4.5) with mesh-sizes $\tau_k = 1/(N_k M)$ for several natural numbers M. The graph of the error $\psi^\tau(1)$ of these solutions is presented in Figure 2.3. To calculate the extrapolated solution U^H, consider the system (4.15). In this case it takes the form

$$\gamma_1 + \gamma_2 = 1,$$

$$\gamma_1 + 2^{-3/2}\gamma_2 = 0$$

and its solution is

$$\gamma_1 = \frac{2^{-3/2}}{2^{-3/2} - 1}, \qquad \gamma_2 = \frac{1}{1 - 2^{-3/2}}.$$

Form the extrapolated solution for several values of M

$$U^H(t) = \gamma_1 u^\tau(t) + \gamma_2 u^{\tau/2}(t), \qquad \tau = 1/M.$$

The error is

$$\eta_3(3M) = |U^H(1) - u(1)|. \tag{4.17}$$

We have used argument $3M$ since the number of operations needed to calculate U^H and the solution $u^{\tau/3}$ approximately coincide.

One can see that the slope of the error η_3 insures a higher order of accuracy for approximate solutions obtained by the extrapolation method.

Integrating a first-order ordinary differential equation is closely related to quadrature; therefore, expansions in fractional powers τ (and sometimes in $\tau^\alpha \ln^p \tau$) are often used to improve on quandrature roles. Results on numerical techniques for solving ordinary differential equations with singularities are presented in [68, 40, 41, 53, 91, 92, 124].

We have described this method because it seems to have great potential for numerical treatments of ordinary differential equations and partial differential equations. Our confidence is based on the fact that experimental results have been encouraging. Positive results have even been obtained for a two-dimensional eigenvalue problem [39]. The rigorous verification of (4.12) unfortunately is a rare occurrence. Another drawback of this method is that it usually requires a great deal of preliminary analytical calculations on the α_i; this is why the method is difficult to automate.

We will study this method as applied to partial differential equations; we will also consider another method based on refining the net near the singularity. This method allows numerical solutions of problems with point singularities, and is presented in Chapter 4.

CHAPTER 3

The One-Dimensional Stationary Diffusion Equation

The diffusion equation is of great interest because of its application to various fields of science. This equation is of special interest in the theory of nuclear reactors, where the diffusion approximation of the transport equation is of great importance. In ecology the computation of the diffusion of industrial aerosols is of great importance. Important diffusion problems occur in physics, chemistry, geophysics, and other fields of science. Because of this we shall focus on various formulations of problems related to the diffusion equation.

It should be emphasized that the most interesting applications of the diffusion equation will have discontinuous coefficients and source functions. This means additional restrictions will be placed on the solution, which will turn out to be continuous but not differentiable. Therefore the method of increasing the accuracy of approximate solutions on different nets set forth in preceding chapters will require considerable modification, since a representation of the solution by a Taylor expansion can be carried out only in a domain where the points of discontinuity of the coefficients and source functions have been excluded. However, problems with continuously differentiable coefficients are also of interest. Thus it is worthwhile to begin with a description of the Richardson extrapolation method for boundary problems associated with the stationary diffusion equation with sufficiently smooth coefficients.

3.1. The Dirichlet Problem

In this section we will consider the Dirichlet problem for the diffusion equation. Since so far we have been interested only in initial-value problems for ordinary differential equations we will now concentrate on boundary-value problems for second-order differential equations, the diffusion equation being of greatest interest to us. First, we will consider equations with sufficiently smooth coefficients. We will argue for the validity of the extrapolation method in this case. In order to obtain an approximate solution of the highest order of accuracy we shall again use approximate solutions for lower-order difference equations on sequence of grids.

Thus let us consider the one-dimensional Dirichlet problem for the stationary diffusion equation

$$-(pu')' + qu = f, \qquad x \in (0, 1), \tag{1.1}$$

$$u(0) = u_0, \qquad u(1) = u_1; \tag{1.2}$$

supposing that the coefficients of the problem satisfy

$$p(x) \geq c_1 > 0, \qquad q(x) \geq 0, \qquad x \in (0, 1), \tag{1.3}$$

where

$$q, f \in C^r[0, 1], \qquad p \in C^{r+1}[0, 1] \tag{1.4}$$

for a natural number $r \geq 2$.

Theorem 1.1. *Suppose conditions* (1.3), (1.4) *hold for* (1.1), (1.2). *Then there is a unique solution* $u \in C^{r+2}[0, 1]$.

PROOF. We will represent the solution of the problems (1.1), (1.2) as

$$u(x) = v(x) + (u_0 + x(u_1 - u_0));$$

where v is the solution of the problem

$$-(pv')' + qv = g \quad \text{on } (0, 1),$$
$$v(0) = v(1) = 0, \tag{1.5}$$

and where $g = f + (u_1 - u_0)p - g(u_0 + x(u_1 - u_0))$. There is [48] a Green's function $G(x, t)$ such that

$$v(x) = \int_0^1 G(x, t)g(t) \, dt.$$

The continuity of function G insures the continuity of the function v on $[0, 1]$. In equation (1.5) let us move the continuous function qv to the right-hand side and integrate from t to x. We have

$$-p(x)v'(x) + p(t)v'(t) = -\int_t^x (q(\xi)v(\xi) - g(\xi)) \, d\xi. \tag{1.6}$$

Divide this equality by $-p(t)$ and again integrate with respect to t from 0 to x. We have

$$p(x)v'(x) \int_0^x p^{-1}(t) \, dt - v(x) = \int_0^x p^{-1}(t) \, dt \int_t^x (q(\xi)v(\xi) - g(\xi)) \, d\xi.$$

From this we find

$$v'(x) = p^{-1}(x)\left[\int_0^x p^{-1}(t)\,dt\right]^{-1}\left\{v(x) + \int_0^x p^{-1}(t)\int_t^x (q(\xi)v(\xi) - g(\xi))\,d\xi\,dt\right\}.$$

Consider the interval $[\frac{1}{3}, 1]$. All integrands are continuous and $p(t)$ is strictly positive, so the expression in braces is continuous. The function

$$\int_0^x p^{-1}(t)\,dt$$

is also continuous with respect to x and strictly positive on $[\frac{1}{3}, 1]$. Consequently, its inverse is also continuous. Thus the derivative v' is continuous on $[\frac{1}{3}, 1]$, as the product of continuous functions.

Let us return now to (1.6). Again we will divide it by $p(t)$ but we will integrate now from x to 1. Then we will obtain an equality from which we have

$$v'(x) = \frac{1}{p(x)}\frac{1}{\int_x^1 dt/p(t)}\left\{-v(x) + \int_x^1 \frac{1}{p(t)}\int_t^x (q(\xi)v(\xi) - g(\xi))\,d\xi\,dt\right\}.$$

From this we get the continuity of the function v' on $[0, \frac{2}{3}]$. Therefore, on the whole interval $[0, 1]$ the function v' is continuous, and $v \in C^1[0, 1]$. Transform (1.5) once more:

$$v'' = -p'v'/p + qv/p - g/p. \tag{1.7}$$

Since all terms on the right-hand side are continuous v'' is also continuous on $[0, 1]$. Therefore $v \in C^2[0, 1]$. This gives us the continuous differentiability of the terms on the right-hand side of (1.7), so that v'' is continuously differentiable on $[0, 1]$, and, consequently, $v \in C^3[0, 1]$. Using this procedure $r - 1$ times we conclude that $v \in C^{r+2}[0, 1]$.

Because the solution u is a sum of the function v and a polynomial we obtain the smoothness result of the theorem. The uniqueness of the solution follows from the maximum modulus principle (e.g., [6]). The theorem is proved. $\square$

To obtain a numerical solution of the problem let us construct a uniform net

$$\bar{\omega}_h = \{x_i = ih, i = 0, 1, \ldots, N\},$$

with mesh size $h = 1/N$, where N is a natural number. Denote the set of midpoints of the net by

$$\check{\omega}_h = \{x_{i+1/2} = (i + 1/2)h, i = 0, 1, \ldots, N - 1\},$$

and denote by ω_h the set of interior points of the net:

$$\omega_h = \{x_i = ih, i = 1, \ldots, N - 1\}.$$

At each point we have the difference equation

$$-(pu^h_{\bar{x}})_{\hat{x}} + qu^h = f \quad \text{on } \omega_h. \tag{1.8}$$

Here, from the definition

$$v_{\hat{x}}(x) = [v(x + h/2) - v(x - h/2)]/h$$

for an arbitrary function v we have

$$(pv_{\bar{x}})_{\hat{x}}|_x = [p(x + h/2)(v(x + h) - v(x)) - p(x - h/2)(v(x) - v(x - h))]/h^2.$$

Adding to equation (1.8) the two boundary conditions:

$$u^h(0) = u_0, \qquad u^h(1) = u_1, \tag{1.9}$$

we obtain $N + 1$ equations for the $N + 1$ values of net function u^h at the knots $\bar{\omega}^h$.

From Lemma 1.2 of §7.1 it follows under the assumptions (1.4) the difference problem will have second-order accuracy. Let us resolve the question of stability.

Lemma 1.2. *Assume that conditions* (1.3) *are valid for* (1.8), (1.9). *Then*

$$\|u^h\|_{C,h} \le \frac{1}{c_1} \max_{\omega_h} |f| + \max\{|u(0)|, |u(1)|\}$$

holds.

PROOF. Divide the problem into two problems:

$$-(pv^h_{\bar{x}})_{\hat{x}} + qv^h = 0 \quad \text{on } \omega_h,$$
$$v^h(0) = u(0), \qquad v^h(1) = u(1)$$

and

$$-(pw^h_{\bar{x}})_{\hat{x}} + qw^h = f \quad \text{on } \omega_h,$$
$$w^h(0) = w^h(1) = 0.$$

The maximum principle for differences is valid for the first (see [114]), from which we have the estimate

$$\|v^h\|_{C,h} \le \max\{|u(0)|, |u(1)|\}.$$

For the second problem let us use the following estimate from [114]:

$$\|w^h\|_{C,h} \le \frac{1}{c_1} \max_{\omega_h} |f|.$$

These estimates and the triangle inequality

$$\|u^h\|_{C,h} \le \|v^h\|_{C,h} + \|w\|_{C,h}$$

give the required estimate. The lemma is proved.   □

Using this estimate and the fact that the difference scheme (1.8), (1.9) is of second-order accuracy one can prove that the solution u^h has accuracy of h^2 order.

Let us construct a corrector to obtain a higher order of accuracy. Assume $l = (r - 1)/2$ and fix natural numbers $M_1 < \cdots < M_{l+1}$. On every net $\bar{\omega}_{h_k}$ with mesh-size $h_k = 1/(M_k N)$ we will solve the difference problem (1.8), (1.9).

Here, as before, N may increase to infinity and our aim is to insure that the accuracy of the corrected solution depends on parameter $h = 1/N$. We should note that all solutions u^{h_k} are defined on the net $\bar{\omega}_h$.

We have mentioned that the system:

$$\sum_{k=1}^{l+1} \gamma_k = 1,$$

$$\sum_{k=1}^{l+1} \gamma_k h_k^{2j} = 0, \qquad j = 1, \ldots, l \quad \text{is regular.} \tag{1.10}$$

Let us take its unique solution $\gamma_1, \ldots, \gamma_{l+1}$ and form a linear combination

$$U^H(x) = \sum_{k=1}^{l+1} \gamma_k u^{h_k}(x), \qquad x \in \bar{\omega}_h. \tag{1.11}$$

Theorem 1.3. *Suppose* (1.3), (1.4) *hold for the problem* (1.1), (1.2). *Then the corrected solution* (1.11) *whose weights are found from* (1.10) *satisfies the estimate*

$$\|U^H - u\|_{c,h} \le c_2 h^r. \tag{1.12}$$

PROOF. Let us check the validity of the conditions of Theorem 2.2 of §12. Assume $M_k(\Omega) = C^k[0, 1]$, $P_k(\bar{\Omega}) = C^{k+2}[0, 1]$, $N_k(D) = R^2$. Then the Condition A of §1.2 is a consequence of Theorem 1.1. For difference problem (2.3) of §1.2 put

$$\bar{\Omega}_h = \bar{\omega}_h, \qquad \check{\Omega}_h = \omega_h, \qquad D_h = \{0, 1\}$$

and

$$\|u\|_{\bar{\Omega}_h} = \|u\|_{c,h}, \qquad \|u\|_{\check{\Omega}_h} = \max_{\omega_h} |u|, \qquad \|u\|_{D_h} = \max\{|u(0)|, |u(1)|\}.$$

Then Condition B of §1.2 coincides with the *a priori* estimate of Lemma 1.2. Now let us prove that it is possible to expand the error of approximation in even powers of h. From Lemma 1.2 of §7.1 it follows that for any function $\varphi \in C^{r-2k+2}[0, 1]$ the expansion

$$-(p\varphi_{\bar{x}})_{\check{x}} + q\varphi = -(p\varphi')' + q\varphi - \sum_{j=1}^{l-k} h^{2j} 4^{-j} \sum_{k+s=j} \frac{(p\varphi^{(2k+1)})^{(2s+1)}}{(2k+1)!\,(2s+1)!}$$

$$+ h^{r-2k} \sigma^h$$

is valid, where $|\sigma^h| \le c_3 \; \forall \; x \in \omega_h$.

Thus Condition D is realized, with $\beta = 0$. All conditions of the Theorem 2.2 of §1.2 are thus satisfied. Now let us use Theorem 3.2 of §1.3. Condition (3.15) in this theorem holds with constant

$$d_3 = \min_{1 \leq k \leq l} (M_{k+1}/M_k) - 1.$$

Therefore from this theorem there follows

$$\max_{\bar{\Omega}_h} |U^H - u| \leq d_4 h_1^r$$

with constant d_4 independent h_k. But the intersection of the $\bar{\Omega}_H$ contains $\bar{\omega}_h$, thus

$$\|U^H - u\|_{C,h} \leq \frac{d_4}{M_1^r} h^r.$$

Assuming $c_2 = d_4/M_1^r$, we arrive at (1.12). Theorem 1.3 is proved. $\qquad\square$

Thus the accuracy of the corrected solution depends only on the degree of smoothness r of the data in (1.1), (1.2).

As an illustration of these results consider the problem of finding a numerical solution of the problem

$$-((1 + x)u')' + xu = \frac{1 + x^2 + x^3}{(1 + x)^2} \quad \text{on } (0, 1),$$

$$u(0) = 0, \qquad u(1) = \tfrac{1}{2}.$$

Its analytic solution is of the form

$$u(x) = x/(1 + x).$$

The dependence of the maximum error on the number of knots $\bar{\omega}_h$ is given in Table 3.1.

Table 3.1

		Maximum error of the extrapolated solution (1.11)		
N	Maximum error of the solution (1.8), (1.9)	$h_1 = 1/N,$ $h_2 = 1/(2N)$	$h_1 = 1/N,$ $h_2 = 1/(2N)$ $h_3 = 1/(3N)$	$h_1 = 1/N,$ $h_2 = 1/(2N)$ $h_3 = 1/(3N),$ $h_4 = 1/(4N)$
10	2.3×10^{-4}	4.2×10^{-7}	5.3×10^{-10}	5.6×10^{-13}
20	5.8×10^{-5}	2.6×10^{-8}	8.4×10^{-12}	4.6×10^{-15}
30	2.6×10^{-5}	5.3×10^{-9}	7.5×10^{-13}	2.1×10^{-14}
40	1.4×10^{-5}	1.7×10^{-9}	1.5×10^{-13}	6.3×10^{-13}

Note that the relative level of round-off error was 10^{-14}. This is clearly seen in two last columns.

3.2. Boundary-Value Problems of the Third Kind

The Dirichlet problem for the diffusion equation has presented no difficulty in approximating the boundary conditions. But when we consider the third boundary-value problem we face this additional problem. This section is devoted to the problem of altering Richardson's method for the third boundary problem.

For equation (1.1), under the assumptions (1.3), (1.4), let us consider the third boundary problem

$$\alpha_0 u(0) - u'(0) = g_0, \tag{2.1}$$

$$\alpha_1 u(1) + u'(1) = g_1, \tag{2.2}$$

with nonnegative constants α_0, α_1 so that

$$\alpha_0 + \alpha_1 = c_2 > 0. \tag{2.3}$$

Theorem 2.1. *Suppose conditions* (1.3), (1.4), *and* (2.3) *are satisfied for* (1.1), (2.1), (2.2). *Then there is a unique solution* $u \in C^{r+2}[0, 1]$.

PROOF. We will represent the solution u as a sum $u(x) = v(x) + ax + b$, where the constants a and b are defined by

$$b = \frac{(1 + \alpha_1)g_0 + g_1}{\alpha_0 + \alpha_1 + \alpha_0\alpha_1}, \qquad a = \alpha_0 b - g_0.$$

This choice provides homogeneous boundary values for the function v:

$$\alpha_0 v - v' = 0 \quad \text{in } x = 0,$$

$$\alpha_1 v + v' = 0 \quad \text{in } x = 1.$$

In addition, v satisfies the equation

$$-(pv')' + qv = g \quad \text{on } [0, 1],$$

where

$$g = f + p'a - q(ax + b).$$

From here, the proof coincides with the proof of Theorem 1.1, and again uses the result from [48], which is also valid for the third boundary-value problem. ☐

If we use (1.8) to construct the difference scheme at the knots ω_h (as has been done, for instance, in [114]) then we have to approximate the boundary conditions by one-side differences. It is easy to understand that we will not have an expansion only in even powers h. Odd powers automatically disappear only when central differences are used. Expansions whose odd powers have nonzero coefficients are not as effective, since when one extrapolates one has to eliminate twice as many terms and consequently to solve twice as many approximate problems.

Therefore, [70] suggested a scheme with net equations

$$-(pu_{\bar{x}}^h)_x + qu^h = f \quad \text{on } \breve{\omega}_h; \tag{2.4}$$

where points not in the internal $[0, 1]$ are excluded from the first and the last equations. This was done by approximating the boundary conditions (2.1), (2.2):

$$\alpha_0 u_x^h - u_{\bar{x}}^h = g_0 \quad \text{in } x = 0, \tag{2.5}$$

$$\alpha_1 u_{\bar{x}}^h + u_{\bar{x}}^h = g_1 \quad \text{in } x = 1. \tag{2.6}$$

The elimination results in an algebraic system with a symmetric $N \times N$ matrix. If one does not eliminate the unknown then the matrix obtained appears to be nonsymmetric (which does not strongly effect the method of solution (factorization method) for tridiagonal matrices). However, if one does not eliminate unknowns then the calculations engendered by the extrapolation method are considerably more illustrative, since the contributions of the approximation to the boundary conditions and of the approximation to the equation itself can be seen separately.

Thus, let us solve the system (2.4)–(2.6) numerically. First we will extend the function $u(x)$ beyond the interval $[0, 1]$, e.g., by polynomials

$$\sum_{k=0}^{r+2} \frac{x^k}{k!} u^{(k)}(0), \qquad \sum_{k=0}^{r+2} \frac{(x-1)^k}{k!} u^{(k)}(1)$$

both to the left and to the right of this interval. As a result, the extended solution will belong to $C^{r+2}[-\frac{1}{2}, \frac{3}{2}]$; we will denote it by $u(x)$ as before.

Let us convince ourselves that the solution of this scheme is stable.

Theorem 2.2. *The solution of system* (2.4)–(2.6) *obeys the estimate*

$$\max_{\breve{\omega}_h} |u^h| \le \frac{2(2 + c_2)}{c_1 c_2} \left(\max_{\breve{\omega}_h} |f| + p(0)|g_0| + p(1)|g_1| \right).$$

PROOF. Multiply each equation of (2.4) for the point $x_{i+1/2} \in \breve{\omega}_h$ by $hu^h(x_{i+1/2})$ and sum up the results. We have

$$-\sum_{\breve{\omega}_h} (pu_{\bar{x}}^h)_{\bar{x}} u^h h + \sum_{\breve{\omega}_h} q(u^h)^2 h = \sum_{\breve{\omega}_h} fu^h h. \tag{2.7}$$

Let us apply the first Green's difference formula to the first term (see [112]), which in our notation is

$$-\sum_{\breve{\omega}_h} v_{\bar{x}} w h = \sum_{\omega_h} v w_{\bar{x}} h + v(0)w(h/2) - v(1)w(1 - h/2),$$

where the function v is defined at the knots $\bar{\omega}_h$, and w is defined at the knots $\breve{\omega}_h$. We obtain

$$-\sum_{\breve{\omega}_h} (pu_{\bar{x}}^h)_{\bar{x}} u^h h = \sum_{\omega_h} p(u_{\bar{x}}^h)^2 h - p(1)u^h(1 - h/2)u_{\bar{x}}^h(1) + p(0)u^h(h/2)u_{\bar{x}}^h(0).$$

$$\tag{2.8}$$

Let us rewrite (2.5), (2.6) in the form

$$-(1 + h\alpha_0/2)u_{\bar{x}}^h(0) + \alpha_0 u^h(h/2) = g_0,$$

$$(1 + h\alpha_1/2)u_{\bar{x}}^h(1) + \alpha_1 u^h(1 - h/2) = g_1$$

and use them in (2.8):

$$-\sum_{\omega_h} (pu_{\bar{x}}^h)_{\bar{x}} u^h h = \sum_{\omega_h} p(u_{\bar{x}}^h)^2 h + \frac{p(0)}{1 + h\alpha_0/2} (\alpha_0 u^h(h/2) - g_0)u^h(h/2)$$

$$+ \frac{p(1)}{1 + h\alpha_1/2} (\alpha_1 u^h(1 - h/2) - g_1)u^h(1 - h/2).$$

Substitute the expression obtained in (2.7) and make the necessary transformations. Then

$$\sum_{\omega_h} p(u_{\bar{x}}^h)^2 h + \frac{\alpha_0 p(0)}{1 + h\alpha_0/2} (u^h(h/2))^2 + \frac{\alpha_1 p(1)}{1 + h\alpha_1/2} (u^h(1 - h/2))^2 + \sum_{\bar{\omega}_h} q(u^h)^2 h$$

$$= \sum_{\bar{\omega}_h} fu^h h + \frac{p(0)g_0}{1 + h\alpha_0/2} u^h(h/2) + \frac{p(1)g_1}{1 + h\alpha_1/2} u^h(1 - h/2). \quad (2.9)$$

Now we estimate every term of (2.9) from below (if it is on the left-hand side) or from above (if it is on the right-hand side). Since $p(x) \geq c_1$ (property (1.3) of the previous section) we have the inequality

$$\sum_{\omega_h} p(u_{\bar{x}}^h)^2 h \geq c_1 \sum_{\omega_h} (u_{\bar{x}}^h)^2 h. \quad (2.10)$$

Since $h < 1$, $\alpha_i \geq 0$ then

$$\frac{\alpha_0 p(0)}{1 + h\alpha_0/2} (u^h(h/2))^2 \geq \frac{\alpha_0 c_1}{1 + \alpha_0/2} (u^h(h/2))^2, \quad (2.11)$$

$$\frac{\alpha_1 p(1)}{1 + h\alpha_1/2} (u^h(1 - h/2))^2 \geq \frac{\alpha_1 c_1}{1 + \alpha_1/2} (u^h(1 - h/2))^2. \quad (2.12)$$

According to property (1.3) $q(x) \geq 0$, so

$$\sum_{\bar{\omega}_h} q(u^h)^2 h \geq 0. \quad (2.13)$$

The estimate

$$\sum_{\bar{\omega}_h} fu^h h \leq \max_{\bar{\omega}_h} |f| \max_{\bar{\omega}_h} |u^h| \quad (2.14)$$

is obtained by replacing every term by its maximum, and taking into account that the number of points in $\bar{\omega}_h$ is N, and $hN = 1$. The simple inequalities

$$\frac{p(0)g_0}{1 + \alpha_0 h/2} u^h(h/2) \leq p(0)|g_0| \max_{\bar{\omega}_h} |u^h|, \quad (2.15)$$

$$\frac{p(1)g_1}{1 + \alpha_1 h/2} u^h(1 - h/2) \leq p(1)|g_1| \max_{\bar{\omega}_h} |u^h| \quad (2.16)$$

follow from $\alpha_i \geq 0$, $h > 0$.

Applying inequalities (2.10)–(2.16) to the expression (2.9) we get

$$c_1 \sum_{\omega_h} (u_{\dot{x}}^h)^2 h + \frac{\alpha_0 c_1}{1 + \alpha_0/2} (u^h(h/2))^2 + \frac{\alpha_1 c_1}{1 + \alpha_1/2} (u^h(1 - h/2))^2$$

$$\leq \left(\max_{\bar{\omega}_h} |f| + p(0)|g_0| + p(1)|g_1| \right) \max_{\bar{\omega}_h} |u^h|. \qquad (2.17)$$

We now use the fact that $\alpha_0 + \alpha_1 \geq c_2$. It is obvious that one of these numbers is not less than $c_2/2$. For simplicity assume that $c_2 \geq \alpha_0 \geq c_2/2$. Thus

$$\frac{\alpha_0}{1 + \alpha_0/2} \geq \frac{c_2}{2 + c_2}.$$

The following estimates also hold:

$$\frac{c_2}{2 + c_2} < 1, \qquad \frac{\alpha_1}{1 + \alpha_1/2} \geq 0.$$

Using them to estimate the left-hand side of (2.17) we obtain

$$\frac{c_1 c_2}{2 + c_2} \left(\sum_{\omega_h} (u_{\dot{x}}^h)^2 h + (u^h(h/2))^2 \right)$$

$$\leq \left(\max_{\bar{\omega}_h} |f| + p(0)|g_0| + p(1)|g_1| \right) \max_{\bar{\omega}_h} |u^h|.$$

To simplify the inequality still further we use a difference estimate (see [112]) which is the analog of a norm estimate in $C[0, 1]$ and $W_2^1(0, 1)$:

$$\max_{\bar{\omega}_h} (u^h)^2 \leq 2 \left(\sum_{\omega_h} (u_{\dot{x}}^h)^2 h + (u^h(h/2))^2 \right).$$

We have

$$\frac{c_1 c_2}{2(2 + c_2)} \max_{\bar{\omega}_h} (u^h)^2 \leq \frac{c_1 c_2}{2 + c_2} \left(\sum_{\omega_h} (u_{\dot{x}}^h)^2 h + (u^h(h/2))^2 \right),$$

which leads to

$$\frac{c_1 c_2}{2(2 + c_2)} \max_{\bar{\omega}_h} |u^h|^2 \leq \max_{\bar{\omega}_h} |u^h| \left(\max_{\bar{\omega}_h} |f| + p(0)|g_0| + p(1)|g_1| \right).$$

The result of the theorem follows from this inequality after dividing both sides by

$$\frac{c_1 c_2}{2(2 + c_2)} \max_{\bar{\omega}_h} |u^h|.$$

The Theorem 2.2 is proved.   □

This estimate allows us to state that for every net $\ddot{\omega}_h$ the problem (2.4)–(2.6) has the unique solution.

We will give a rule for constructing a corrector from several solutions. Assume $l = [(r - 1)/2]$ and fix odd numbers $0 < M_1 < \cdots < M_{l+1}$. On each net $\ddot{\omega}_{h_k}$ with mesh size $h_k = 1/(M_k N)$ we will solve the difference problem (2.4)–(2.6). We point out that for any odd numbers M_k and any natural number N the inclusion $\ddot{\omega}_{h_k} \supset \ddot{\omega}_h$ $(h = 1/N)$ holds. So all solutions are indeed defined on the net $\ddot{\omega}_h$.

Consider the system

$$\sum_{k=1}^{l+1} \gamma_k = 1,$$

$$\sum_{k=1}^{l+1} \gamma_k h_k^{2j} = 0, \qquad j = 1, \ldots, l. \tag{2.18}$$

It has a unique solution $\gamma_1, \ldots, \gamma_{l+1}$. Using it we form the linear combination

$$U^H = \sum_{k=1}^{l+1} \gamma_k u^{h_k} \quad \text{on } \ddot{\omega}_h \tag{2.19}$$

and establish its degree of accuracy.

Theorem 2.3. *Suppose that for* (1.1), (2.1), (2.2) *the conditions* (1.3), (1.4), (2.3) *hold. Then the corrected solution* (2.19), *with weights obtained from* (2.18), *obeys the estimate*

$$\max_{\ddot{\omega}_h} |U^H - u| \leq c_3 h^r. \tag{2.20}$$

PROOF. Let us check the conditions of Theorem 2.2 of §1.2. Assume $M_k(\Omega) = C^k[0, 1]$, $P_k(\bar{\Omega}) = C^{k+2}[0, 1]$ and for any k $N_k(D) = R^2$. Then Condition A of §1.2 is seen to hold from Theorem 2.1. For the difference problem (2.3) of §1.2 we should take $\bar{\Omega}_h = \ddot{\omega}_h$, $\dot{\Omega}_h = \ddot{\omega}_h$, $D_h = \{0, 1\}$

$$\|u\|_{\bar{\Omega}_h} = \|u\|_{\dot{\Omega}_h} = \max_{\ddot{\omega}_h} |u|,$$

$$\|u\|_{D_h} = \max\{|u(0)|, |u(1)|\}.$$

In this notation Condition B of §1.2 coincides with the estimate of Theorem 2.2. It remains to prove that one can expand the approximation error in even powers of h. From Lemma 1.2 of §7.1 we have that for any function $\varphi \in C^{r-2k+2}[0, 1]$ there is the expansion

$$-(p\varphi_{\hat{x}})_{\hat{x}} + q\varphi = -(p\varphi')' + q\varphi$$

$$- \sum_{j=1}^{l-k} h^{2j} 4^{-j} \sum_{k+s=j} \frac{(p\varphi^{(2k+1)})^{(2s+1)}}{(2k+1)! \, (2s+1)!} + h^{r-2k}\sigma^h \quad \text{on } \ddot{\omega}_h,$$

where

$$|\sigma^h| \leq c_4 \quad \forall \, x \in \ddot{\omega}_h.$$

At boundary points we have

$$\alpha_0 \varphi_{\bar{x}}(0) - \varphi_{\bar{x}}(0) = \alpha_0 \varphi(0) - \varphi'(0)$$

$$+ \sum_{j=1}^{l-k} h^{2j} 4^{-j} (\alpha_0 \varphi^{(2j)}(0)/(2j)! - \varphi^{(2j+1)}(0)/(2j+1)!) + h^{r-2k}\rho_0,$$

where $|\rho_0| \le c_5$, and

$$\alpha_1 \varphi_{\bar{x}}(1) + \varphi_{\bar{x}}(1) = \alpha_1 \varphi(1) + \varphi'(1)$$

$$+ \sum_{j=1}^{l-k} h^{2j} 4^{-j} (\alpha_1 \varphi^{(2j)}(1)/(2j)! + \varphi^{(2j+1)}(1)/(2j+1)!) + h^{r-2k}\rho_1,$$

where $|\rho_1| \le c_6$.

Thus Condition D holds with $\beta = 0$. So all conditions of Theorem 2.2 of §1.2 are satisfied. Let us verify that the conditions of Theorem 3.2 of §1.3 hold. It is not difficult to see that only condition (3.15) requires checking, and if we choose

$$d_3 = \min_{1 \le k \le l} (M_{k+1}/M_k) - 1$$

it follows immediately.

Thus from Theorem 3.2 it follows that

$$\max_{\bar{\Omega}_H} |U^H - u| \le d_4 h_1^r,$$

where the constant d_4 does not depend on h_k. Since the numbers M_k are odd the intersection $\bar{\Omega}_H$ contains $\breve{\omega}_h$, and thus

$$\max_{\breve{\omega}_h} |U^H - u| \le \frac{d_4}{M_1^r} h^r.$$

Choosing the constant c_3 equal to d_4/M_1^r, we arrive at the estimate (2.20). Theorem 2.3 is proved.   □

So, for the third boundary-value problem we have established that the accuracy of the linear corrector depends only on the degree of smoothness r of coefficients and the right-hand side of (1.1).

To demonstrate the effect of the corrector let us consider the problem

$$-((1 + x)u')' + xu = \frac{1 + x^2 + x^3}{(1 + x)^2} \quad \text{on } (0, 1),$$

$$u(0) - u'(0) = -1, \qquad 2u(1) + u'(1) = \tfrac{5}{4}.$$

Its analytic solution has the form

$$u(x) = x/(1 + x).$$

We present a table of maximum errors versus the number of knots $\breve{\omega}_h$.

Table 3.2

		Maximal error of the extrapolated solution (2.19)		
		$h_1 = 1/N$ $h_2 = 1/(3N)$	$h_1 = 1/N,$ $h_2 = 1/(3N)$ $h_3 = 1/(5N)$	$h_1 = 1/N,$ $h_2 = 1/(3N)$ $h_3 = 1/(5N),$ $h_4 = 1/(7N)$
N	Maximal error of the solution of (2.4)–(2.6)			
10	3.5×10^{-3}	7.7×10^{-7}	8.6×10^{-10}	7.6×10^{-13}
20	8.7×10^{-4}	4.8×10^{-8}	1.3×10^{-11}	1.6×10^{-14}
30	3.9×10^{-4}	9.6×10^{-9}	1.2×10^{-12}	2.6×10^{-14}
40	2.2×10^{-4}	3.0×10^{-9}	2.0×10^{-13}	1.7×10^{-12}

The relative level of the round-off error was 10^{-14}. This contributed considerably to values in the last column.

3.3. Equations with Discontinuous Coefficients

In §3.1 we considered the Dirichlet problem for a diffusion equation with smooth coefficients. Now we will study a diffusion equation with discontinuous coefficients. As has already been mentioned in the introduction to this chapter problems of this type are of particular interest for applications. Let us also note that the approach of the previous sections was based on expanding the smooth solution in powers of a small parameter—the mesh-size. Such an expansion is meaningless for discontinuous coefficients, where even the first derivative of the solution will be discontinuous. The present section treats modifications of this method and its application to problems with discontinuous coefficients.

Let us consider the Dirichlet problem

$$-(pu')' + qu = f, \tag{3.1}$$

$$u(0) = u_0, \qquad u(1) = u_1, \tag{3.2}$$

where $p(x) \geq c_1 > 0$, $q(x) \geq 0$.

Now we allow the functions p, q, f to have the discontinuities of the first kind at a finite number of points of the interval $[0, 1]$. For simplicity let us examine the case when these functions are discontinuous at only one point $\xi \in (0, 1)$. We will assume equation (3.1) holds the intervals $(0, \xi)$, $(\xi, 1)$, and impose consistency conditions at ξ:

$$u(\xi + 0) = u(\xi - 0), \tag{3.3}$$

$$p(\xi + 0)u'(\xi + 0) - p(\xi - 0)u'(\xi - 0) = g, \tag{3.4}$$

with a fixed constant g.

Here v $(\xi \pm 0)$, for an arbitrary function v, denotes the limiting expressions.

$$\lim_{\substack{\delta \to 0 \\ \delta > 0}} v(\xi + \delta), \qquad \lim_{\substack{\delta \to 0 \\ \delta > 0}} v(\xi - \delta),$$

respectively.

Let us introduce the function classes Q_ξ^k, k a natural number. We will say that $v \in Q_\xi^k$ if it is defined on $[0, 1]$ and has piecewise continuous derivatives up to degree k, and that the function itself and its derivatives can have discontinuities only of the first kind and only at the point ξ.

Theorem 3.1. *If the coefficients of the problem* (3.1)–(3.4) *satisfy*

$$p \in Q_\xi^{r+1}, \qquad f, q \in Q_\xi^r, \qquad r \geq 2 \tag{3.5}$$

then there is a unique solution u such that

$$u \in Q_\xi^{r+2}, \qquad u \in C[0, 1]. \tag{3.6}$$

PROOF. We will look for a solution u in the form $u = w_1 + w_2$. w_1 is defined as follows:

$$w_1(x) = \begin{cases} u_0 + ax & \text{if } x \in [0, \xi], \\ u_1 + (1 - x)b & \text{if } x \in [\xi, 1], \end{cases}$$

$$a = \frac{p(\xi + 0)(u_1 - u_0) - (1 - \xi)g}{\xi p(\xi + 0) + (1 - \xi)p(\xi - 0)},$$

$$b = \frac{p(\xi - 0)(u_0 - u_1) - \xi g}{\xi p(\xi + 0) + (1 - \xi)p(\xi - 0)}.$$

This function has the following properties:

$$w_1(0) = u_0, \qquad w(1) = u_1$$

$$w_1(\xi + 0) - w_1(\xi - 0) = 0,$$

$$p(\xi + 0)w_1'(\xi + 0) - p(\xi - 0)w_1'(\xi - 0) = g,$$

$$-(pw_1')' + qw_1 = z_1,$$

where

$$z_1(x) = \begin{cases} -ap'(x) + q(x)(u_0 + ax) & \text{if } x \in (0, \xi), \\ bp'(x) + q(x)(u_1 + (1 - x)b) & \text{if } x \in (\xi, 1). \end{cases}$$

We will define w_2 as the solution of the problem with homogeneous conditions on the boundary of $[0, 1]$, and at ξ:

$$w_2(0) = 0, \qquad w_2(1) = 0,$$

$$w_2(\xi + 0) - w_2(\xi - 0) = 0,$$

$$p(\xi + 0)w_2'(\xi + 0) - p(\xi - 0)w_2'(\xi - 0) = 0, \tag{3.7}$$

$$-(pw_2')' + qw_2 = f - z_1.$$

It is known (see [6]), that there is a unique solution for (3.7) which is continuous on $[0, 1]$. It is this solution that we will use as the function w_2.

It is apparent that the function $u = w_1 + w_2$ is a solution of the problem (3.1)–(3.4). There can be no other solution for (3.1)–(3.4). Indeed, let us assume the contrary: let u_1 be another solution of (3.1)–(3.4). Then the difference $u_1 - u$ satisfies the homogeneous problem (3.7), where $f - z_1 = 0$. But this problem has only the trivial solution, and therefore the functions u and u_1 coincide.

Thus the solution u is unique and continuous on $[0, 1]$ since both w_1 and w_2 are continuous on $[0, 1]$. Hence it follows that $u(\xi)$ is finite-valued. Let us consider two boundary-value problems:

$$-(pv_1')' + qv_1 = f \quad \text{on } (0, \xi),$$

$$v_1(0) = u_0, \qquad v_1(\xi) = u(\xi)$$

and

$$-(pv_2')' + qv_2 = f \quad \text{on } (\xi, 1),$$

$$v_2(\xi) = u(\xi), \qquad v_2(1) = u_1.$$

The coefficients p, q, f have no discontinuities in these domains, and therefore according to Theorem 1.1 both problems have unique solutions, and $v_1 \in C^{r+2}[0, \xi]$, $v_2 \in C^{r+2}[\xi, 1]$. But the function u also satisfies these equations. Thus

$$u(x) = \begin{cases} v_1(x) & \text{if } x \in [0, \xi], \\ v_2(x) & \text{if } x \in [\xi, 1], \end{cases}$$

has properties (3.6). The theorem is proved. $\qquad\qquad\qquad\qquad\qquad\square$

Assume $h_1 = \xi/N_1$, $h_2 = (1 - \xi)/N_2$ and construct the net so that it contains the point of discontinuity:

$$\omega_{h_1} = \{x_i = ih_1; i = 1, \dots, N_1 - 1\},$$

$$\omega_{h_2} = \{x_i = \xi + (i - N_1)h_2; i = N_1 + 1, \dots, N_1 + N_2 - 1\}, \quad (3.8)$$

$$\omega_h = \omega_{h_1} \cup \omega_{h_2} \cup \{x_{N_1} = \xi\}.$$

We should note that within each interval where the coefficients are smooth the mesh-sizes are uniform. Let us introduce a set of midpoints

$$\breve{\omega}_h = \{x_{i+1/2} = (x_i + x_{i+1})/2; i = 0, 1, \dots, N_1 + N_2 - 1\}. \quad (3.9)$$

Consider the difference equations

$$-(pu_{\bar{x}}^h)_{\hat{x}} + qu^h = f \quad \text{on } \omega_{h_1}, \tag{3.10}$$

$$-(pu_{\bar{x}}^h)_{\hat{x}} + qu^h = f \quad \text{on } \omega_{h_2}. \tag{3.11}$$

Since the nets ω_{h_1} and ω_{h_2} are uniform, with mesh-sizes h_1 and h_2, respectively, the meaning of the difference derivatives is simple. Because the functions q and f are defined ambiguously at the point ξ we will transform equations (3.10), (3.11) as follows:

$$-(pu_{\bar{x}}^h)_{\hat{x}}|_\xi + \frac{h_1 q(\xi - 0) + h_2 q(\xi + 0)}{h_1 + h_2} u^h(\xi)$$

$$= \frac{h_1 f(\xi - 0) + h_2 f(\xi + 0)}{h_1 + h_2} - \frac{2g}{h_1 + h_2}. \qquad (3.12)$$

Here, from the definition of difference derivatives for nonuniform nets, we have

$$(pu_{\bar{x}}^h)_{\hat{x}}|_\xi = (pu_{\bar{x}}^h)_{\hat{x}}|_{x_{N_1}} = [(pu_{\bar{x}}^h)|_{x_{N_1+1/2}} - (pu_{\bar{x}}^h)|_{x_{N_1-1/2}}]/(x_{N_1+1/2} - x_{N_1-1/2})$$

$$= \frac{2}{h_2(h_1 + h_2)} p(\xi + h_2/2)[u^h(\xi + h_2) - u^h(\xi)]$$

$$- \frac{2}{h_1(h_1 + h_2)} p(\xi - h_1/2)[u^h(\xi) - u^h(\xi - h_1)].$$

The two boundary conditions

$$u^h(0) = u_0, \qquad u^h(1) = u_1 \qquad (3.13)$$

complete the description of the system of difference equations. The scheme we have constructed is uniquely solvable and stable (see [112]).

The principles used in constructing equations of type (3.12) are described in [11, 87, 114], where difference equations for discontinuous coefficients are also examined.

We should note that in order to successfully use the extrapolation procedure we need that there be only one net parameter, while there are two of them in this difference scheme. To eliminate this unnecessary degree of freedom we impose the restriction

$$h_1 = c_2 h_2, \qquad (3.14)$$

where c_2 remains constant independent of all changes in the mesh-sizes. Now it is possible to introduce a single parameter $\hbar = \sqrt{h_1 h_2}$ which characterizes the net:

$$h_1 = \sqrt{c_2}\,\hbar, \qquad h_2 = \hbar/\sqrt{c_2}. \qquad (3.15)$$

Theorem 3.2. *Suppose that for* (3.1)–(3.4) *the smoothness condition* (3.5) *holds, and that for* (3.10)–(3.13) *condition* (3.14) *is satisfied. Then the approximate solution can be expanded as*

$$u^h = u + \sum_{j=1}^{l} h^{2j}v_j + h^r\eta^h \quad \text{on } \bar\omega_h. \tag{3.16}$$

Here $l = [(r - 1)/2]$, *the functions* v_j *belong to* Q_ξ^{r+2-2j} *and the modulus of the net function* η^h *is uniformly bounded on* $\bar\omega_h$ *independent of* h.

PROOF. Assume

$$v_0 = u$$

and consider the set of differential equations

$$-(pv_j')' + qv_j = \sum_{1 \le s+k \le j} \frac{(pv_{j-s-k}^{(2s+1)})^{(2k+1)}c_2^{s+k}}{4^{s+k}(2s+1)!\,(2k+1)!} \quad \text{on } (0, \xi), \tag{3.17}$$

$$-(pv_j')' + qv_j = \sum_{1 \le s+k \le j} \frac{(pv_{j-s-k}^{(2s+1)})^{(2k+1)}c_2^{-s-k}}{4^{s+k}(2s+1)!\,(2k+1)!} \quad \text{on } (\xi, 1), \tag{3.18}$$

$$v_j(0) = 0, \qquad v_j(1) = 0, \tag{3.19}$$

$$v_j(\xi + 0) - v_j(\xi - 0) = 0, \tag{3.20}$$

$$p(\xi + 0)v_j'(\xi + 0) - p(\xi - 0)v_j'(\xi - 0) = - \sum_{1 \le s+k \le j} [c_2^{-s-k}(pv_{j-s-k}^{(2k+1)})^{(2s)}|_{\xi+0}$$

$$- c_2^{s+k}(pv_{j-s-k}^{(2k+1)})^{(2s)}|_{\xi-0}]/[(2s)!\,(2k+1)!\,4^{s+k}] \qquad j = 1, \ldots, l. \tag{3.21}$$

Let us assume that the functions $v_0, \ldots, v_{j-1}$ are known and $v_k \in Q_\xi^{r+2-2k}$, $0 \le k \le j - 1$. Then one can consider the equalities (3.17)–(3.20) as the problem which defines the function v_j. If we extend right-hand sides of (3.17) and (3.18) by their left and right limits at the point ξ, then they become $r - 2j$ times continuously differentiable on the intervals $[0, \xi]$ and $[\xi, 1]$, correspondingly. The constants on the right-hand side (3.21) are finite, because they are limits of functions having a discontinuity of the first kind. Therefore on the basis of Theorem 3.1 the function v_j is defined uniquely and belongs to Q_ξ^{r-2j+2}.

Let us prove that the net function

$$\eta^h = h^{-r}\left(u^h - \sum_{j=1}^{l} h^{2j}v_j\right) \quad \text{on } \bar\omega_h \tag{3.22}$$

is bounded as $h \to 0$. To do this we solve equation (3.22) with respect to u^h and put the result in (3.10). Then

$$-\sum_{j=0}^{l} h^{2j}(p(v_j)_{\bar x})_{\hat x} - h^r(p\eta_{\bar x}^h)_{\hat x} + \sum_{j=0}^{l} h^{2j}qv_j + h^r q\eta^h = f. \tag{3.23}$$

We apply Lemma 1.2 of §7.1 to the terms as follows:

$$(p(v_j)_{\hat{x}})_{\hat{x}} = \sum_{k=0}^{l-j} h_1^{2k} \frac{4^{-k}}{(2k+1)!} \sum_{s=0}^{l-j-k} h_1^{2s} \frac{4^{-s}}{(2s+1)!} (pv_j^{(2k+1)})^{(2s+1)} + h_1^{r-2j} \varkappa_j^h,$$

(3.24)

where

$$|\varkappa_j^h| \le c_3 \quad \text{on } \omega_{h_1}.$$

Substitute the expansion obtained into (3.23). Taking into consideration that $h_1 = \sqrt{c_2}\, h$ we combine similar terms and obtain the equality

$$\sum_{j=0}^{l} h^{2j}\left(- \sum_{0 \le k+s \le j} \frac{c_2^{k+s}(pv_{j-s-k}^{(2k+1)})^{(2s+1)}}{(2k+1)!\,(2s+1)!\,4^{k+s}} + qv_j\right)$$

$$- h^r(p\eta_{\hat{x}}^h)_{\hat{x}} - h^r \sum_{j=0}^{l} c_2^{r/2-j}\varkappa_j^h + h^r q\eta^h = f.$$

On the basis of equation (3.1) for v_0, and using the definition the functions v_j, we eliminate all terms of $h^0, \ldots, h^{2l}$ orders. We divide all remaining terms by h^r and finally obtain

$$-(p\eta_{\hat{x}}^h)_{\hat{x}} + q\eta^h = \sum_{j=0}^{l} c_2^{r/2-j}\varkappa_j^h \quad \text{on } \omega_{h_1},$$

(3.25)

where

$$\max_{\omega_{h_1}} \sum_{j=0}^{l} c_2^{r/2-j}|\varkappa_j^h| \le c_4 = c_3 \max\{1, c_2^{r/2}\}.$$

(3.26)

Carrying out a similar procedure on the net ω_{h_2}, we have

$$-(p\eta_{\hat{x}}^h)_{\hat{x}} + q\eta^h = -\sum_{j=0}^{l} c_2^{-r/2+j}\varkappa_j^h \quad \text{on } \omega_{h_2},$$

(3.27)

where

$$\max_{\omega_{h_2}} \sum_{j=0}^{l} c_2^{-r/2+j}|\varkappa_j^h| \le c_5 = c_3 \max\{1, c_2^{-r/2}\}.$$

(3.28)

Now substitute the expression for u^h into the equation at the point ξ:

$$-\sum_{j=0}^{l} h^{2j}(p(v_j)_{\hat{x}})_{\hat{x}} - h^r(p\eta_{\hat{x}}^h)_{\hat{x}} + \sum_{j=0}^{l} h^{2j}v_j[h_1 q(\xi - 0) + h_2 q(\xi + 0)]/(h_1 + h_2)$$

$$+ h^r \eta^h[h_1 q(\xi - 0) + h_2 q(\xi + 0)]/(h_1 + h_2)$$

$$= [h_1 f(\xi - 0) + h_2 f(\xi + 0)]/(h_1 + h_2) - 2g/(h_1 + h_2). \quad (3.29)$$

Apply Lemma 1.1 of §7.1 to the divided difference

$$(v_j(\xi + h_2/2))_{\hat{x}} = \sum_{k=0}^{l-j} h_2^{2k} \frac{4^{-k}}{(2k+1)!} v^{(2k+1)}(\xi + h_2/2) + h_2^{r-2j+1}\sigma_j,$$

where

$$|\sigma_j| \le c_6.$$

Multiply this equality by $p(\xi + h_2/2)$ and expand each term $pv_j^{(2k+1)}$ by Taylor's formula, taking into account that the derivatives have finite limits from the right at the point ξ:

$$p(v_j)_{\hat{x}}|_{\xi+h_2/2} = \sum_{k=0}^{l-j} \frac{h_2^{2k}4^{-k}}{(2k+1)!} \sum_{s=0}^{r-2j-2k} h_2^s \frac{2^{-s}}{s!} (pv_j^{(2k+1)})^{(s)}|_{\xi+0} + h_2^{r-2j+1}\rho_j^+,$$

(3.30)

where

$$|\rho_j^+| \le c_7, \qquad j = 0, 1, \ldots, l.$$

(3.31)

We have a similar expression on the left:

$$p(v_j)_{\hat{x}}|_{\xi-h_1/2}$$
$$= \sum_{k=0}^{l-j} \frac{h_1^{2k}4^{-k}}{(2k+1)!} \sum_{s=0}^{r-2j-2k} (-h_1)^s \frac{2^{-s}}{s!} (pv_j^{(2k+1)})^{(s)}|_{\xi-0} + h^{r-2j+1}\rho_j^-, \quad (3.32)$$

where

$$|\rho_j^-| \le c_8, \qquad j = 0, 1, \ldots, l.$$

(3.33)

Subtract (3.32) from (3.30) and divide both sides of the resulting equality by $(h_1 + h_2)/2$:

$$(p(v_j)_{\hat{x}})_{\hat{x}}|_\xi = \sum_{k=0}^{l-j} \sum_{s=0}^{l-j-k} \frac{4^{-s-k}}{(2s+1)!\,(2k+1)!} [h_2^{2k+2s+1}(pv_j^{(2k+1)})^{(2s+1)}|_{\xi+0}$$
$$+ h_1^{2k+2s+1}(pv_j^{(2k+1)})^{(2s+1)}|_{\xi-0}]/(h_1 + h_2)$$
$$+ \sum_{k=0}^{l-j} \sum_{s=0}^{l-j-k} \frac{2^{-2s-2k+1}}{(2s)!\,(2k+1)!} [h_2^{2k+2s}(pv_j^{(2k+1)})^{(2s)}|_{\xi+0}$$
$$- h_1^{2k+2s}(pv_j^{(2k+1)})^{(2s)}|_{\xi-0}]/(h_1 + h_2) + h^{r-2j}\mu_j/(h_1 + h_2),$$

(3.34)

where

$$|\mu_j| \le c_9, \qquad j = 0, 1, \ldots, l$$

(3.35)

from (3.31), (3.33).

On the right-hand side of (3.34) we have divided the terms into two groups so that in the first group we have those terms which are nonzero on the uniform part of the net ω_h, while in the second group we have those which vanish on the uniform part of the net ω_h. Substituting this expansion into (3.29) and using the relations (3.15) between the mesh sizes h_i and $\hbar$ we have

$$-\sum_{j=0}^{l} \hbar^{2j} \Bigg\{ \sum_{0 \le k+s \le j} \frac{4^{-s-k}}{(2s+1)!\,(2k+1)!}$$

$$\times \left[h_2 c_2^{-s-k}(pv_{j-s-k}^{(2k+1)})^{(2s+1)}|_{\xi+0} + h_1 c_2^{s+k}(pv_{j-s-k}^{(2k+1)})^{(2s+1)}|_{\xi-0} \right]/(h_1+h_2)$$

$$+ v_j(\xi)[h_2 q(\xi+0) + h_1 q(\xi-0)]/(h_1+h_2) \Bigg\}$$

$$-\sum_{j=0}^{l} \hbar^{2j} \sum_{0 \le k+s \le j} \frac{2^{-2s-2k+1}}{(2s)!\,(2k+1)!}$$

$$\times \left[c_2^{-k-s}(pv_{j-s-k}^{(2k+1)})^{(2s)}|_{\xi+0} - c_2^{k+s}(pv_{j-s-k}^{(2k+1)})^{(2s)}|_{\xi-0} \right]/(h_1+h_2)$$

$$- \hbar^r(p\eta_{\hat{x}}^\hbar)_\xi|_\xi + \hbar^r\eta^\hbar(\xi)[h_1 q(\xi-0) + h_2 q(\xi+0)]/(h_1+h_2)$$

$$+ \hbar^r v^\hbar/(h_1+h_2)$$

$$= [h_1 f(\xi-0) + h_2 f(\xi+0)]/(h_1+h_2) - 2g/(h_1+h_2), \qquad (3.36)$$

where $v^\hbar$ (from (3.35)) satisfies the condition

$$|v^\hbar| \le c_{10}. \qquad (3.37)$$

In order to simplify (3.36) we use $v_0 = u$. Therefore the equalities derived from equation (3.2) and the relations for the corresponding limits:

$$-(pv_0')'|_{\xi+0} + q(\xi+0)v_0(\xi) = f(\xi+0),$$

$$-(pv_0')'|_{\xi-0} + q(\xi-0)v_0(\xi) = f(\xi-0)$$

hold.

Combining these relations results in the equality

$$-[h_1(pv_0')'|_{\xi-0} + h_2(pv_0')'|_{\xi+0}]/(h_1+h_2)$$

$$+ v_0(\xi)[h_1 q(\xi-0) + h_2 q(\xi+0)]/(h_1+h_2)$$

$$= [h_1 f(\xi-0) + h_2 f(\xi+0)]/(h_1+h_2). \qquad (3.38)$$

Dividing both sides of (3.4) by $(h_1+h_2)/2$ we have

$$2[(pv_0')|_{\xi+0} - (pv_0')|_{\xi-0}]/(h_1+h_2) = 2g/(h_1+h_2). \qquad (3.39)$$

Equalities (3.38) and (3.39) allow us to eliminate all terms on the right-hand side of (3.36), and those terms on the left-hand side, corresponding to the index $j = 0$.

As above from (3.17), (3.18) we obtain

$$-\sum_{0 \le k+s \le j} [h_1 c_2^{k+s}(pv_{j-s-k}^{(2k+1)})^{(2s+1)}|_{\xi-0}$$

$$+ h_2 c_2^{-k-s}(pv_{j-s-k}^{(2k+1)})^{(2s+1)}|_{\xi+0}]/[(h_1+h_2)4^{k+s}(2s+1)!\,(2k+1)!]$$

$$+ v_j(\xi)[h_1 q(\xi-0) + h_2 q(\xi+0)]/(h_1+h_2) = 0. \qquad (3.40)$$

Let us transpose all terms in the consistency condition (3.21) to the left-hand side and divide by $(h_1 + h_2)/2$. Hence we have

$$\sum_{0 \le k+s \le j} [c_2^{-k-s}(pv_{j-k-s}^{(2k+1)})^{(2s)}|_{\xi+0} - c_2^{k+s}(pv_{j-k-s}^{(2k+1)})^{(2s)}|_{\xi-0}]/$$

$$[(h_1 + h_2)(2s)!\,(2k + 1)!\,2^{2k+2s+1}] = 0. \quad (3.41)$$

Using (3.40), (3.41) in (3.36) many terms drop out. Dividing the remained terms by h^r we have

$$-(p\eta_{\bar{x}}^h)_{\hat{x}}|_\xi + \eta^h(\xi)[h_1 q(\xi - 0) + h_2 q(\xi + 0)]/(h_1 + h_2) = -v^h/(h_1 + h_2). \quad (3.42)$$

There is nothing unexpected in the fact that all terms of less than h^2 order have disappeared, because condition (3.17), (3.18), (3.21) were chosen so as to accomplish this.

From equations (3.25), (3.27), (3.42) we can estimate the solution η^h, if we take into account that corresponding to the boundary conditions (3.2), (3.13) and (3.19) we have

$$\eta^h(0) = h^{-r}\left(u^h(0) - \sum_{j=0}^l h^{2j}v_j(0)\right) = 0,$$

$$\eta^h(1) = h^{-r}\left(u^h(1) - \sum_{j=0}^l h^{2j}v_j(1)\right) = 0. \quad (3.43)$$

Indeed, for functions vanishing at the ends of the interval $[0, 1]$ and satisfying the equations

$$-(p\eta_{\bar{x}}^h)_{\hat{x}} + q\eta^h = w \quad \text{on } \omega_{h_1} \cup \omega_{h_2},$$

$$-(p\eta_{\bar{x}}^h)_{\hat{x}}|_\xi + \eta^h(\xi)[h_1 q(\xi - 0) + h_2 q(\xi + 0)]/(h_1 + h_2) = w(\xi)$$

there is a Green's difference function $G(x, t)$ such that (see [14])

$$\eta^h(x) = \sum_{t \in \omega_{h_1}} G(x, t)w(t)h_1 + G(x, \xi)w(\xi)(h_1 + h_2)/2 + \sum_{t \in \omega_{h_2}} G(x, t)w(t)h_2.$$

Moreover, the Green's function is uniformly bounded:

$$|G(x, t)| \le 1/c_1 \quad \forall\, x \in \bar{\omega}_h, \forall\, t \in \bar{\omega}_h.$$

Hence

$$|\eta^h(x)| \le \left(\sum_{\omega_{h_1}}|w|h_1 + |w(\xi)|(h_1 + h_2)/2 + \sum_{\omega_{h_2}}|w|h_2\right)\bigg/c_1.$$

Using inequalities (3.26), (3.28) and (3.37) to estimate $w(t)$ we have:

$$|\eta^h(x)| \le \left(\sum_{\omega_{h_1}}c_4 h_1 + |v^h|/2 + \sum_{\omega_{h_2}}c_5 h_2\right)\bigg/c_1$$

$$\le (c_4\xi + c_{10}/2 + c_5(1 - \xi))/c_1 = c_{11}.$$

Since the point $x \in \bar{\omega}_h$ was arbitrary we have proved the estimate

$$\max_{\bar{\omega}_h} |\eta^h| \leq c_{11} \tag{3.44}$$

completing the proof of the validity of the expansion (3.16). Theorem 3.2 is proved. $\qquad\square$

Having obtained (3.16) we can extrapolate using the $l + 1$ solutions of the difference problem (3.10)–(3.13) for nets with different parameters h_i; the accuracy of the corrected solution turns out to be $O(h^r)$. We should note that when we extrapolate we must not use a succession of the nets with different ratios N_1/N_2 because from $h_1 = c_2 h_2$ and $h_1 = \xi/N_1$, $h_2 = (1 - \xi)/N_2$ we have

$$N_1/N_2 = \xi/(c_2(1 - \xi)).$$

Thus the ratio N_1/N_2, determined by the first net, should remain constant for all other nets.

Now we choose the initial values $N_1 = K$ and $N_2 = L$ which define the net $\bar{\omega}_{h_1}$. Then construct the nets $\bar{\omega}_{h_i}$ with integers $N_1 = iK$ and $N_2 = iL$. Having solved the difference problem (3.10)–(3.13) on each of the nets, we will obtain solutions u^{h_i}. All these functions are defined on the net $\bar{\omega}_{h_1}$. Let us form the linear combination at its knots

$$U = \sum_{k=1}^{l+1} \gamma_k u^{h_k} \tag{3.45}$$

with weights

$$\gamma_k = \frac{2(-1)^{l+k+1} k^{2l+2}}{(l - k + 1)! \, (l + k + 1)!}. \tag{3.46}$$

Theorem 3.3. *When the assumptions of Theorem 3.2 for the corrected solutions (3.45) are satisfied then estimate*

$$\max_{\bar{\omega}_{h_1}} |U - u| \leq h_1^r c_{12} \tag{3.47}$$

with constant independent of h_i, holds.

PROOF. From Theorem 3.2 the expansions

$$u^{h_k} = u + \sum_{j=1}^{l} h^{2j} v_j + h_k^r \eta^{h_k} \quad \text{on } \bar{\omega}_{h_1}$$

obtain, with functions v_j independent of k. Hence

$$U = \sum_{k=1}^{l+1} \gamma_k u + \left(\sum_{j=1}^{l} \sum_{k=1}^{l+1} \gamma_k h_k^{2j} \right) v_j + \sum_{k=1}^{l+1} \gamma_k h_k^r \eta^{h_k}. \tag{3.48}$$

From Lemma 2.2 of §7.2 the weights γ_k satisfy the equalities

$$\sum_{k=1}^{l+1} \gamma_k = 1,$$

$$\sum_{k=1}^{l+1} \gamma_k/k^{2j} = 0, \qquad j = 1, \ldots, l. \tag{3.49}$$

Multiplying the above equality by h_1^{2j} and using

$$h_1/k = \sqrt{\xi(1 - \xi)}/\sqrt{k^2 KL} = h_k,$$

we have

$$\sum_{k=1}^{l+1} \gamma_k h_k^{2j} = 0. \tag{3.50}$$

Using (3.49) and (3.50) we transform (3.48) to

$$U = u + \sum_{k=1}^{l+1} \gamma_k h_k^r \eta^{h_k}.$$

We have

$$|U - u| \le \sum_{k=1}^{l+1} |\gamma_k| |\eta^{h_k}| h_k^r \quad \text{on } \bar{\omega}_{h_1}.$$

Using (3.44) and the explicit form of the weights γ_k we obtain (3.47), where

$$c_{12} = c_{11} \sum_{k=1}^{l+1} \frac{2k^{2l+2-r}}{(l - k + 1)! \, (l + k + 1)!}.$$

Thus Theorem 3.3 is proved. $\qquad\qquad\qquad\qquad\qquad\qquad\square$

In the case where it is necessary to determine the value of the function at a point z not coinciding with the knot of any net ω_{h_i} we interpolate. In order to obtain accuracy $O(h_i^r)$ it is useful to construct the Lagrange interpolation polynomial for r knots. To obtain the maximal expected accuracy (see [14]) we used the r knots nearest to z in §2.1. However in this case, because we have only piecewise smoothness this rule is not valid and must be reformulated so that we interpolate on knots in a region in which the function is smooth. Therefore if $z \in (0, \xi)$ then we should interpolate with respect to the r nearest knots from the set $\omega_{h_1} \cup \{0, \xi\}$. If $z \in (\xi, 1)$ then the Lagrange interpolation polynomial is constructed with respect to the r nearest knots of the set $\omega_{h_2} \cup \{\xi, 1\}$. Other than this there are no other major differences from §2.1. In the case when there is more than one point of discontinuity the algorithm will differ from that described above only in the construction of the sequence of nets. Let

$$0 = z_0 < z_1 < \cdots < z_m = 1$$

be the set of points of discontinuity (of the first kind) of the functions p, q, f and their derivatives. Let us choose m integers $N_i \geq 2$ and construct the difference net $\bar{\omega}_{h_1}$ uniformly within each interval (z_{i-1}, z_i). To do this we divide each interval into N_i equal parts and assign all the points of this subdivision to the net $\bar{\omega}_{h_1}$, including the points $z_0, \ldots, z_m$ themselves. Subsequent nets are constructed similarly, except that one makes kN_i subdivisions for each net $\bar{\omega}_{h_k}$, with

$$\hbar_k = 1/(k(N_1 \cdots N_m)^{1/m})$$

defining the parameter $\hbar_k$. Further details of the construction do not differ from those for the construction (3.45), (3.46).

3.4. The Sturm–Liouville Problem

In solving problems in mathematical physics one often has to deal with eigenvalue problems. The simplest problem of this kind is the Sturm–Liouville problem. The solution of this problem yields a family of eigenfunctions and the corresponding eigenvalues. In the majority of cases of practical interest an analytic solution of this problem appears to be difficult or completely impossible. Therefore an effective solution of the problem is found using numerical methods. These methods, however, always entail the introduction of critical approximation errors. Even if such errors are negligible for the first eigenfunctions and eigenvalues, for higher harmonics they usually turn out to be important. Thus, if it is necessary to obtain results for a wide range of eigenfunctions and eigenvalues then the approximate solutions obtained will have to be further improved. This appears to be possible using the Richardson extrapolation method. It is thus necessary to have a family of approximate solutions for different nets. An appropriate linear combination of these solutions with given coefficients will yield a solution with greater accuracy. In this section we will construct such an algorithm.

The simplest Sturm–Liouville problem asks for the values of parameter λ (eigenvalues) at which nontrivial solutions (called eigenfunctions) of the equation

$$-(py')' + qy = \lambda y,$$

$$p(x) \geq c_1 > 0, \qquad q(x) \geq 0, \qquad x \in (0, 1), \tag{4.1}$$

will exist with homogeneous boundary conditions

$$y(0) = y(1) = 0. \tag{4.2}$$

Let us require that the coefficients of the solution satisfy the smoothness conditions

$$p \in C^{r+1}[0, 1], \qquad q \in C^r[0, 1] \tag{4.3}$$

for a natural number r.

Since equation (4.1) defines the function only up to an aribtrary constant multiplier we will further assume that the eigenfunction satisfies the normalization conditions

$$\int_0^1 y^2 \, dx = 1 \qquad (4.4)$$

and

$$y'(0) > 0. \qquad (4.5)$$

Let us recall some properties of the eigenvalues and eigenfunctions of (4.1)–(4.5) (see [27, 114, 48]).

There is a countable set eigenvalues $0 < \lambda_1 < \cdots < \lambda_n < \cdots$ with corresponding eigenfunctions $y_1(x), \ldots, y_n(x), \ldots$; only one eigenfunction corresponds to each eigenvalue. The eigenfunctions form an orthonormal system, that is

$$\int_0^1 y_n y_m \, dx = 0$$

if $n \neq m$. For the eigenvalues we have the estimate

$$c_2 n^2 - c_3 \leq \lambda_n \leq c_2 n^2 + c_3, \qquad (4.6)$$

where c_2, c_3 do not depend on n.

We also have for the eigenfunctions that:

$$|y_n| \leq c_4, \qquad |y'_n| \leq c_4 n, \qquad (4.7)$$

where c_4 is a constant independing of n, x. With the help of these inequalities and (4.1) the moduli of the second derivatives can be estimated. Estimates of the higher derivatives can be obtained by differentiating equation (4.1) and using these inequalities. We get

$$\max_{[0, 1]} |y_n^{(k)}| \leq c_5 n^k, \qquad k = 0, 1, \ldots, r + 2, \qquad (4.8)$$

where the constant c_5 does not depend on n, k, x.

Let us now formulate the difference eigenvalue problem. In accordance with our usual notation we will introduce the uniform net $\bar{\omega}_h$ and consider the system of difference equations

$$-(p y_{\bar{x}}^h)_{\hat{x}} + q y^h = \lambda^h y^h \quad \text{on } \omega_h \qquad (4.9)$$

together with boundary conditions

$$y^h(0) = y^h(1) = 0. \qquad (4.10)$$

Then the Sturm–Liouville difference problem can be stated as follows: find the values of the parameter λ^h (the difference eigenvalues) to which the non-zero solutions (the difference eigenfunctions) of the system of equations (4.9), (4.10) correspond.

If we eliminate $y^h(0)$, and $y^h(1)$ from (4.9) and define

$$A = \begin{bmatrix} b_1 & e_1 & & & 0 \\ a_2 & b_2 & \ddots & & \\ & \ddots & \ddots & \ddots & \\ & & \ddots & \ddots & e_{N-2} \\ 0 & & & a_{N-1} & b_{N-1} \end{bmatrix}, \qquad Y^h = \begin{bmatrix} y^h(x_1) \\ y^h(x_2) \\ \vdots \\ y^h(x_{N-1}) \end{bmatrix},$$

where

$$a_i = e_{i-1} = -p(x_{i-1/2})/h^2, \qquad b_i = -a_i - e_i + q(x_i),$$

then the system obtained will have the form of the usual algebraic spectral problem

$$AY^h = \lambda^h Y^h \tag{4.11}$$

with a tridiagonal symmetric matrix A.

Therefore, the difference eigenfunctions are defined up to an arbitrary constant multiplier. To insure uniqueness we will assume the normalization relations

$$\sum_{\omega_h} (y^h)^2 h = 1, \tag{4.12}$$

$$y^h(x_1) > 0. \tag{4.13}$$

Let us note some properties of the Sturm–Liouville difference problem (see [114, 34]).

There are $N - 1$ difference eigenvalues $0 < \lambda_1^h < \cdots < \lambda_{N-1}^h$ with $N - 1$ corresponding difference eigenfunctions $y_1^h(x), \ldots, y_{N-1}^h(x)$ defined on ω_h. Only one eigenfunction correspond to each eigenvalue. The eigenfunctions form an orthonormal system:

$$\sum_{\omega_h} y_i^h y_j^h h = \delta_{ij},$$

where δ_{ij} is the Kronecker symbol.

We have the following estimates for the difference eigenvalues

$$c_6 n^2 \leq \lambda_n^h \leq c_7 n^2, \qquad 1 \leq n \leq N - 1, \tag{4.14}$$

where c_6, c_7 do not depend on n or h.

If the condition of smoothness (4.3) is satisfied then the estimate

$$\max_{\omega_h} |y_n^h| \leq c_8 n^{1/2} \tag{4.15}$$

will hold, with c_8 independent of n and h.

Let us fix n and expand the difference eigenfunction and eigenvalue in powers of h

$$y_n^h(x) = \sum_{s=0}^{l} h^{2s} v_s(x) + h^r \eta^h(x) \quad \text{on } \bar{\omega}_h, \tag{4.16}$$

$$\lambda_n^h = \sum_{s=0}^{l} h^{2s} \sigma_s + h^r \rho^h, \tag{4.17}$$

where $l = [(r - 1)/2]$, the functions v_s and the constants σ_s are independent of h, and ρ^h and the net function η^h are uniformly bounded by constants independent of h.

Assume that $v_j \in C^{r-2j+2}[0, 1]$ and substitute (4.16), (4.17) into (4.9). Taking into account

$$-(p(v_j)_{\hat{x}})_{\hat{x}} = - \sum_{0 \le k+s \le l-j} h^{2k+2s} \frac{(pv_j^{(2s+1)})^{(2k+1)}}{4^{k+s}(2s+1)!\,(2k+1)!} + h^{r-2j}\mu_j^h \quad \text{on } \omega_h,$$

(4.18)

where

$$\max_{\omega_h} |\mu_j^h| \le c_{11},$$

(4.19)

after canceling similar terms we obtain

$$\sum_{j=0}^{l} h^{2j}\left(- \sum_{0 \le k+s \le j} \frac{(pv_{j-k-s}^{(2k+1)})^{(2s+1)}}{(2k+1)!\,(2s+1)!\,4^{k+s}} + qv_j \right)$$

$$- h^r(p\eta_{\hat{x}}^h)_{\hat{x}} - h^r \sum_{j=0}^{l} \mu_j^h + h^r q\eta^h$$

$$= \sum_{j=0}^{l} h^{2j} \sum_{k=0}^{j} \sigma_k v_{j-k} + \sum_{j=l+1}^{2l} h^{2j} \sum_{k=j-l}^{l} \sigma_k v_{j-k}$$

$$+ h^r \lambda_n^h \eta^h + h^r \rho^h \sum_{j=0}^{l} h^{2j} v_j.$$

(4.20)

Here for convenience we use the notation $\sigma_0 = \lambda_n$, $v_0 = y_n$.

In the previous sections, by special choice of the functions v_j, we managed to eliminate all terms of lower order. After dividing the remaining terms by h^r we obtained a net equation for the function η^h, the right-hand side of which contained only bounded terms .We estimated η^h based on the stability of the difference problem.

Using this same principle let us begin to construct sufficient conditions for the defining constants σ_j and functions v_j. Equating the coefficients in h^0 on the both sides of the equation we have

$$-(pv_0')' + qv_0 = \sigma_0 v_0.$$

(4.21)

Note that it suffices to use $\sigma_0 = \lambda_n$ and $v_0 = y_n$ in order that the equality is satisfied at all points of ω_h.

Comparing the coefficients of h^2 we have

$$-(pv_1')' + qv_1 - \sum_{k+s=1} \frac{(pv_0^{(2k+1)})^{(2s+1)}}{(2k+1)!\,(2s+1)!\,4} = \sigma_0 v_1 + \sigma_1 v_0.$$

Let us use this to determine the values v_1 and σ_1, and rewrite it in the form

$$-(pv_1')' + qv_1 - \sigma_0 v_1 = (pv_0')'''/24 + (pv_0''')'/24 + \sigma_1 v_0 \qquad (4.22)$$

and use the boundary conditions

$$v_1(0) = v_1(1) = 0. \qquad (4.23)$$

These boundary conditions are a consequence of the assumption that the expansion (4.16) exists. Indeed, comparing the coefficients in even powers of h in this expansion at $x = 0$ and $x = 1$ yields

$$v_j(0) = v_j(1) = 0, \qquad j = 0, 1, \ldots, l \qquad (4.24)$$

(the first of which is automatically satisfied when $v_0 = y_n$).

   Note that one may extend equation (4.22) to the whole interval $[0, 1]$. But in this case we obtain a problem with a singular operator, since $\sigma_0 = \lambda_n$ is its eigenvalue.

Lemma 4.1. *Let $f \in C^k[0, 1]$ $(k \le r)$, λ_n an eigenvalue and y_n the corresponding eigenfunction, for (4.1)–(4.5). Then the problem*

$$-(pu')' + qu - \lambda_n u = f \quad \text{on } [0, 1], \quad u(0) = u(1) = 0 \qquad (4.25)$$

is solvable only if

$$\int_0^1 f y_n \, dx = 0;$$

if this holds then there is a unique solution $z(x)$ satisfying the condition

$$\int_0^1 z y_n \, dx = 0 \qquad (4.26)$$

and belonging to $C^{k+2}[0, 1]$.

PROOF. From [48] it follows that there is a unique solution z of (4.25), which is continuous on $[0, 1]$, all other solutions being of the form

$$u = z + \alpha y_n.$$

The smoothness of z is proved by the standard recurrence procedure. Indeed, let us rewrite equation (4.25) as follows:

$$-(pz')' = (\lambda_n - q)z + f. \qquad (4.27)$$

Since the functions z, q, and f are continuous the left-hand side of (4.27) is also continuous. Therefore by Theorem 1.1 $z \in C^2[0, 1]$. If we now assume that $p \in C^3[0, 1]$, $q \in C^2[0, 1]$ and $f \in C^2[0, 1]$, then again by Theorem 1.1, $z \in C^4[0, 1]$. Iterating, we get the stated result on smoothness.
   The lemma is proved.   □

From this lemma we see that the system (4.22), (4.23) will be solvable if the right-hand side is orthogonal to the given eigenfunction; this forces the choice of σ_1:

$$\sigma_1 = - \int_0^1 [(pv_0')'''/24 + (pv_0''')'/24]y_n \, dx,$$

where the relation $v_0 = y_n$ has been used. It is clear that the value σ_1 is independent of h, and bounded if $v_0 \in C^4[0, 1]$.

Now the question of choosing the function v_1 uniquely arises, since there are an infinite number of solutions of (4.22), (4.23); all solutions are given by

$$z_1 + \alpha_1 y_n,$$

where z_1 is the solution orthogonal to y_n and α_1 is a real parameter.

To uniquely determine the constant α_1 we use the normalization (4.12) for the difference eigenfunction. Substitute (4.16) into (4.12):

$$\sum_{\omega_h} \left[\sum_{s=0}^l h^{2s}v_s \cdot \sum_{j=0}^l h^{2j}v_j + h^r y_n^h \eta^h + h^r \eta^h \sum_{s=0}^l h^{2s}v_s \right] h = 1. \qquad (4.28)$$

We will now transform the terms on the left-hand side of (4.28) using a well-known result on the trapezoidal quadrature formula (see [68]).

Lemma 4.2. *If* $\varphi \in C^k[0, 1]$ ($k \geq 3$) *then the expansion*

$$\frac{h}{2} f(0) + \sum_{\omega_h} fh + \frac{h}{2} f(1) = \int_0^1 f \, dx + \sum_{j=1}^m h^{2j} \frac{(-1)^j B_j}{(2j)!} \int_0^1 f^{(2j)} \, dx + h^k g^h$$

is valid, where $m = [(k - 1)/2]$; B_j *are Bernoulli numbers (see* [37]):

$$B_0 = 1, \quad B_1 = \tfrac{1}{6}, \quad B_2 = \tfrac{1}{30}, \quad B_3 = \tfrac{1}{42}, \ldots$$

and the constant g^h *is uniformly bounded as* $h \to 0$.

Using Lemma 4.2, (4.28) can be rewritten

$$\sum_{j=0}^l h^{2j} \left[\sum_{s=0}^j \sum_{k=0}^{l-j} h^{2k}(-1)^k B_k/(2k)! \int_0^1 (v_s v_{j-s})^{(2k)} \, dx + h^{r-2j}v_j^h \right]$$

$$+ \sum_{\omega_h} \left[\sum_{j=l+1}^{2l} h^{2j} \sum_{k=j-l}^l v_k v_{j-k} + h^r y_n^h \eta^h + h^r \eta^h \sum_{s=0}^l h^{2s}v_s \right] h = 1,$$

$$\qquad (4.29)$$

where

$$|v_j^h| \leq c_{12}, \quad j = 0, \ldots, l.$$

Compare coefficients at like powers of h on both sides of (4.29). Equality at h^0

$$\int_0^1 v_0^2 \, dx = 1$$

is automatically realized, since $v_0 = y_n$. Comparison of the coefficients in (4.29) for h^2 yields

$$2 \int_0^1 v_0 v_1 \, dx - \frac{1}{12} \int_0^1 (v_0^2)'' \, dx = 0.$$

If one chooses v_1 as $z_1 + \alpha_1 y_n$ then this equality and the condition that z_1 and y_n be orthogonal allows us to find the coefficient α_1:

$$\alpha_1 = \frac{1}{24} \int_0^1 (v_0^2)'' \, dx. \tag{4.30}$$

Assume

$$v_1 = z_1 + \alpha_1 y_n, \tag{4.31}$$

where z_1 is orthogonal to y_n and solves (4.22), (4.23) and α_1 is defined in (4.30).

Comparing coefficients of h^4 in (4.20), we have

$$-\sum_{0 \le k+s \le 2} \frac{(p v_{2-k-s}^{(2k+1)})^{(2s+1)}}{(2k+1)! \, (2s+1)! \, 4^{k+s}} + q v_2 = \sigma_0 v_2 + \sigma_1 v_1 + \sigma_2 v_0 \quad \text{on } \omega_h.$$

From this equality we can construct an equation for function v_2:

$$-(p v_2')' + q v_2 - \sigma_0 v_2 = \sum_{1 \le k+s \le 2} \frac{(p v_{2-k-s}^{(2k+1)})^{(2s+1)}}{(2k+1)! \, (2s+1)! \, 4^{k+s}}$$

$$+ \sigma_1 v_1 + \sigma_2 v_0 \quad \text{on } [0, 1], \tag{4.32}$$

to which we add the boundary conditions

$$v_2(0) = v_2(1) = 0 \tag{4.33}$$

(see (4.24)). The system (4.32), (4.33) will have a solution if the right-hand side of (4.32) is orthogonal to y_n. To insure this we need:

$$\sum_{1 \le k+s \le 2} \int_0^1 \frac{(p v_{2-k-s}^{(2k+1)})^{(2s+1)} y_n}{(2k+1)! \, (2s+1)! \, 4^{k+s}} \, dx + (\sigma_1 \alpha_1 + \sigma_2) \int_0^1 y_n^2 \, dx = 0,$$

hence

$$\sigma_2 = -\sigma_1 \alpha_1 - \sum_{1 \le k+s \le 2} \int_0^1 \frac{(p v_{2-k-s}^{(2k+1)})^{(2s+1)} y_n}{(2k+1)! \, (2s+1)! \, 4^{k+s}} \, dx. \tag{4.34}$$

Now we write out the set of all solutions as

$$v_2 = z_2 + \alpha_2 y_n \tag{4.35}$$

where z_2 is the solution orthogonal to y_n. We choose the coefficient α_2 so that the coefficients of h^4 in (4.29) are equal

$$\int_0^1 (2 v_0 v_2 + v_1^2) \, dx - \frac{1}{6} \int_0^1 (v_1 v_0)'' \, dx + \frac{1}{720} \int_0^1 (v_0^2)^{(IV)} \, dx = 0.$$

Taking (4.35) into account we obtain a relation for α_2:

$$\alpha_2 = -\frac{1}{2} \int_0^1 v_1^2 \, dx + \frac{1}{12} \int_0^1 (v_1 v_0)'' \, dx - \frac{1}{1440} \int_0^1 (v_0^2)^{(IV)} \, dx.$$

It is easily seen that this process can be continued.

So far we have assumed that the expansions (4.16), (4.17) exist *a priori*. Under this assumption we then calculate the terms.

Let us now prove that the expansions (4.16), (4.17) really do converge.

Theorem 4.3. *Let n be a fixed integer and assume that conditions (4.3) hold for the Sturm–Liouville problem (4.1), (4.2), (4.4), (4.5). Then an $h_0 > 0$ exists such that the nth eigenvalue λ_n^h of the Sturm–Liouville difference problem (4.9), (4.10), (4.12), (4.13) and the corresponding net eigenfunction $y_n^h(x)$, $h \le h_0$ can be expanded:*

$$y_n^h = y_n + \sum_{s=1}^l h^{2s} v_s + h^r \eta^h \quad \text{on } \bar\omega_h, \tag{4.36}$$

$$\lambda_n^h = \lambda_n + \sum_{s=1}^l h^{2s} \sigma_s + h^r \rho^h, \tag{4.37}$$

where λ_n is the eigenvalue of the Sturm–Liouville differential equation and y_n the corresponding eigenfunction. Here $l = [(r-1)/2]$, the functions v_s and the constants σ_s are independent of h, $v_s \in C^{r-2s+2}[0, 1]$ and the reminders satisfy the estimates

$$|\rho^h| \le c_9, \tag{4.38}$$

$$\max_{\bar\omega_h} |\eta^h| \le c_{10} \tag{4.39}$$

with constants independent of h.

PROOF. Assume $v_0 = y_n$, $\sigma_0 = \lambda_h$ and construct the system of differential equations

$$-(p v_j')' + q v_j - \sigma_0 v_j = f_j + \sum_{k=1}^{j-1} \sigma_k v_{j-k} + \sigma_j v_0 \quad \text{on } [0, 1], \tag{4.40}$$

$$v_j(0) = v_j(1) = 0, \tag{4.41}$$

where

$$f_j = \sum_{1 \le k+s \le j} \frac{(p v_{j-k-s}^{(2s+1)})^{(2k+1)}}{4^{s+k}(2k+1)!\,(2s+1)!}, \tag{4.42}$$

$$\sigma_j = -\int_0^1 f_j y_n \, dx - \sum_{k=1}^{j-1} \sigma_k \int_0^1 v_{j-k} y_n \, dx, \qquad j = 1, 2, \ldots, l. \tag{4.43}$$

Find a set of functions v_j satisfying the additional condition

$$\int_0^1 v_j y_n \, dx = - \sum_{\substack{1 \le k+s \le j \\ s \ne j}} \frac{(-1)^k B_k}{2(2k)!} \int_0^1 (v_s v_{j-k-s})^{(2k)} \, dx, \qquad j = 1, 2, \ldots, l.$$

(4.44)

We have already shown how to find the functions v_1, v_2. Assume that functions $v_0, \ldots, v_{j-1}$ satisfying equations (4.40)–(4.44) have been found with $v_k \in C^{r+2-2k}[0, 1]$, $k = 0, \ldots, j - 1$. Find a function v_j satisfying equations (4.40)–(4.44). Note that by assumption the right-hand side of (4.40) is a linear combination of the known functions $v_0, \ldots, v_{j-1}$ and belongs to $C^{r-2j}[0, 1]$. It is also orthogonal to y_n since from (4.43) and the normalization of y_n and v_0 we have

$$\int_0^1 \left(f_j + \sum_{k=1}^{j-1} \sigma_k v_{j-k} + \sigma_j v_0 \right) y_n \, dx = 0.$$

Therefore from Lemma 4.1 there is a unique solution of (4.40)–(4.43) in $C^{r-2j+2}[0, 1]$ if we require it to be orthogonal to y_n. Denote this solution by z_j. We will seek for function v_j among the solutions $z_j + \alpha_j y_n$ with real α_j defined (from the additional condition (4.44)) by

$$\alpha_j = - \sum_{\substack{1 \le k+s \le j \\ s \ne j}} \frac{(-1)^k B_k}{2(2k)!} \int_0^1 (v_s v_{j-k-s})^{(2k)} \, dx.$$

(4.45)

From the continuity of v_k, $k = 0, \ldots, j - 1$ we conclude that α_j is finite. Therefore this

$$v_j = z_j + \alpha_j y_n$$

is the function sought for, and belongs to $C^{r-2j+2}[0, 1]$.

All the functions $v_0, v_1, \ldots, v_l$ are defined in the same way.

Let λ_n^h be the nth eigenvalue of (4.9), (4.10) and let y_n^h be the corresponding difference eigenfunction satisfying the normalization conditions (4.12), (4.13). It is obvious that h should be less than $1/n$. Assuming that the functions y_n^h, v_j and constants λ_n^h, σ_j are known we define the functions η^h and ρ^h by

$$\eta^h = h^{-r}\left(y_n^h - \sum_{j=1}^l h^{2j} v_j \right) \qquad \text{on } \bar{\omega}_h,$$

(4.46)

$$\rho^h = h^{-r}\left(\lambda_n^h - \sum_{j=1}^l h^{2j} \sigma_j \right)$$

(4.47)

and show that they are bounded when $h \to 0$.

Solve for y_n^h and λ_n^h in (4.46), (4.47) and substitute them into equation (4.9). It is evident that we again get (4.20). Using equation (4.1) we eliminate terms of h^0 order in (4.20). From the definitions of the functions v_j (4.40) and constants σ_j (4.43) we eliminate all $h^2, \ldots, h^{2l}$ order terms as follows:

$$-(p\eta^h_{\hat{x}})_{\hat{x}} + q\eta^h - \lambda^h_n \eta^h = \sum_{j=0}^{l} \mu^h_j + h^{2l+2-r} \sum_{j=0}^{l-1} h^{2j} \sum_{k=j+1}^{l} \sigma_k v_{j-k} + \rho^h \sum_{j=0}^{l} h^{2j} v_j.$$

(4.48)

Note that the right-hand side is independent of η^h. Since this identity must hold for the real function η^h it must be solvable with respect to η^h. The condition for solvability is given by the following lemma.

Lemma 4.4. Let λ^h_n be a difference eigenvalue and y^h_n the corresponding difference eigenfunction. Then the difference problem

$$-(pw^h_{\hat{x}})_{\hat{x}} + qw^h - \lambda^h_n w^h = f \quad on \ \omega_h, \quad w^h(0) = w^h(1) = 0 \quad (4.49)$$

is solvable only if

$$\sum_{\omega_h} f y^h_n h = 0; \quad (4.50)$$

in this case the set of all solutions can be represented in the form

$$w^h = z^h + \alpha y^h_n \quad on \ \bar{\omega}_h, \quad (4.51)$$

where α is an arbitrary real parameter, and z^h is the solution of (4.49), (4.50) with smallest norm

$$\sum_{\omega_h} (z^h)^2 h.$$

In addition, from the estimate

$$\max_{\omega_h} |f| \leq c_{12} \quad (4.52)$$

it follows that

$$\max_{\bar{\omega}_h} |z^h| \leq c_{13}. \quad (4.53)$$

PROOF. Write the problem in matrix form:

$$(A - \lambda^h_n I)W^h = F, \quad (4.54)$$

where I is the identity matrix, and the symmetric matrix A is as in (4.11). Here the vectors W^h and F are defined by

$$W^h = \begin{bmatrix} w^h(x_1) \\ \vdots \\ w^h(x_{N-1}) \end{bmatrix}, \quad F^h = \begin{bmatrix} f(x_1) \\ \vdots \\ f(x_{N-1}) \end{bmatrix}.$$

Since λ^h_n is an eigenvalue of A then

$$\det(A - \lambda^h_n I) = 0.$$

The system (4.54) has a solution only when the right-hand side F is orthogonal to the kernel of $A - \lambda_n^h I$ as an operator in $(N - 1)$-dimensional Euclidean space. The kernel is one-dimensional and consists of the vectors αY_n^h, where α is an arbitrary real parameter, and Y_n^h is the eigenfunction of matrix A corresponding to λ_n^h. Thus for a solution of (4.49) to exist it is necessary and sufficient to have $F^T Y_n^h = 0$. Multiplying this relation by h we get (4.50). Therefore (4.54) is solvable, and any solution can be written as

$$W^h = Z^h + \alpha Y_n^h,$$

where Z^h is a normal solution of the system (see [134]), i.e., the solution with the smallest norm $(Z^h)^T Z^h$. The expression

$$\sum_{\omega_h} (z^h)^2 h$$

coincides with this value except for a factor h if we define the net function $z^h(x)$ by

$$Z^h = \begin{bmatrix} z^h(x_1) \\ \vdots \\ z^h(x_{N-1}) \end{bmatrix}.$$

Thus the representation (4.51) is valid. To prove the (4.53) we write the right-hand side of (4.49) as

$$f = \sum_{\substack{i=1 \\ i \neq n}}^{N-1} \beta_i y_i^h.$$

This is possible from (4.50). The real parameters β_i will thus satisfy

$$\sum_{\substack{i=1 \\ i \neq n}}^{N-1} \beta_i^2 = \sum_{\omega_h} f^2 h \le c_{12}^2. \tag{4.55}$$

Thus the solution $z^h(x)$ has the form

$$z^h = \sum_{\substack{i=1 \\ i \neq n}}^{N-1} \frac{\beta_i}{\lambda_i^h - \lambda_n^h} y_i^h.$$

We will now estimate the error in our approximate eigenvalues. For this purpose we will use the fact that the difference eigenvalues and eigenfunctions converge to the true ones.

Lemma 4.5 (see [112]). *Suppose the coefficients p, q in (4.1), (4.2) belong to $C^2[0, 1]$. Then the nth solution of the difference equation (4.9), (4.10) converges to the solution of the corresponding differential equation as h^2:*

$$\max_{[0, 1]} |y_n - y_n^h| \le c_{14} h^2,$$

$$|\lambda_n - \lambda_n^h| \le c_{15} h^2,$$

where the constants c_{14} and c_{15} are independent of h (but dependent on n).

Consider the eigenvalues of (4.1), (4.2). Since they increase in n then either λ_{n+1} or λ_{n-1} is the nearest to λ_n. Denote the least distance by a:

$$a = \min\{|\lambda_n - \lambda_{n-1}|, |\lambda_n - \lambda_{n+1}|\}.$$

Note that

$$|\lambda_n - \lambda_i| \geq a > 0 \quad \forall\, i \neq n.$$

Let us now consider the eigenvalues of the difference equation. The eigenvalue nearest to the eigenvalue λ_n^h is either λ_{n+1}^h or λ_{n-1}^h. For simplicity assume it is λ_{n+1}^h. Then

$$|\lambda_n^h - \lambda_i^h| \geq |\lambda_n^h - \lambda_{n+1}^h| \quad \forall\, i \neq n. \tag{4.56}$$

By Lemma 4.5 each of λ_n^h and λ_{n+1}^h converge to λ_n and λ_{n+1}, respectively, when $h \to 0$. Therefore one can find $\varepsilon_1 > 0$ such that $\forall\, h < \varepsilon_1$

$$|\lambda_n - \lambda_n^h| \leq a/4, \qquad |\lambda_{n+1} - \lambda_{n+1}^h| \leq a/4.$$

Thus

$$
\begin{aligned}
|\lambda_n^h - \lambda_{n+1}^h| &= |\lambda_n^h - \lambda_n + \lambda_n - \lambda_{n+1} + \lambda_{n+1} - \lambda_{n+1}^h| \\
&\geq |\lambda_n - \lambda_{n+1}| - |\lambda_n^h - \lambda_n| - |\lambda_{n+1} - \lambda_{n+1}^h| \geq a/2.
\end{aligned}
$$

Taking into account (4.46) we have

$$|\lambda_n^h - \lambda_i^h| \geq a/2 \quad \forall\, i \neq n, \forall\, h \leq \varepsilon_1.$$

This relation and (4.55) lead to

$$\sum_{\omega_h} (z^h)^2 h \leq \sum_{\substack{i=1 \\ i \neq n}}^{N-1} 4\beta_i^2/a^2 \leq 4c_{12}^2/a^2.$$

Note that z^h satisfies

$$-(pz_{\bar{x}}^h)_{\hat{x}} + qz^h = f + \lambda_n^h z^h \quad \text{on } \omega_h,$$
$$z^h(0) = z^h(1) = 0 \tag{4.57}$$

with bounded right-hand side $g = f + \lambda_n^h z^h$. From inequality (4.14) we have

$$
\left(\sum_{\omega_h} g^2 h\right)^{1/2} \leq \left(\sum_{\omega_h} f^2 h\right)^{1/2} + |\lambda_n^h|\left(\sum_{\omega_h} (z^h)^2 h\right)^{1/2}
$$
$$
\leq c_{12} + 2c_{12}c_7 n^2/a.
$$

Use the following estimate (from [114]) for z^h

$$\max_{[0,1]} |z^h| \leq \frac{1}{c_1}\left(\sum_{\omega_h} g^2 h\right)^{1/2}.$$

The estimate (4.53) follows from the last two relations if we assume

$$c_{13} = c_{12}(1 + 2c_7 n^2/a)/c_1.$$

Lemma 4.4 is proved.

Let us again consider equation (4.48).

Let us make sure that the net function η^h is zero at the endpoints of the interval $[0, 1]$. Indeed

$$\eta^h(b) = h^{-r}\left(y_n^h(b) - \sum_{j=0}^{l} h^{2j} v_j(b)\right) = 0 \qquad (4.58)$$

if $b = 0$ or $b = 1$ due to the homogeneous boundary conditions (4.2), (4.10), and (4.41). From Lemma 4.4 the right-hand side of (4.48) should be orthogonal (on the net) to y_n^h, thus

$$\rho^h = -\left[\sum_{\omega_h}\left(\sum_{j=0}^{l} \mu_j^h + h^{2l+2-r}\sum_{j=0}^{l-1} h^{2j}\sum_{k=j+1}^{l} \sigma_k v_{j-k}\right)y_n^h h\right] \Big/ \left[\sum_{\omega_h} y_n^h \sum_{j=0}^{l} h^{2j} v_j h\right].$$

The modulus of the numerator is estimated using the Cauchy–Schwartz–Bunyakovsky inequality:

$$\left(\sum_{\omega_h}\left(\sum_{j=0}^{l} \mu_j^h + h^{2l+2-r}\sum_{j=0}^{l-1} h^{2j}\sum_{k=j+1}^{l-1} \sigma_k v_{j-k}\right)^2 h\right)^{1/2} \leq c_{11}(l+1) + c_{16}c_{17}\, l^2,$$

where the constant c_{11} is gotten from (4.19),

$$c_{16} = \max_{0 \leq k \leq l} |\sigma_k|, \qquad c_{17} = \max_{\substack{[0,1] \\ 0 \leq j \leq l}} |v_j|. \qquad (4.59)$$

Let us estimate the denominator. From Lemma 4.5 it follows that y_n^h converges to y_n uniformly over $\bar{\omega}_h$ when $h \to 0$. Thus for some $\varepsilon_2 > 0$

$$\max_{\omega_h} |y_n^h - y_n| \leq \tfrac{1}{4} \quad \forall\, h \leq \varepsilon_2.$$

The functions v_j are continuous on $[0, 1]$, and consequently are bounded. Thus

$$\lim_{h \to 0}\max_{[0,1]}\left|\sum_{j=1}^{l} h^{2j} v_j\right| = 0.$$

Choose $\varepsilon_3 > 0$ such that

$$\max_{[0,1]}\left|\sum_{j=1}^{l} h^{2j} v_j\right| \leq \tfrac{1}{4} \quad \forall\, h \leq \varepsilon_3.$$

Hence

$$\sum_{\omega_h} y_n^h \sum_{j=0}^{l} h^{2j} v_j h = \sum_{\omega_h} y_n^h\left(y_n^h - y_n^h + v_0 + \sum_{j=1}^{l} h^{2j} v_j\right)h$$

$$= 1 - \sum_{\omega_h} y_n^h(y_n^h - v_0)h + \sum_{\omega_h} y_n^h \sum_{j=1}^{l} h^{2j} v_j h \geq 1 - \frac{1}{2}\sum_{\omega_h} |y_n^h| h$$

$$\geq 1 - \frac{1}{2}\left(\sum_{\omega_h} (y_n^h)^2 h\right)^{1/2} = \frac{1}{2}, \qquad h \leq \min\{\varepsilon_2, \varepsilon_3\}.$$

Therefore for $h \leq \min\{\varepsilon_1, \varepsilon_2, \varepsilon_3\}$ we have

$$|\rho^h| \leq c_{18} = 2(l + 1)c_{11} + 2l^2 c_{16} c_{17}. \tag{4.60}$$

For such a ρ^h the right-hand side of (4.48) is bounded uniformly by the constant

$$c_{18}/2 + c_{18}(l + 1)c_{17}$$

and consequently the solution of (4.48), (4.58) orthogonal to y_n^h is also bounded, by Lemma 4.4:

$$\max_{\omega_h} |z^h| \leq c_{19}, \qquad h \leq \min\{\varepsilon_1, \varepsilon_2, \varepsilon_3\}. \tag{4.61}$$

However, the general solution of this problem is

$$z^h + \alpha^h y_n^h;$$

thus it is necessary to find α^h and prove that it is bounded. We will determine it from (4.12). Solve (4.46) for y_n^h and substitute it into (4.12). Obviously we get (4.29). Using the normalization condition (4.4), which is valid for v_0, we can eliminate the terms in h^0 in (4.29), and rearrange the rest:

$$\sum_{j=1}^{l} h^{2j} \left(\sum_{0 \leq k+s \leq j} \frac{(-1)^k B_k}{2(2k)!} \int_0^1 (v_s v_{j-k-s})^{(2k)} \, dx \right)$$

$$+ h^r \sum_{j=0}^{l} v_j^h + \sum_{\omega_h} \left(\sum_{j=l+1}^{2l} h^{2j} \sum_{k=j-l}^{l} v_k v_{j-k} \right) h + h^r \sum_{\omega_h} \eta^h \left(y_n^h + \sum_{s=0}^{l} h^{2s} v_s \right) h = 0.$$

One can see from this that condition (4.44) leads to cancellation of all items in $h^2, h^4, \ldots, h^{2l}$. Divide the remaining terms by h^r, to get

$$\sum_{\omega_h} \eta^h \left(y_n^h + \sum_{s=0}^{l} h^{2s} v_s \right) h = \varkappa^h, \tag{4.62}$$

where

$$\varkappa^h = - \sum_{j=0}^{l} v_j^h - h^{2l+2-r} \sum_{\omega_h} \sum_{j=0}^{l-1} h^{2j} \sum_{k=j+1}^{l} v_k v_{j-k} h$$

is bounded value because

$$|\varkappa^h| \leq c_{20} = (l + 1)c_{12} + c_{17}^2 l^2. \tag{4.63}$$

Here the constant c_{12} is derived from (4.29) and the constant c_{17} is defined by (4.59).

Using the definition of the function η^h (4.46) we can transform (4.62) to read:

$$2 \sum_{\omega_h} \eta^h y_n^h h - h^r \sum_{\omega_h} (\eta^h)^2 h - \varkappa^h = 0.$$

Substituting $\eta^h = z^h + \alpha^h y_n^h$ we get an equation for α^h:

$$2\alpha^h - h^r \sum_{\omega_h} (z^h)^2 h - (\alpha^h)^2 h^r + \varkappa^h = 0.$$

Note that this equation has two roots:

$$\alpha_1^h = \left[1 + \sqrt{1 - h^r\left(\varkappa^h + h^r \sum_{\omega_h} (z^h)^2 h\right)}\right]\Bigg/ h^r,$$

$$\alpha_2^h = \left[1 - \sqrt{1 - h^r\left(\varkappa^h + h^r \sum_{\omega_h} (z^h)^2 h\right)}\right]\Bigg/ h^r.$$

Since the function η^h is defined unambiguously only one root can be used as α^h. The function z^h is uniformly bounded (see (4.61)) by the constant c_{19} and $\varkappa^h$ is uniformly bounded by the constant c_{20} (see (4.63)); therefore

$$\left| \varkappa^h + h^r \sum_{\omega_h} (z^h)^2 h \right| \le c_{20} + c_{19}^2$$

and with

$$\varepsilon_4 = [0.75/(c_{20} + c_{19}^2)]^{1/r}$$

we have

$$\alpha_1 \ge 1.5 h^{-r} \quad \forall\, h \le \varepsilon_4. \tag{4.64}$$

Assume $\alpha^h = \alpha_1^h$. From the definition of the function η^h it follows that

$$y_n^h(x_1) = y_n(x_1) + \sum_{s=1}^{l} h^{2s} v_s(x_1) + h^r \eta^h(x_1),$$

and from the relation

$$\eta^h = z^h + \alpha_1^h y_n^h$$

we have

$$(1 - h^r \alpha_1^h) y_n^h(x_1) = y_n(x_1) + \sum_{s=1}^{l} h^{2s} v_s(x_1) + h^r z^h(x_1). \tag{4.65}$$

Taking into account the boundedness of $v_s(x_1)$ and $z^h(x_1)$ we face the contradiction with the normalization conditions (4.5) and (4.13). Indeed

$$y_n(x_1) = y_n(0) + h y_n'(0) + h^2 y_n''(\xi)/2,$$

where $\xi \in (0, x_1)$. Since $y_n'(0) > 0$ and $y_n(0) = 0$ then for some ε_5 we can guarantee that $y_n(x_1) > 0$ when $h \to 0$. Thus for some ε_6 when $h \le \varepsilon_6$ the term $y_n(x_1)$ becomes the principle term on the right-hand side of (4.65) and determines its sign. The left-hand side (4.65) is negative because from (4.64) it follows that $(1 - h^r \alpha_1^h) \le -\frac{1}{2}$, and by the normalization condition $y_n^h(x_1) > 0$. The incompatability in signs is caused by the fact that we have taken $\alpha^h = \alpha_1^h$.

Therefore we will assume $\alpha^h = \alpha_2^h$. Let us make sure that this quantity is bounded. We have

$$|\alpha_2^h| = \begin{cases} [\sqrt{1-\beta}-1]/h^r & \text{if } \beta \geq 0, \\ [1-\sqrt{1+\beta}]/h^r & \text{if } \beta \in (-1, 0), \end{cases}$$

where

$$\beta = -h^r\left(\varkappa^h + h^r \sum_{\omega_h} (z^h)^2 h\right).$$

Let us consider the case $\beta \geq 0$ first. Then from the inequality

$$\sqrt{1+\beta} \leq 1 + \beta/2$$

it follows that

$$|\alpha_2^h| \leq -\tfrac{1}{2}\varkappa^h - \frac{h^r}{2}\sum_{\omega_h}(z^h)^2 h \leq (c_{20} + c_{19}^2)/2.$$

If $\beta < 0$, then

$$\beta \geq -\tfrac{3}{4} \quad \forall\, h \leq \varepsilon_4$$

and so

$$\sqrt{1+\beta} \geq 1 + 2\beta/3.$$

Therefore

$$|\alpha_2^h| \leq -2\beta/(3h^r) = 2\left(\varkappa^h + h^r\sum_{\omega_h}(z^h)^2 h\right)/3 \leq 2(c_{20} + c_{19}^2)/3.$$

Thus in both cases we get the same estimate

$$|\alpha_2^h| \leq 2(c_{20} + c_{19}^2)/3.$$

Bearing in mind (4.15) we obtain the required inequality

$$|\eta^h| \leq c_8 n^{1/2} 2(c_{20} + c_{19}^2)/3 + c_{19} \quad \forall\, h \leq h_0,$$

where

$$h_0 = \min_{1 \leq i \leq s} \varepsilon_i.$$

The proof of the theorem is thus completed. $\qquad\square$

Remark. The constants c_9 and c_{10} in (4.38) and (4.39) depend greatly on the index n of the eigenvalue used (they increase with increasing n). This dependence of the constants on n can be explicitly written down if one takes more care in the proof, but it is best illustrated by a numerical example to be presented at the end of this section.

It should be noted that both (4.36) and (4.37) admit Richardson extrapolation. The linear combinations used will have the same weights for both. Let us formulate an extrapolation algorithm for the nth eigenvalue and the corresponding nth eigenfunction.

Assume the smoothness conditions (4.3) hold. Assume $s = [(r - 1)/2]$. Let us build the nets $\bar{\omega}_{h_k}$ with the mesh-sizes $h_k = 1/N_k, k = 1, \ldots, s + 1$ for $s + 1$ integers $N_k = kL$ and construct the Sturm–Liouville difference equations (4.9)–(4.11). In each such equation find the nth eigenvalue $\lambda_n^{h_k}$ and the corresponding eigenfunction. We normalize every function in such a way as to satisfy (4.12), (4.13). All $s + 1$ of the difference eigenfunctions $y_n^{h_k}$ obtained will be defined on the net $\bar{\omega}_{h_1}$. Let us construct the linear combination

$$Y_n = \sum_{k=1}^{s+1} \gamma_k y_n^{h_k} \quad \text{on } \bar{\omega}_{h_1} \tag{4.66}$$

with weights

$$\gamma_k = \frac{2(-1)^{s-k+1} k^{2s+2}}{(s - k + 1)! \, (s + k + 1)!}. \tag{4.67}$$

Let us also construct corrected difference eigenvalues

$$\Lambda_n = \sum_{k=1}^{s+1} \gamma_k \lambda_n^{h_k} \tag{4.68}$$

with the same weights.

Theorem 4.6. *Suppose the conditions of Theorem 4.3 hold for a certain fixed integer* n. *Then the corrected difference eigenvalues* (4.68) *and difference eigenfunctions* (4.66) *satisfy the estimates*

$$\max_{\bar{\omega}_{h_1}} |Y_n - y_n| \le h_1^r c_{21}, \tag{4.69}$$

$$|\Lambda_n - \lambda_n| \le h_1^r c_{22}. \tag{4.70}$$

PROOF. According to Theorem 4.3 we can expand the eigenvalues as

$$\lambda_n^{h_k} = \lambda_n + \sum_{j=1}^{s} h_k^{2j} \sigma_j + h_k^r \rho^{h_k}, \qquad k = 1, 2, \ldots, s,$$

Summing them up with weights γ_k we have

$$\Lambda_n = \sum_{k=1}^{s+1} \gamma_k \lambda_n + \sum_{j=1}^{s} \left(\sum_{k=1}^{s+1} \gamma_k h_k^{2j} \right) \sigma_j + \sum_{k=1}^{s+1} \gamma_k h_k^r \rho^{h_k}. \tag{4.71}$$

By Lemma 2.2 of §7.2 the weights γ_k satisfy

$$\sum_{k=1}^{s+1} \gamma_k = 1, \qquad \sum_{k=1}^{s+1} \gamma_k / k^{2j} = 0, \qquad j = 1, \ldots, s,$$

which is equivalent to the relations

$$\sum_{k=1}^{s+1} \gamma_k h_k^{2j} = 0, \qquad j = 1, \ldots, s.$$

Taking these relations into account in (4.71), we obtain

$$\Lambda_n = \lambda_n + \sum_{k=1}^{s+1} \gamma_k h_k^r \rho^{h_k}$$

which results in

$$|\Lambda_n - \lambda_n| \leq \sum_{k=1}^{s+1} |\gamma_k| |h_k^r| |\rho^{h_k}|. \tag{4.72}$$

The quantity $|\rho^{h_k}|$ is bounded by the constant c_9 from (4.38); and $|\gamma_k| |h_k^r|$ can be put in the form:

$$|\gamma_k| |h_k^r| = h_1^r 2k^{2s+2-r}/[(s-k+1)!\,(s+k+1)!].$$

Therefore (4.72) can be written as

$$|\Lambda_n - \lambda_n| \leq c_9 h_1^r \sum_{k=1}^{s+1} \frac{2k^{2s+2-r}}{(s-k+1)!\,(s+k+1)!},$$

which is equivalent to (4.70) with

$$c_{22} = c_9 \sum_{k=1}^{s+1} \frac{2k^{2s+2-r}}{(s-k+1)!\,(s+k+1)!}.$$

The analogous proof using (4.36) leads to (4.69). Theorem 4.6 is proved. $\square$

Let us illustrate these results with some numerical examples. Assume $\alpha = (e^2 - 1)/2$ and consider the problem of finding the eigenvalues and eigenfunctions of

$$-((2\alpha x + 1)^2 u')' = \lambda u \quad \text{on } (0, 1),$$

$$u(0) = u(1) = 0. \tag{4.73}$$

The solution of this problem is a sequence of eigenvalues

$$\lambda_n = (1 + \pi^2 n^2)\alpha^2, \qquad n = 1, 2, \ldots,$$

and a corresponding orthonormal set of eigenfunctions

$$y_n(x) = \sqrt{\frac{2\alpha}{2\alpha x + 1}} \sin\left(\frac{n\pi}{2} \ln(2\alpha x + 1)\right).$$

Fix n and graph the error in logarithmic coordinates

$$\xi_\lambda(N) = |\lambda_n^h - \lambda_n|, \qquad \xi_y(N) = \max_{\bar{\omega}_h} |y_n^h - y_n|.$$

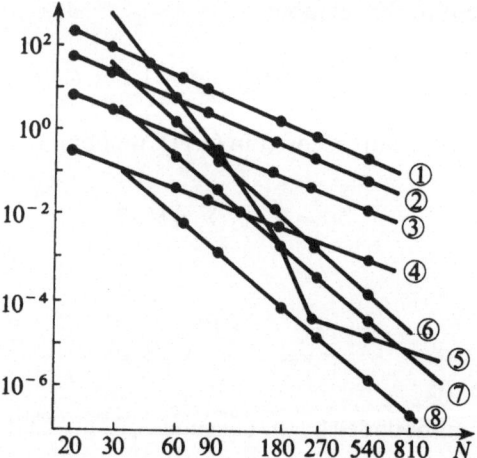

Figure 3.1. Errors in the difference eigenvalues and extrapolated values in (4.73).

The error in the difference eigenvalues for (4.9), (4.10): (1) for the fourth eigenvalue; (2) for the third eigenvalue; (3) for the second eigenvalue; (4) for the first eigenvalue. The errors in the extrapolated eigenvalues (4.68): (5) for the fourth eigenvalue; (6) for the third eigenvalue; (7) for the second eigenvalue; (8) for the first eigenvalue.

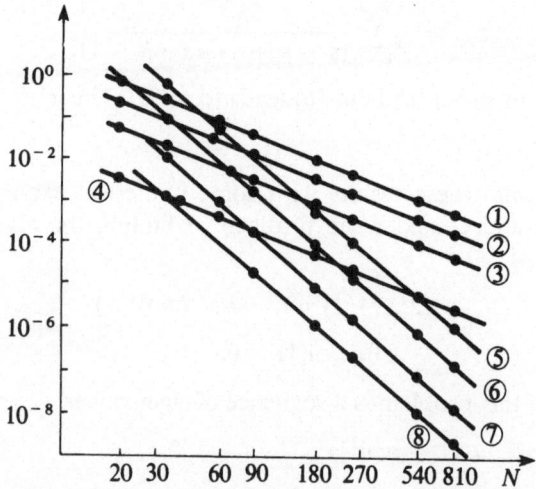

Figure 3.2. Maximum errors of the difference eigenfunctions and the extrapolated functions (4.66) for (4.73).

Maximum errors of the difference eigenfunctions of (4.9), (4.10): (1) for the fourth eigenfunction; (2) for the third eigenfunction; (3) for the second eigenfunction; (4) for the first eigenfunction. Maximum errors of the extrapolated eigenfunctions (4.66): (5) for the fourth eigenfunction; (6) for the third eigenfunction; (7) for the second eigenfunction, (8) for the first eigenfunction.

Then let us construct extrapolated values of Λ_n and Y_n using (4.66), (4.68) at $s = 1$ for several N with the help of nets $\bar{\omega}_h$ and $\bar{\omega}_{h/2}$ where $h = 1/N$ and find the error

$$\xi_\lambda(3N) = |\Lambda_n - \lambda_n|, \qquad \xi_y(3N) = \max_{\bar{\omega}_h} |Y_n - y_n|.$$

These errors have argument $3N$ since the number of calculations necessary to find Λ_h and Y_n and to solve (4.9), (4.10) on a net with mesh-size $1/(3N)$ are of the same order.

The graphs of these errors are shown in Figure 3.1 (for the first four eigenvalues) and in Figure 3.2 (for the first four eigenfunctions). The graphs clearly show the dependence of the accuracy of the approximate solutions on n.

3.5. Improving the Accuracy of the Finite Elements Method

In recent years variational-difference approachs to solving certain problems of mathematical physics have attracted great attention. Among these the finite elements method is the most widely used (see, for example, [32, 87, 100, 125], etc.) For selfadjoint equations a certain variational functional is minimized and the solution is then given as a superposition of certain special test functions—the "finite elements". The simplest choice of finite elements (in the form of "wedges") will lead, as a rule, to the simplest type of three point difference equation for the unknown coefficients of the solution, typically yielding first- or second-order accuracy. Using more complex (or smoother) finite element functions will result in more complex multipoints difference equations of greater accuracy. It should be noted that when one uses smooth finite elements the algorithm used to construct the difference equations will be far more complex when the initial difference equation has discontinuous coefficients. Therefore, it is preferable to use the simplest finite elements, and then take a linear combination of the solutions of such variational difference equations. This leads to a solution with greater accuracy. Let us note that such an approach affords us the advantage of describing the solution set in terms of a single parameter (mesh-size). The present section deals with this problem.

First let us consider a problem simplier than (1.1):

$$-u'' + qu = f \quad \text{on } (0, 1), \tag{5.1}$$

$$u(0) = u_0, \qquad u(1) = u_1 \tag{5.2}$$

with

$$f, q \in C^r[0, 1], \qquad q \geq 0 \quad \text{on } [0, 1]. \tag{5.3}$$

where the integer $r \geq 2$. Some methods by which (1.1) can be brought to the form (5.1) are given in [27].

To obtain a numerical solution of this problem we associate each net point of ω_h with a test function defined on the interval $[0, 1]$ according to the following formula

$$\varphi_i(x) = \begin{cases} (x - x_{i-1})/h & \text{if } x - x_i \in (-h, 0], \\ (x_{i+1} - x)/h & \text{if } x - x_i \in (0, h), \\ 0 & \text{in other cases.} \end{cases} \tag{5.4}$$

Fix i, $1 \leq i \leq N - 1$. Taking the product (5.1) and the function φ_i and integrating it over $[0, 1]$ we have:

$$\int_0^1 (-u''\varphi_i + qu\varphi_i) \, dx = \int_0^1 f\varphi_i \, dx.$$

Integrating the above equality by parts and keeping in mind that $\varphi_i(x_i \pm h) = 0$ we have:

$$\int_0^1 (u'\varphi_i' + qu\varphi_i) \, dx = \int_0^1 f\varphi_i \, dx. \tag{5.5}$$

For the sake of simplicity we introduce the following notation:

$$(v, w) = \int_0^1 vw \, dx, \qquad [v, w] = \int_0^1 (v'w' + qvw) \, dx.$$

Then it follows from (5.5) that

$$[u, \varphi_k] = (f, \varphi_k), \qquad k = 1, \ldots, N - 1. \tag{5.6}$$

The idea of the finite elements method is to search for a solution u^h in the form

$$u^h = \sum_{i=0}^N \alpha_i \varphi_i. \tag{5.7}$$

Here α_i are certain constants which are determined by substituting u^h in place of u in (5.6). We have

$$[u^h, \varphi_k] = (f, \varphi_k) \tag{5.8}$$

or

$$\sum_{i=0}^N \alpha_i [\varphi_i, \varphi_k] = (f, \varphi_k), \qquad k = 1, \ldots, N - 1.$$

In order to satisfy the boundary conditions we assume

$$\alpha_0 = u_0, \qquad \alpha_N = u_1. \tag{5.9}$$

The system of equations we get for α_i can be expressed in matrix form:

$$A\alpha = \begin{bmatrix} b_1 & e_2 & & & & 0 \\ a_2 & b_2 & & & & \\ & & \ddots & & & \\ & & & \ddots & & \\ & & & & e_{N-2} & \\ 0 & & & & a_{N-1} & b_{N-1} \end{bmatrix} \begin{bmatrix} \alpha_1 \\ \alpha_2 \\ \vdots \\ \alpha_{N-2} \\ \alpha_{N-1} \end{bmatrix} = \begin{bmatrix} g_1 - a_1 u_0 \\ g_2 \\ \vdots \\ g_{N-2} \\ g_{N-1} - e_{N-1} u_1 \end{bmatrix}, \quad (5.10)$$

where

$$a_i = [\varphi_{i-1}, \varphi_i], \qquad e_i = a_{i+1} = [\varphi_i, \varphi_{i+1}],$$

$$b_i = [\varphi_i, \varphi_i] = \frac{2}{h} + \int_{x_{i-1}}^{x_i} \frac{(x - x_{i-1})^2}{h^2} q(x) \, dx + \int_{x_i}^{x_{i+1}} \frac{(x_{i+1} - x)^2}{h^2} q(x) \, dx,$$

$$g_i = (f, \varphi_i) = \int_{x_{i-1}}^{x_i} \frac{(x - x_{i-1})}{h} f(x) \, dx + \int_{x_i}^{x_{i+1}} \frac{(x_{i+1} - x)}{h} f(x) \, dx.$$

We will show that system (5.10) is solvable.

Theorem 5.1. *The matrix of system* (5.10) *is positive definite.*

PROOF. Let Y be a nonzero vector with the components $y_i, i = 1, \ldots, N - 1$.
Then

$$Y^T A Y = \sum_{j=1}^{N-1} \sum_{i=1}^{N-1} y_i [\varphi_i, \varphi_j] y_j.$$

Assume

$$\tilde{y} = \sum_{i=1}^{N-1} y_i \varphi_i.$$

Then

$$Y^T A Y = [\tilde{y}, \tilde{y}] \geq \int_0^1 (\tilde{y}')^2 \, dx.$$

Using the Cauchy–Schwartz–Bungakovsky we have

$$\int_0^x (\tilde{y}')^2 \, dx \geq \left(\int_0^x \tilde{y}' \, dx \right)^2.$$

Since $\tilde{y}(0) = 0$ the expression on the right-hand side of this inequality is equal
to $\tilde{y}^2(x)$. Combining both inequalities we have

$$Y^T A Y \geq (\tilde{y}(x))^2 \quad \forall \, x \in [0, 1]$$

and in particular

$$Y^T A Y \geq y_i^2, \qquad i = 1, \ldots, N - 1,$$

since $\tilde{y}(ih) = y_i$. From this there follows

$$Y^T A Y \geq h \sum_{i=1}^{N-1} y_i^2,$$

which ensures the positive definiteness and regularity of the matrix A.   □

Thus for any N there is a unique set of constants α_i which gives an approximate solution from (5.7).

Theorem 5.2. *If the conditions* (5.3) *for* (5.1), (5.2) *are satisfied then the expansion*

$$u^h = u + \sum_{j=1}^{l} h^{2j} v_j + h^{r+2} \eta^h \quad on \; \bar{\omega}_h \qquad (5.11)$$

is a valid solution of the approximate system (5.7)–(5.9). *Here* $l = [(r+1)/2]$, *the functions* v_j *belong to* $C^{r+4-2j}[0, 1]$ *and are independent of* h, *and the net function* η^h *is bounded*:

$$|\eta^h| \leq c_2 \quad on \; \bar{\omega}_h. \qquad (5.12)$$

PROOF. Let us note that to any function w defined at the net points of ω_h one may associate the continuous function

$$\tilde{\omega}(x) = \sum_{i=0}^{N} \omega(x_i) \varphi_i(x),$$

on the interval $[0, 1]$ (its "piecewise linear extension"). Note $\tilde{u}^h = u^h$. Therefore (5.11) leads to

$$u^h = \tilde{u} + \sum_{j=1}^{l} h^{2j} \tilde{v}_j + h^{r+2} \tilde{\eta}^h \quad on \; [0, 1].$$

We assume

$$v_0 = u \qquad (5.13)$$

and will search for the functions v_j as solutions to

$$-v_j'' + q v_j = R_j \quad on \; (0, 1), \quad v_j(0) = v_j(1) = 0, \quad j = 1, \dots, l, \qquad (5.14)$$

where

$$R_1 = -q v_0''/24,$$

$$R_j = -\sum_{s=1}^{j-1} \frac{2}{(2s+2)!} R_{j-s}^{(2s)}$$

$$- \sum_{s=1}^{j} \sum_{\substack{k+m=2s \\ k \geq 2, m \geq 0}} \frac{(m+1)(k^2 + 2km + 3k - 2m - 4) v_{j-s}^{(k)} q^{(m)}}{(m+3)! \, k! \, (k+m+1)(k+m+2)},$$

$$j = 2, \dots, l. \qquad (5.15)$$

It is seen from formulae (5.15) that $R_1 \in C^r[0, 1]$; therefore by Theorem 1.1 there is a unique solution of (5.14) which belongs to $C^{r+2}[0, 1]$. Assume that the functions $R_1, \ldots, R_j$ have been defined and $R_k \in C^{r+2-2r}[0, 1]$. Using (5.14) we find the functions $v_1, \ldots, v_j$, which, by theorem 1.1, have two derivatives more than $R_1, \ldots, R_j$, i.e.,

$$v_k \in C^{r+4-2k}[0, 1], \qquad k = 1, \ldots, j.$$

Checking how many derivatives each term on the right-hand side of (5.15) has with $k = j + 1$ results in

$$R^{(2s)}_{j+1-s} \in C^{r-2j}[0, 1],$$
$$v^{(k)}_{j+1-s} \in C^{r+2-2j}[0, 1], \qquad s \geq 1,$$
$$q^{(m)} \in C^{r-2j}[0, 1]$$

(the last is because $m \leq j$). Thus $R_{j+1} \in C^{r-2j}[0, 1]$. By Theorem 1.1 there is a unique solution of (5.14) with $k = j + 1$; it belongs to $C^{r+2-2j}[0, 1]$. Therefore all the R_j and v_j are defined unambiguously and

$$v_j \in C^{r-2j+4}[0, 1], \qquad R_j \in C^{r-2j+2}[0, 1]. \tag{5.16}$$

Now since the functions $v_j, u,$ and u^h are known let us define the net function

$$\eta^h = \left(u^h - \sum_{j=0}^{l} v_j\right)\bigg/ h^{r+2} \qquad \text{on } \bar{\omega}_h \tag{5.17}$$

which is equivalent to

$$\tilde{\eta}^h = \left(u^h - \sum_{j=0}^{l} h^{2j}\tilde{v}_j\right)\bigg/ h^{r+2} \qquad \text{on } [0, 1]. \tag{5.18}$$

Equation (5.8)

$$[u^h, \varphi_i] = (f, \varphi_i), \qquad i = 1, \ldots, N - 1,$$

together with (5.18) yields

$$[u^h, \varphi_i] = \sum_{j=0}^{l} h^{2j}[\tilde{v}_j, \varphi_i] + h^{r+2}[\tilde{\eta}^h, \varphi_i] = (f, \varphi_i), \qquad i = 1, \ldots, N - 1.$$

$$\tag{5.19}$$

Before simplifying these expressions let us consider several auxiliary expansions.

Lemma 5.3. *Let t be a natural number and let $v \in C^t[0, 1]$; then*

$$(v, \varphi_i) = hv(x_i) + \sum_{j=1}^{[(t-1)/2]} h^{2j+1} 2v^{(2j)}(x_i)/(2j + 2)! + h^{t+1}\sigma_i, \tag{5.20}$$

where

$$|\sigma_i| \le 2 \max_{[0, 1]} |v^{(t)}|/(t + 2)!$$

PROOF. Let the function g be such that $g'' = v$. It is sufficient to assume, for example, that

$$g(x) = \int_0^x \int_0^t v(z) \, dz \, dt.$$

Then, as it is not difficult to see, we have

$$(g'', \varphi_i) = [g(x_{i-1}) - 2g(x_i) + g(x_{i+1})]/h. \tag{5.21}$$

From Lemma 1.2 of §7.1 it follows that

$$(g'', \varphi_i) = h(g_{\hat{x}}(x_i))_{\hat{x}}$$

$$= \sum_{j=0}^{[(t-1)/2]} h^{2j+1} 2g^{(2j+2)}(x_i)/(2j + 2)! + 2h^{t+1} g^{(t+2)}(\xi_i)/(t + 2)!.$$

Since $g^{(k+2)} = v^{(k)}$ we arrive at the statement of the lemma.   □

Lemma 5.4. *Let $v \in C^t[0, 1]$, where t is an integer and $1 \le t \le r + 2$. Then*

$$(q\tilde{v}, \varphi_i) = (qv, \varphi_i)$$

$$+ \sum_{j=1}^{[(t-1)/2]} h^{2j+1} \sum_{\substack{k+m=2j \\ k \ge 2, m \ge 0}} \frac{(m + 1)(k^2 + 2km + 3k - 2m - 4)}{(m + 3)! \, k!} v^{(k)}(x_i) q^{(m)}(x_i)$$

$$+ h^{t+1} \varkappa^h, \tag{5.22}$$

where

$$|\varkappa_i^h| \le c_3. \tag{5.23}$$

PROOF. Let us rewrite the left-hand side of (5.22). The support of the function φ_i is concentrated on the interval $[x_{i-1}, x_{i+1}]$, and thus

$$(q\tilde{v}, \varphi_i) = \int_{x_{i-1}}^{x_{i+1}} q\tilde{v}\varphi_i \, dx.$$

Since the functions $\tilde{v}$, φ_i piecewise coincide with polynomials it is convenient to divide the interval of integration into two parts. First consider

$$\int_{x_i}^{x_{i+1}} q\tilde{v}\varphi_i \, dx = \frac{1}{h^2} \int_{x_i}^{x_{i+1}} q(x)[v(x_i)(x_{i+1} - x)$$

$$+ v(x_{i+1})(x - x_i)](x_{i+1} - x) \, dx.$$

In the last integral let us take the Taylor series expansion of the function $v(x_{i+1})$ with respect to x_i. We have

$$\int_{x_i}^{x_{i+1}} q\tilde{v}\varphi_i \, dx = \frac{1}{h} \int_{x_i}^{x_{i+1}} q(x)v(x_i)(x_{i+1} - x) \, dx$$

$$+ \frac{1}{h} \int_{x_i}^{x_{i+1}} q(x)v'(x_i)(x - x_i)(x_{i+1} - x) \, dx$$

$$+ \sum_{k=2}^{r+1} \frac{h^{k-2}}{k!} v^{(k)}(x_i) \int_{x_i}^{x_{i+1}} (x - x_i)(x_{i+1} - x)q(x) \, dx$$

$$+ \frac{h^{r+3}}{(r+2)!} v^{(r+2)}(\xi_i)v_i^h, \tag{5.24}$$

where

$$|v_i^h| \le \tfrac{1}{6} \max_{[0,1]} |q|.$$

In the expression under the summation replace the function $q(x)$ by its Taylor series expansion with respect to x_i. Then, after integrating, we have

$$\frac{h^{k-2}}{k!} v^{(k)}(x_i) \int_{x_i}^{x_{i+1}} (x - x_i)(x_{i+1} - x)$$

$$\times \left[\sum_{m=0}^{r+1-k} \frac{(x - x_i)^m}{m!} q^{(m)}(x_i) + \frac{(x - x_i)^{r+2-k}}{(r+2-k)!} q^{(r+2-k)}(\xi_i) \right] dx$$

$$= \sum_{m=0}^{r+1-k} \frac{h^{k+m+1}(m+1)}{k!\,(m+3)!} v^{(k)}(x_i)q^{(m)}(x_i) + h^{r+3} \mu_{k,i}^h,$$

where

$$|\mu_{k,i}^h| \le \frac{r+3-k}{k!\,(r+5-k)!} \max_{[0,1]} |q^{(r+2-k)}| \max_{[0,1]} |v^{(k)}|.$$

Therefore, finally we have

$$\int_{x_i}^{x_{i+1}} q\tilde{v}\varphi_i \, dx = \frac{1}{h} \int_{x_i}^{x_{i+1}} q(x)v(x_i)(x_{i+1} - x) \, dx$$

$$+ \frac{1}{h} \int_{x_i}^{x_{i+1}} q(x)v'(x_i)(x_{i+1} - x)(x - x_i) \, dx$$

$$+ \sum_{k=2}^{r+1} \sum_{m=0}^{r+1-k} \frac{h^{k+m+1}(m+1)}{(m+3)!\,k!} v^{(k)}(x_i)q^{(m)}(x_i) + h^{r+3} \mu_i^h,$$

$$\tag{5.25}$$

where

$$|\mu_i^h| \leq c_4, \qquad i = 1, \ldots, N - 1.$$

Analogous manipulations of the integral

$$\int_{x_i}^{x_{i+1}} qv\varphi_i \, dx$$

yield

$$\int_{x_i}^{x_{i+1}} qv\varphi_i \, dx = \frac{1}{h} \int_{x_i}^{x_{i+1}} q(x)v(x_i)(x_{i+1} - x) \, dx$$

$$+ \frac{1}{h} \int_{x_i}^{x_{i+1}} q(x)v'(x_i)(x_{i+1} - x)(x - x_i) \, dx$$

$$+ \sum_{k=2}^{r+1} \sum_{m=0}^{r+1-k} \frac{h^{k+m+1}}{m!\, k!\, (k + m + 1)(k + m + 2)} v^{(k)}(x_i)q^{(m)}(x_i)$$

$$+ h^{r+3}\rho_i^h, \tag{5.26}$$

where

$$|\rho_i^h| \leq c_5, \qquad i = 1, \ldots, N - 1.$$

Substracting (5.26) from (5.25) we obtain

$$\int_{x_i}^{x_{i+1}} q\tilde{v}\varphi_i \, dx - \int_{x_i}^{x_{i+1}} qv\varphi_i \, dx$$

$$= \sum_{k=2}^{r+1} \sum_{m=0}^{r+1-k} \frac{h^{k+m+1}(m + 1)(k^2 + 2km + 3k - 2m - 4)}{(m + 3)!\, k!\, (k + m + 1)(k + m + 2)}$$

$$\times v^{(k)}(x_i)q^{(m)}(x_i) + h^{r+3}(\mu_i^h - \rho_i^h). \tag{5.27}$$

Now replace x_{i+1} by x_{i-1} in the above (this will change h to $-h$ on the right-hand side of (5.27) and change the limits of integration (which changes a sign on the right-hand side of (5.27)). Then we have

$$\int_{x_{i-1}}^{x_i} q\tilde{v}\varphi_i \, dx - \int_{x_{i-1}}^{x_i} qv\varphi_i \, dx$$

$$= \sum_{k=2}^{r+1} \sum_{m=0}^{r+1-k} -\frac{(-h)^{k+m+1}(m + 1)(k^2 + 2km + 3k - 2m - 4)}{(m + 3)!\, k!\, (k + m + 1)(k + m + 2)}$$

$$\times v^{(k)}(x_i)q^{(m)}(x_i) + h^{r+3}\delta_i^h,$$

where

$$|\delta_i^h| \leq c_4 + c_5.$$

Adding this to (5.27) gives us the statement of Lemma 5.4, where $c_3 = 2c_4 + 2c_5$. $\qquad\qquad\qquad\qquad\qquad\qquad\qquad\qquad\qquad\qquad\qquad\square$

It should be noted that this proof could not be simplified by a simultaneous expansion of the functions v and q since the function q is not sufficiently smooth.

Lemma 5.5. *For any function* $v \in C^1[0, 1]$

$$(v', \varphi_i') = (\tilde{v}', \varphi_i')$$

is an identity.

PROOF. Transform the left-hand side to read

$$(v', \varphi_i') = -\frac{1}{h} \int_{x_i}^{x_{i+1}} v' \, dx + \frac{1}{h} \int_{x_{i-1}}^{x_i} v' \, dx$$

$$= -[v(x_{i+1}) - v(x_i)]/h + [v(x_i) - v(x_{i+1})]/h. \quad (5.28)$$

Now

$$\tilde{v}'(x) = [v(x_{i+1}) - v(x_i)]/h \quad \text{when } x \in [x_i, x_{i+1}],$$

$$\tilde{v}'(x) = [v(x_i) - v(x_{i-1})]/h \quad \text{when } x \in [x_{i-1}, x_i].$$

Therefore

$$(\tilde{v}', \varphi_i') = -\frac{1}{h^2} \int_{x_i}^{x_{i+1}} [v(x_{i+1}) - v(x_i)] \, dx + \frac{1}{h^2} \int_{x_{i-1}}^{x_i} [v(x_i) - v(x_{i-1})] \, dx.$$

comparing this equation with (5.28) proves the lemma.

Let us now return to (5.19). Using Lemma 5.5 we can rewrite its left-hand side and get

$$[u^h, \varphi_i] = \sum_{j=0}^{l} h^{2j}\{(\tilde{v}_j', \varphi_i') + (q\tilde{v}_j, \varphi_i)\} + h^{r+2}[\tilde{\eta}^h, \varphi_i]$$

$$= \sum_{j=0}^{l} h^{2j}\{(v_j', \varphi_i') + (q\tilde{v}_j, \varphi_i)\} + h^{r+2}[\tilde{\eta}^h, \varphi_i].$$

We now use Lemma 5.4 to rewrite appropriate terms:

$$[u^h, \varphi_i] = \sum_{j=0}^{l} h^{2j}\{(v_j', \varphi_i') + (qv_j, \varphi_i)\}$$

$$+ \sum_{j=0}^{l-1} \sum_{s=1}^{l-j} h^{2j+2s+1} \sum_{\substack{k+m=2s \\ k \geq 2, m \geq 0}} \frac{(m+1)(k^2 + 2km + 3k - 2m - 4)}{(m+3)! \, k! \, (k+m+1)(k+m+2)}$$

$$\times v_j^{(k)}(x_i)q^{(m)}(x_i) + h^{r+3}\beta_i^h + h^{r+2}[\tilde{\eta}^h, \varphi_i], \quad (5.29)$$

where

$$|\beta_i^h| \leq c_6.$$

Due to the fact that the functions v_j solve (5.14), (5.15) we have

$$[v_j, \varphi_i] = (v_j', \varphi_i') + (qv_j, \varphi_i) = (R_j, \varphi_i) \tag{5.30}$$

which we obtain by integrating the product of equation (5.14) and φ_i by part. By Lemma 5.3 we have

$$(R_j, \varphi_i) = \sum_{s=0}^{l-j} h^{2s+1} 2R_j^{(2s)}(x_i)/(2s+2)! + h^{r+3-2j} S_{j,i}^h,$$

where

$$|S_{j,i}^h| \le c_7, \qquad i = 1, \ldots, N-1; \quad j = 1, \ldots, l.$$

Now transform (5.29), keeping in mind (5.30), (5.8), and the last equation

$$[u^h, \varphi_i] = \sum_{j=1}^{l} h^{2j} \sum_{s=0}^{l-j} h^{2s+1} 2R_j^{(2s)}(x_i)/(2s+2)!$$

$$+ \sum_{j=0}^{l} \sum_{s=1}^{l-j} h^{2j+2s+1} \sum_{\substack{k+m=2s \\ k \ge 2, m \ge 0}} \frac{(m+1)(k^2 + 2km + 3k - 2m - 4)}{(m+3)! \, k! \, (k+m+1)(k+m+2)}$$

$$\times v_j^{(k)}(x_i) q^{(m)}(x_i) + h^{r+3} T_i^h + h^{r+2}[\tilde{\eta}^h, \varphi_i] + (f, \varphi_i).$$

Here

$$|T_i^h| \le c_8, \qquad i = 1, \ldots, N-1.$$

Change the order of summation in the double sums:

$$[u^h, \varphi_i] = \sum_{j=1}^{l} h^{2j+1} \left\{ \sum_{s=0}^{j-1} 2R_{j-s}^{(2s)}(x_i)/(2s+2)! \right.$$

$$+ \sum_{s=1}^{j} \sum_{\substack{k+m=2s \\ k \ge 2, m \ge 0}} \frac{(m+1)(k^2 + 2km + 3k - 2m - 4)}{(m+3)! \, k! \, (k+m+1)(k+m+2)}$$

$$\left. \times v_{j-s}^{(k)}(x_i) q^{(m)}(x_i) \right\} + h^{r+3} T_i^h + h^{r+2}[\tilde{\eta}^h, \varphi_i] + (f, \varphi_i).$$

It is obvious that the double sum is zero with the choice of functions (formula (5.15)):

$$[u^h, \varphi_i] = h^{r+3} T_i^h + h^{r+2}[\tilde{\eta}^h, \varphi_i] + (f, \varphi_i).$$

From the definition u^h it follows that

$$[u^h, \varphi_i] = (f, \varphi_i),$$

so that

$$[\tilde{\eta}^h, \varphi_i] = -hT_i^h, \qquad i = 1, \ldots, N - 1.$$

Multiply each equation by $\eta^h(x_i)$ and sum over $i, i = 1, \ldots, N - 1$. Then from $\tilde{\eta}^h(0) = \tilde{\eta}^h(1) = 0$ (from the definition of $\tilde{\eta}^h$ at the points 0 and 1) we have

$$[\tilde{\eta}^h, \tilde{\eta}^h] = -\sum_{i=1}^{N-1} T_i^h \eta^h(x_i) h.$$

Theorem 5.1 shows that

$$(\tilde{\eta}^h(x))^2 \le [\tilde{\eta}^h, \tilde{\eta}^h] \quad \forall \, x \in \omega_h;$$

and thus

$$\max_{\omega_h} |\tilde{\eta}^h|^2 \le \sum_{i=1}^{N-1} |T_i^h| |\eta^h(x_i)| h \le \max_{\omega_h} |\tilde{\eta}^h| \cdot c_8.$$

After dividing this inequality by $\max_{\omega_h} |\tilde{\eta}^h|$ we have

$$\max_{\omega_h} |\tilde{\eta}^h| \le c_8.$$

Thus Theorem 5.2 is proved, and $c_2 = c_8$. $\qquad\qquad\square$

Let us now formulate the extrapolation algorithm which uses the expansion of Theorem 5.2.

Let $l = [(r + 1)/2]$, where the integer r is taken from (5.3). For the first $l + 1$ values of the integer $N_k = kN$ we construct the nets $\bar{\omega}_{h_k}$ with mesh-sizes $h_k = 1/N_k$, $k = 1, \ldots, l + 1$. On each net $\bar{\omega}_{h_k}$ we formulate the system (5.8), (5.9) and solve it by finding the weights α_j in (5.7). As a result we obtain $l + 1$ functions u^{h_k}. We are interested only in their values at the points of $\bar{\omega}_{h_1}$. Let us form the linear combination

$$U = \sum_{k=1}^{l+1} \gamma_k u^{h_k} \quad \text{on } \bar{\omega}_{h_1} \tag{5.31}$$

with weights

$$\gamma_k = \frac{2(-1)^{l-k+1} k^{2l+2}}{(l - k + 1)! \, (l + k + 1)!}. \tag{5.32}$$

Theorem 5.6. *Suppose the hypotheses of Theorem 5.2 holds. Then the solution* (5.31), (5.32) *obey the estimate*

$$\max_{\bar{\omega}_{h_1}} |U - u| \le h_1^{r+2} c_9. \tag{5.33}$$

The proof is analogous to that of Theorem 4.6, the constant c_9 being equal to

$$2c_2 \sum_{k=1}^{l+1} \frac{k^{2l-r}}{(l+1-k)!\,(l+1+k)!},$$

where the constant c_2 is taken from Theorem 5.2.

Now consider the more general equation

$$-(pu')' + qu = f \qquad \text{on } (0, 1),$$
$$p(x) \geq c_1 > 0, \qquad q(x) \geq 0, \tag{5.34}$$
$$p \in C^{r+1}[0, 1], \qquad f, q \in C^r[0, 1].$$

Let us note two possible ways of generalizing the results obtained earlier.

We can leave the basic functions (5.4) unchanged and redefine the expression $[v, w]$ in (5.7)–(5.10):

$$[v, w] = \int_0^1 (pv'w' + qvw)\, dx. \tag{5.35}$$

The algebraic system of equations (5.10) will still have a tridiagonal structure, though coefficients a_i, b_i, e_i, g_i will differ from the above:

$$a_i = [\varphi_{i-1}, \varphi_i] = -\frac{1}{h^2} \int_{x_{i-1}}^{x_i} p(x)\, dx + \frac{1}{h^2} \int_{x_{i-1}}^{x_i} (x_i - x)(x - x_{i-1})q(x)\, dx,$$

$$e_i = a_{i+1} = [\varphi_i, \varphi_{i+1}],$$

$$b_i = [\varphi_i, \varphi_i] = \frac{1}{h^2} \int_{x_{i-1}}^{x_i} p(x)\, dx + \frac{1}{h^2} \int_{x_i}^{x_{i+1}} p(x)\, dx$$

$$+ \frac{1}{h^2} \int_{x_{i-1}}^{x_i} (x - x_{i-1})^2 q(x)\, dx + \frac{1}{h^2} \int_{x_i}^{x_{i+1}} (x_{i+1} - x)^2 q(x)\, dx, \tag{5.36}$$

$$g_i = (f, \varphi_i) = \frac{1}{h} \int_{x_{i-1}}^{x_i} (x - x_{i-1})f(x)\, dx + \frac{1}{h} \int_{x_i}^{x_{i+1}} (x_{i+1} - x)f(x)\, dx.$$

In this case Lemmas 5.3 and 5.4 remain valid but Lemma 5.5 does not. We should instead expand the difference $(p\tilde{v}', \varphi_i') - (pv', \varphi_i')$ with respect to h. This leads to the remainder of order h^{r+2}. Finally, the remainder in (5.11) is a quantity of order h^{r+1} with the same requirements on q and f.

One can instead increase the complexity of the trial functions φ_i:

$$\varphi_i(x) = \begin{cases} \displaystyle \int_{x_{i-1}}^x p^{-1}(t)\, dt \Big/ \int_{x_{i-1}}^{x_i} p^{-1}(t)\, dt & \text{if } x \in (x_{i-1}, x_i], \\[2ex] \displaystyle \int_x^{x_{i+1}} p^{-1}(t)\, dt \Big/ \int_{x_i}^{x_{i+1}} p^{-1}(t)\, dt & \text{if } x \in (x_i, x_{i+1}), \\[2ex] 0 & \text{in all other cases.} \end{cases} \tag{5.37}$$

In this case the coefficients of (5.10) look more complex:

$$a_i = [\varphi_{i-1}, \varphi_i] = -\left(\int_{x_{i-1}}^{x_i} p^{-1}(t)\, dt\right)^{-1}$$

$$+ \int_{x_{i-1}}^{x_i} \left(\int_x^{x_i} p^{-1}(t)\, dt\right)\left(\int_{x_{i-1}}^x p^{-1}(t)\, dt\right)\left(\int_{x_{i-1}}^{x_i} p^{-1}(t)\, dt\right)^{-2} q(x)\, dx,$$

$$e_i = a_{i+1} = [\varphi_i, \varphi_{i+1}],$$

$$b_i = \left(\int_{x_{i-1}}^{x_i} p^{-1}(t)\, dt\right)^{-1} + \left(\int_{x_i}^{x_{i+1}} p^{-1}(t)\, dt\right)^{-1}$$

$$+ \int_{x_{i-1}}^{x_i} \left(\int_{x_{i-1}}^x p^{-1}(t)\, dt\right)^2 \left(\int_{x_{i-1}}^{x_i} p^{-1}(t)\, dt\right)^{-2} q(x)\, dx$$

$$+ \int_{x_i}^{x_{i+1}} \left(\int_x^{x_{i+1}} p^{-1}(t)\, dt\right)^2 \left(\int_{x_i}^{x_{i+1}} p^{-1}(t)\, dt\right)^{-2} q(x)\, dx,$$

$$g_i = \int_{x_{i-1}}^{x_i} \left(\int_{x_{i-1}}^x p^{-1}(t)\, dt\right)\left(\int_{x_{i-1}}^{x_i} p^{-1}(t)\, dt\right)^{-1} f(x)\, dx \qquad (5.38)$$

$$+ \int_{x_i}^{x_{i+1}} \left(\int_x^{x_{i+1}} p^{-1}(t)\, dt\right)\left(\int_{x_i}^{x_{i+1}} p^{-1}(t)\, dt\right)^{-1} f(x)\, dx.$$

For such trial functions Lemma 5.3 and the identity $(pv', \varphi_i') = (p\tilde{v}', \varphi_i')$ are still valid, as is Lemma 5.4. Lemma 5.4 is now proved only with considerable difficulty; the remainder is of the same order as before, however,

The trial functions (5.37) allow us to insure that the order of the remainder in (5.11) is as before. Let us note that for $q \equiv 0$ the last identity guarantees that (5.34) is exactly integrable, as before.

Note that the trial functions (5.37), for real mesh-size h, give greater accuracy than the trial functions (5.4). This is especially noticeable when the coefficient p varies greatly.

This is illustrated by the equation

$$-\frac{d}{dx}\left(e^{-4x}\frac{du}{dx}\right) + 4e^{-4x}u = 8 - 4x + 4xe^{-4x},$$

$$u(0) = u(1) = 0. \qquad (5.39)$$

Its exact solution is the function

$$u(x) = (e^{4x} - 1)(1 - x).$$

At the beginning of this section we solved some variational-difference problems (5.10) using the simplest trial functions (5.4). In this case the coefficients of the system (5.10) are given by (5.36). For each solution (5.7) the maximal deviation $\xi(N)$ from the exact solution was found, where N is the number of net points of $\bar{\omega}_h$. The graph of this quantity is shown in Figure 3.3.

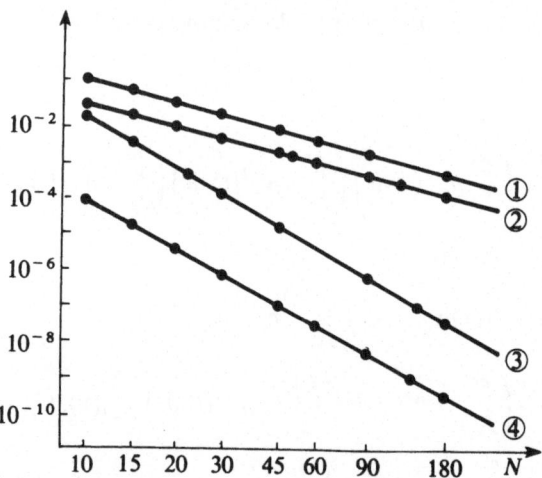

Figure 3.3. Maximal errors of variational-difference solutions and the extrapolated solutions.

Maximal error of variational difference solution: (1) with piecewise linear trial functions; (2) with special trial functions (5.37). Maximal error of the extrapolated solutions: (3) with piecewise linear trial functions; (4) with trial functions (5.37).

We then solved an approximate problem (5.10) using the trial functions (5.37). In this case the coefficients of the system (5.10) are given by (5.38), using these weights we constructed approximate solutions of (5.7). The maximal error $\xi(N)$ was again found where N is again the number of net points of $\overline{\omega}_h$. The graph of this quantity is also shown in Figure 3.3. To clarify the effect of extrapolation corrected solutions were constructed for $l = 1$ by (5.31); the extrapolations were carried out over meshes where mesh-sizes were in a ratio of $2:1$. The extrapolation was applied to solutions using the trial functions (5.4) and (5.37) on two nets with mesh-sizes $h = 1/N$ and $h/2$. The graphs of the maximal errors of these corrected solutions are also shown in Figure 3.3.

3.6. The Quasilinear Problem

Let us consider the Dirichlet problem for a second-order quasilinear equation. We assume that the equation can be solved explicitly for the second derivative, where the right-hand side will be a function of independent variable, the unknown function itself, and its first derivative. It is known that for the solution to exist and to be unique the right-hand side must be sufficiently smooth, and certain additional conditions must be satisfied. In the present section we will prove that the solution of an approximate version of the non-linear problem can be written as a series expansion in integer powers of the mesh-size parameter. This theorem gives us a constructive method for using Richardson extrapolation. When we have a set of approximate solutions with different mesh-size parameters we can again obtain a solution with maximal accuracy.

Let us consider the problem

$$u'' = f(x, u, u') \quad \text{on } [0, 1], \tag{6.1}$$

$$u(0) = u_0, \qquad u(1) = u_1. \tag{6.2}$$

Assume that $f(x, u, v)$, as a function of its three arguments, belongs to $C^r([0, 1] \times (-\infty, \infty) \times (-\infty, \infty))$ where $r \geq 2$. Let us also assume that the system (6.1), (6.2) is solvable in $C^{r+2}[0, 1]$.

Since we are only going to use Richardson extrapolation for nonlinear equations let us consider only one condition which insures unique solvability, e.g.:

$$\frac{1}{2} - \frac{1}{16} \left| \frac{\partial f}{\partial v}(x, u, v) \right|^2 + \frac{1}{8} \frac{\partial f}{\partial u}(x, u, v) \geq c_1,$$

$$c_1 \in (0, \tfrac{1}{2}), \quad x \in [0, 1], \quad u, v \in (-\infty, \infty). \tag{6.3}$$

To completely side-step the question of the existence and uniqueness of the solutions is not wholly satisfactory, since the assumptions of existence and uniqueness underlies many of the schemes we use.

We will show that the inequality characterizing strong monotonicity (see [73], [101]) of (6.1), (6.2) follows from condition (6.3):

$$\int_0^1 \{u'(u' - v') + f(x, u, u')(u - v)\} \, dx - \int_0^1 \{v'(u' - v')$$

$$+ f(x, v, v')(u - v)\} \, dx \geq c_1 \int_0^1 (u' - v')^2 \, dx \tag{6.4}$$

for any functions $u, v \in C^1[0, 1]$ which are equal at the endpoints of the interval: $u(0) = v(0), u(1) = v(1)$.

Assume

$$q_0 = \tau u + (1 - \tau)v, \qquad q_1 = \tau u' + (1 - \tau)v';$$

then

$$f(x, u, u') - f(x, v, v') = \int_0^1 \frac{d}{d\tau} f(x, q_0, q_1) \, d\tau$$

$$= \int_0^1 \frac{\partial f}{\partial q_0} \, d\tau(u - v) + \int_0^1 \frac{\partial f}{\partial q_1} \, d\tau(u' - v').$$

Denote the expression on the left-hand side of (6.4) by $J(u, v)$; by the previous relation we have

$$J(u, v) = \int_0^1 \left\{ \int_0^1 \frac{\partial f}{\partial q_0}(x, q_0, q_1) \, d\tau (u - v)^2 \right.$$

$$\left. + \int_0^1 \frac{\partial f}{\partial q_1}(x, q_0, q_1) \, d\tau (u - v)(u' - v') + (u' - v')^2 \right\} dx.$$

Subtract $c_1 \int_0^1 (u' - v')^2 \, dx$ from both sides and apply the inequality $ab \leq a^2 \varepsilon/2 + b^2/(2\varepsilon)$, $\varepsilon > 0$. We have

$$J(u, v) - c_1 \int_0^1 (u' - v')^2 \, dx \geq \int_0^1 \left\{ \int_0^1 \frac{\partial f}{\partial q_0}(x, q_0, q_1) \, d\tau (u - v)^2 \right.$$

$$- \int_0^1 \left| \frac{\partial f}{\partial q_1}(x, q_0, q_1) \right| d\tau \left(\frac{\varepsilon}{2}(u - v)^2 + \frac{1}{2\varepsilon}(u' - v')^2 \right)$$

$$\left. + (1 - c_1)(u' - v')^2 \right\} dx. \tag{6.5}$$

For any two functions u and v, which attain equal values at the endpoints of the interval $[0, 1]$ we have the inequality (see [114])

$$\frac{1}{8} \int_0^1 (u' - v')^2 \, dx \geq \int_0^1 (u - v)^2 \, dx.$$

Applying it to the right-hand side of the (6.5) and assuming

$$\varepsilon = \int_0^1 \left| \frac{\partial f}{\partial q_1}(x, q_0, q_1) \right| d\tau,$$

we have

$$J(u, v) - c_1 \int_0^1 (u' - v')^2 \, dx$$

$$\geq \int_0^1 \left\{ \int_0^1 \left(4 - 8c_1 - \frac{1}{2} \left| \frac{\partial f}{\partial q_1}(x, q_0, q_1) \right|^2 + \frac{\partial f}{\partial q_0}(x, q_0, q_1) \right) d\tau \right\} (u - v)^2 \, dx.$$

The right-hand side of this relation is nonnegative from (6.3). Therefore

$$J(u, v) \geq c_1 \int_0^1 (u' - v')^2 \, dx,$$

which is equivalent to (6.4).

The unique solvability of the problem follows from its strong monotonicity. Suppose for example, that u and v are two different solutions of (6.1), (6.2) in $C^{r+2}[0, 1]$. Then $u(0) = v(0)$, $u(1) = v(1)$, (6.4) is valid, and its left-hand side is zero.

But, from the embedding theorem for $\mathring{W}_2^1[0, 1]$ to $C[0, 1]$ it follows that [114]

$$\max_{[0, 1]} |u - v| \leq \frac{1}{2} \left(\int_0^1 (u' - v')^2 \, dx \right)^{1/2}.$$

Hence $u = v$ on $[0, 1]$, and we obtain a contradiction.

To obtain a numerical solution of (6.1), (6.2) we use the difference scheme

$$u_{\bar{x}\bar{x}}^h(x) = f_{\bar{x}}(x, u_{\bar{x}}^h(x), u_{\bar{x}}^h(x)), \qquad x \in \omega_h, \tag{6.6}$$

$$u^h(0) = u_0, \qquad u^h(1) = u_1 \tag{6.7}$$

The equation (6.6) the explicit form

$$[u(x - h) - 2u(x) + u(x + h)]/h^2$$
$$= \tfrac{1}{2} f(x + h/2, [u(x + h) + u(x)]/2, [u(x + h) - u(x)]/h)$$
$$+ \tfrac{1}{2} f(x - h/2, [u(x) + u(x - h)]/2, [u(x) - u(x - h)]/h).$$

Let us prove that the system (6.6), (6.7) is uniquely solvable. The first step is to show that the difference analog of strong monotonicity is valid if h is sufficiently small:

$$\sum_{\omega_h} \{-u_{\bar{x}\bar{x}} + f_{\bar{x}}(x, u_{\bar{x}}, u_{\bar{x}})\} (u - v)h$$

$$- \sum_{\omega_h} \{-v_{\bar{x}\bar{x}} + f_{\bar{x}}(x, v_{\bar{x}}, v_{\bar{x}})\} (u - v)h \geq c_1 \sum_{\hat{\omega}_h} (u_{\bar{x}} - v_{\bar{x}})^2 h, \tag{6.8}$$

where u and v are arbitrary functions, defined on $\bar{\omega}_h$ and attaining equal values at the points 0 and 1. Assume as before that

$$q_0 = \tau u_{\bar{x}} + (1 - \tau)v_{\bar{x}} \quad \text{on } \check{\omega}_h,$$

$$q_1 = \tau u_{\bar{x}} + (1 - \tau)v_{\bar{x}} \quad \text{on } \check{\omega}_h.$$

Then

$$f(x, u_{\bar{x}}, u_{\bar{x}}) - f(x, v_{\bar{x}}, v_{\bar{x}}) = \int_0^1 \frac{d}{d\tau} f(x, q_0, q_1) \, d\tau$$

$$= \int_0^1 \frac{\partial f}{\partial q_0} \, d\tau (u_{\bar{x}} - v_{\bar{x}}) + \int_0^1 \frac{\partial f}{\partial q_1} \, d\tau (u_{\bar{x}} - v_{\bar{x}}) \quad \forall x \in \check{\omega}_h.$$

Using Green's formula for the second divided difference and for the remaining terms in the last identity, we rewrite the left-hand side of (6.8) in the form:

$$J^h(u, v) = \sum_{\bar{\omega}_h} \left\{ (u_{\bar{x}} - v_{\bar{x}})^2 + \int_0^1 \frac{\partial f}{\partial q_0} (x, q_0, q_1) \, d\tau (u_{\bar{x}} - v_{\bar{x}})^2 \right.$$

$$\left. + \int_0^1 \frac{\partial f}{\partial q_1} (x, q_0, q_1) \, d\tau (u_{\bar{x}} - v_{\bar{x}})(u_{\bar{x}} - v_{\bar{x}}) \right\} h.$$

Here the fact that the values of u and v coincide at the endpoints of the interval $[0, 1]$ is used in the form of the following difference identity, which is valid for any function w defined at the knots $\breve{\omega}_h$:

$$\sum_{\omega_h} w_{\bar{x}}(u - v) = \sum_{\breve{\omega}_h} w(u_x - v_x).$$

Transforming $J^h(u, v)$ as in continuous case, we have

$$J^h(u, v) - c_1 \sum_{\breve{\omega}_h} (u_{\bar{x}} - v_{\bar{x}})^2 h \geq \sum_{\breve{\omega}_h} \left\{ \int_0^1 \frac{\partial f}{\partial q_0} (x, q_0, q_1) \, d\tau (u_{\bar{x}} - v_{\bar{x}})^2 \right.$$

$$- \frac{1}{2} \int_0^1 \left| \frac{\partial f}{\partial q_1} (x, q_0, q_1) \right|^2 d\tau (u_{\bar{x}} - v_{\bar{x}})^2$$

$$\left. + (\tfrac{1}{2} - c_1)(u_{\bar{x}} - v_{\bar{x}})^2 \right\} h.$$

We use now two inequalities valid for any function w, defined on $\bar{\omega}_h$ and equal to zero at $x = 0$ and $x = 1$:

$$\sum_{\breve{\omega}_h} (w_{\bar{x}})^2 h \geq 8 \sum_{\breve{\omega}_h} w^2 h,$$

$$\sum_{\omega_h} w^2 h \geq \sum_{\breve{\omega}_h} (w_{\bar{x}})^2 h.$$

The first of these inequalities is the difference analog of the norm estimate for the embedding of $\mathring{W}_2^1[0, 1]$ into $L_2[0, 1]$. It is proved in [114]. The second inequality follows almost immediately from the relation

$$w(x)w(x + h) \leq (w^2(x) + w^2(x + h))/2.$$

Applying these inequalities we have

$$J^h(u, v) - c_1 \sum_{\breve{\omega}_h} (u_{\bar{x}} - v_{\bar{x}})^2 h$$

$$\geq \sum_{\breve{\omega}_h} \left\{ \int_0^1 \left(4 - 8c_1 - \frac{1}{2} \left| \frac{\partial f}{\partial q_1} (x, q_0, q_1) \right|^2 + \frac{\partial f}{\partial q_0} (x, q_0, q_1) \right) d\tau \right\}$$

$$\times (u_{\bar{x}} - v_{\bar{x}})^2 h.$$

Since from (6.3) we know that the right-hand side of this relation is non-negative we obtain

$$J^h(u, v) \geq c_1 \sum_{\breve{\omega}_h} (u_{\bar{x}} - v_{\bar{x}})^2 h.$$

From the definition of $J^h(u, v)$ this inequality is seen to be equivalent to (6.8).

To use the difference analog of strong monotonicity it is necessary to pass from the boundary conditions (6.7) to homogeneous ones. Consider the linear function

$$F(x) = u_0 + x(u_1 - u_0)$$

and substitute

$$u^h = F + \xi \quad \text{on } \omega_h.$$

Then ξ will satisfy

$$\xi_{\hat{x}\hat{x}} = f_{\hat{x}}(x, \xi_{\hat{x}} + F_{\hat{x}}, \xi_{\hat{x}} + F_{\hat{x}}) \quad \text{on } \omega_h, \quad \xi(0) = \xi(1) = 0, \qquad (6.9)$$

which is solvable if (6.6), (6.7) is solvable.

We proceed as in [101]. We need a lemma from [84]:

Lemma 6.1. Let $\xi \to P(\xi)$ be a continuous mapping of E^m into itself, so that for suitable $p > 0 \, (P(\xi), \xi) \geq 0$ for all ξ with $\|\xi\| = p$. Then we can find some $\mathring{\xi}$ with $\|\mathring{\xi}\| \leq p$ so that $P(\mathring{\xi}) = 0$.

In order to use this lemma consider the mapping of $(N - 1)$-dimensional space into itself which sends the vector ξ with components $\xi_i = \xi(x_i)$, $x_i \in \omega_h$ to the vectors P_ξ with components

$$P_\xi(x_i) = -\xi_{\hat{x}\hat{x}}(x_i) + f_{\hat{x}}(x_i, \xi_{\hat{x}}(x_i) + F_{\hat{x}}(x_i), \xi_{\hat{x}}(x_i) + F_{\hat{x}}(x_i)).$$

Here for the sake of simplicity we have assumed $\xi(0) = 0$, $\xi(1) = 0$. The continuity of mapping $\xi \to P_\xi$ follows from the continuity of the function f.

Compute

$$\sum_{\omega_h} P_\xi \xi h = \sum_{\omega_h} \{-\xi_{\hat{x}\hat{x}} + f_{\hat{x}}(x, \xi_{\hat{x}} + F_{\hat{x}}, \xi_{\hat{x}} + F_{\hat{x}})\} \xi h.$$

Rewrite the sum on the right-hand side:

$$\sum_{\omega_h} P_\xi \xi h = \sum_{\omega_h} \{-\xi_{\hat{x}\hat{x}} + f_{\hat{x}}(x, \xi_{\hat{x}} + F_{\hat{x}}, \xi_{\hat{x}} + F_{\hat{x}})$$

$$- f_{\hat{x}}(x, F_{\hat{x}}, F_{\hat{x}})\} \xi h + \sum_{\omega_h} f_{\hat{x}}(x, F_{\hat{x}}, F_{\hat{x}}) \xi h.$$

Since $\xi(0) = \xi(1) = 0$, strong monotonicty and the Cauchy–Schwartz–Bunyakovsky inequality give us

$$\sum_{\omega_h} P_\xi \xi h \geq c_1 \sum_{\bar{\omega}_h} \xi_{\hat{x}}^2 h - c_2 \left(\sum_{\omega_h} \xi^2 h \right)^{1/2},$$

where

$$c_2 = \max_{x \in [0, 1]} |f(x, F(x), F'(x)|.$$

Here we used the fact that F is a linear function, so that:

$$F_{\hat{x}}(x) = F(x) \quad \text{and} \quad F_{\hat{x}}(x) = F'(x).$$

Since $\xi(0) = \xi(1) = 0$ we have

$$\sum_{\bar{\omega}_h} \xi_{\hat{x}}^2 h \geq 8 \sum_{\omega_h} \xi^2 h.$$

Therefore

$$\sum_{\bar\omega_h} P_\xi \xi h \geq 8c_1 \sum_{\omega_h} \xi^2 h - c_2 \left(\sum_{\omega_h} \xi^2 h \right)^{1/2}.$$

Hence, it is obvious that if we choose σ so that $8c_1\sigma^2 - c_2\sigma \geq 0$, then

$$\sum_{\omega_h} P_\xi \xi \geq 0$$

if

$$\left(\sum_{\omega_h} \xi^2 h \right)^{1/2} \leq \sigma.$$

Therefore according to Lemma 6.1 a solution of (6.9) can surely be found, satisfying

$$\left(\sum_{\omega_h} \xi^2 h \right)^{1/2} \leq \sigma.$$

Theorem 6.2. *When* (6.3), *which guarantees the unique solvability of* (6.1), (6.2) *in* $C^{r+2}[0, 1]$, *is satisfied the expansion*

$$u^h = u + \sum_{j=1}^{l} h^{2j} v_j + h^r \eta^h \quad \text{on } \bar\omega_h \qquad (6.10)$$

will give a solution of the approximate system (6.6), (6.7). *Here* $l = [(r - 1)/2]$, *the functions* v_j *belong to* $C^{r+2-2j}[0, 1]$ *and do not depend on* h, *and the net function* η^h *is uniformly bounded:*

$$|\eta^h| \leq c_3 \quad \text{on } \bar\omega_h. \qquad (6.11)$$

PROOF. Let us assume $v_0 = u$ on $[0, 1]$ and try to find l functions v_j from

$$-v_j'' + \frac{\partial f}{\partial v}(x, u, u')v_j' + \frac{\partial f}{\partial u}(x, u, u')v_j$$

$$= -\frac{1}{4^j(2j)!}\frac{d^{2j}}{dx^{2j}}f(x, u, u') + \sum_{k=1}^{j}\frac{2v_{j-k}^{(2k+2)}}{(2k+2)!}$$

$$- \frac{\partial f}{\partial v}(x, u, u')\sum_{k=1}^{j}\frac{v_{j-k}^{(2k+1)}}{4^k(2k+1)!} - \frac{\partial f}{\partial u}(x, u, u')\sum_{k=1}^{j}\frac{v_{j-k}^{(2k)}}{4^k(2k)!}$$

$$- \sum_{\substack{2 \leq s+m+p \leq j \\ s+m \geq 1}}\frac{1}{4^p(2p)!}\frac{d^{2p}}{dx^{2p}}\left\{ \frac{1}{s!\,m!}\frac{\partial^{s+m}f}{\partial u^s\,\partial v^m}(x, u, u') \right.$$

$$\times \sum_{t_1 + \cdots + t_{s+m} = j-p}\prod_{i=1}^{s}\left(\sum_{k=0}^{t_i}\frac{v_{t_i-k}^{(2k)}}{4^k(2k)!} \right)$$

$$\times \left. \prod_{i=s+1}^{s+m}\left(\sum_{k=0}^{t_i}\frac{v_{t_i-k}^{(2k+1)}}{4^k(2k+1)!} \right) \right\} \quad \text{on } [0, 1], \qquad (6.12)$$

$$v_j(0) = 0, \qquad v_j(1) = 0; \qquad j = 1, 2, \ldots, l. \qquad (6.13)$$

The first step is to find the function v_1. Equation (6.12) for this function has the simple form:

$$-v_1'' + \frac{\partial f}{\partial v}(x, u, u')v_1' + \frac{\partial f}{\partial u}(x, u, u')v_1$$

$$= -\frac{\partial f}{\partial v}(x, u, u')\frac{v_0'''}{24} - \frac{\partial f}{\partial u}(x, u, u')\frac{v_0''}{8} - \frac{1}{8}\frac{d^2}{dx^2} f(x, u, u') + \frac{v_0^{(4)}}{12}.$$

The right-hand side of this equation contains only the function v_0 and is $r - 2$ times continuously differentiable. Since the equation is linear and the coefficients are $r - 2$ times continuously differentiable there is a solution, satisfying the boundary condition (6.13). It is unique and belongs to $C^r[0, 1]$. Let us now assume, that the functions $v_0, \ldots, v_{j-1}$ (with $v_k \in C^{r+2-2k}[0, 1]$) have been already determined. From (6.12) we see that the right-hand side of the equation for v_j does not contain functions v_k whose index k is more than $j - 1$, and is $r - 2j$ times continuously differentiable on the interval $[0, 1]$. The latter follows from a simple calculation on the smoothness of v_k and the number of derivatives one may take. The coefficients of the linear equation obtained are functions of $C^{r-1}[0, 1]$; therefore there is a unique solution satisfying the boundary conditions (6.13) as well as equation (6.12). This solution will belong to $C^{r+2-2j}[0, 1]$. Thus, all $l + 1$ functions v_j have been found and satisfy the condition

$$v_j \in C^{r-2j+2}[0, 1], \qquad j = 0, 1, \ldots, l.$$

Using these functions we define

$$w = \sum_{j=0}^{l} h^{2j}v^j \quad \text{on } \bar\omega_h \tag{6.14}$$

and substitute this w into the difference operator of (6.6):

$$-\sum_{j=0}^{l} h^{2j}(v_j)_{\bar x \hat x} + f_{\hat x}\left(x, \sum_{j=0}^{l} h^{2j}(v_j)_{\hat x}, \sum_{j=0}^{l} h^{2j}(v_j)_{\hat x}\right) = -w_{\bar x \hat x} + f_{\hat x}(x, w_{\hat x}, w_{\hat x}).$$

$$\tag{6.15}$$

Applying Lemma 1.1 of §7.1, we have

$$-\sum_{j=0}^{l} h^{2j}(v_j)_{\bar x \hat x} = -2\sum_{j=0}^{l} h^{2j}\sum_{k=0}^{l-j} h^{2k}\frac{v_j^{(2k+2)}}{(2k+2)!} + h^r\theta_1$$

$$= -2\sum_{j=0}^{l} h^{2j}\sum_{k=0}^{l} \frac{v_{j-k}^{(2k+2)}}{(2k+2)!} + h^r\theta_1 \quad \text{on } \omega_h,$$

$$\sum_{j=0}^{l} h^{2j}(v_j)_{\hat x} = \sum_{j=0}^{l} h^{2j}\sum_{k=0}^{j} \frac{v_{j-k}^{2k}}{4^k(2k)!} + h^r\theta_2 \quad \text{on } \ddot\omega_h, \tag{6.16}$$

$$\sum_{j=0}^{l} h^{2j}(v_j)_{\hat x} = \sum_{j=0}^{l} h^{2j}\sum_{k=0}^{j} \frac{v_{j-k}^{(2k+1)}}{4^k(2k+1)!} + h^r\theta_3 \quad \text{on } \ddot\omega_h,$$

where

$$|\theta_i| \le c_4 \quad \text{for } i = 1, 2, 3.$$

Let us now expand the function $f(x, w_{\bar{x}}, w_{\overset{\circ}{x}})$ by Taylor's formula:

$$f\left(x, \sum_{j=0}^{l} h^{2j} \sum_{k=0}^{j} \frac{v_{j-k}^{(2k)}}{4^k (2k)!} + h^r \theta_2, \sum_{j=0}^{l} h^{2j} \sum_{k=0}^{j} \frac{v_{j-k}^{(2k+1)}}{4^k (2k+1)!} + h^r \theta_3\right)$$

$$= f(x, u, u') + \sum_{1 \le s+m \le l} \frac{1}{s!\, m!} \frac{\partial^{s+m}}{\partial u^s\, \partial v^m} f(x, u, u')$$

$$\times \left(\sum_{j=1}^{l} h^{2j} \sum_{k=0}^{j} \frac{v_{j-k}^{(2k)}}{4^k (2k)!} + h^r \theta_2\right)^s \left(\sum_{j=1}^{l} h^{2j} \sum_{k=0}^{j} \frac{v_{j-k}^{(2k+1)}}{4^k (2k+1)!} + h^r \theta_3\right)^m$$

$$+ h^{2l+2} \theta_4 \quad \text{on } \breve{\omega}_h, \tag{6.17}$$

where

$$\theta_4 = \sum_{s+m=l+1} \frac{1}{s!\, m!} \frac{\partial^{l+1} f}{\partial u^s\, \partial v^m} (x, \xi_s, \eta_m)$$

$$\times \left(\sum_{j=1}^{l} h^{2j} \sum_{k=0}^{j} \frac{v_{j-k}^{(2k)}}{4^k (2k)!} + h^r \theta_2\right)^s \left(\sum_{j=1}^{l} h^{2j} \sum_{k=0}^{j} \frac{v_{j-k}^{(2k+1)}}{4^k (2k+1)!} + h^r \theta_3\right)^m$$

is uniformly bounded

$$|\theta_4| \le c_5$$

since at every point $x \in \breve{\omega}_h$ the functions v_j and their derivatives in (6.17) are bounded (because they are continuous on the interval $[0, 1]$).

Expand the expressions in brackets in (6.17), and write them as polynomials in h:

$$f(x, w_{\bar{x}}, w_{\overset{\circ}{x}}) = f(x, u, u')$$

$$+ \sum_{1 \le s+m \le l} \frac{1}{s!\, m!} \frac{\partial^{s+m} f}{\partial u^s\, \partial v^m} (x, u, u') \sum_{j=s+m}^{l} h^{2j}$$

$$\times \left\{ \sum_{\substack{t_1 + \cdots + t_{s+m} = j \\ t_i \ge 1}} \prod_{i=1}^{s} \left(\sum_{k=0}^{t_i} \frac{v_{t_i-k}^{(2k)}}{4^k (2k)!}\right) \prod_{i=s+1}^{s+m} \left(\sum_{k=0}^{t_i} \frac{v_{t_i-k}^{(2k+1)}}{4^k (2k+1)!}\right) \right\} + h^r \theta_5$$

and change the order of summation

$$f(x, w_{\bar{x}}, w_{\overset{\circ}{x}}) = f(x, u, u')$$

$$+ \sum_{1 \le s+m \le l} \frac{1}{s!\, m!} \frac{\partial^{s+m} f}{\partial u^s\, \partial v^m} (x, u, u') \sum_{j=1}^{l} h^{2j}$$

$$\times \left\{ \sum_{t_1 + \cdots + t_{s+m} = j} \prod_{i=1}^{s} \left(\sum_{k=0}^{t_i} \frac{v_{t_i-k}^{(2k)}}{4^k (2k)!}\right) \right.$$

$$\left. \times \prod_{i=s+1}^{s+m} \left(\sum_{k=0}^{t_i} \frac{v_{t_i-k}^{(2k+1)}}{4^k (2k+1)!}\right) \right\} + h^r \theta_5 \quad \text{on } \breve{\omega}_h.$$

Again let us use formula (1.2) of §7.1:

$$f_{\bar{x}}(x, w_{\bar{x}}, w_{\bar{x}}) = f(x, u, u') + \sum_{j=1}^{l} h^{2j}\left\{\frac{1}{4^j(2j)!}\frac{d^{2j}}{dx^{2j}} f(x, u, u')\right.$$

$$+ \sum_{p=0}^{l-j} h^{2p}\frac{1}{4^p(2p)!}\frac{d^{2p}}{dx^{2p}}\left(\sum_{1 \le s+m \le j}\frac{1}{s!\,m!}\frac{\partial^{s+m}f}{\partial u^s\,\partial v^m}(x, u, u')\right)$$

$$\times \sum_{t_1+\cdots+t_s+m=j}\prod_{i=1}^{s}\left(\sum_{k=0}^{t_i}\frac{v_{t_i-k}^{(2k)}}{4^k(2k)!}\right)$$

$$\times \prod_{i=s+1}^{s+m}\left(\sum_{k=0}^{t_i}\frac{v_{t_i-k}^{(2k+1)}}{4^k(2k+1)!}\right)\right\} + h^r\theta_6.$$

This equality is valid on the entire net region ω_h. Let us arrange the terms of this expansion in increasing powers of h and substitute it in (6.15), and using (6.16), this results in:

$$-w_{\bar{x}\bar{x}} + f_{\bar{x}}(x, w_{\bar{x}}, w_{\bar{x}}) = -v_0'' + f(x, u, u')$$

$$+ \sum_{j=1}^{l} h^{2j}\left\{\frac{1}{4^j(2j)!}\frac{d^{2j}}{dx^{2j}} f(x, u, u') - 2\sum_{k=0}^{j}\frac{v_{j-k}^{(2k+2)}}{(2k+2)!}\right.$$

$$+ \sum_{p=0}^{j}\frac{1}{4^p(2p)!}\frac{d^{2p}}{dx^{2p}}\left(\sum_{1 \le s+m \le j-p}\frac{1}{s!\,m!}\frac{\partial^{s+m}f}{\partial u^s\,\partial v^m}(x, u, u')\right)$$

$$\times \sum_{t_1+\cdots+t_s+m=j-p}\prod_{i=1}^{s}\left(\sum_{k=0}^{t_i}\frac{v_{t_i-k}^{(2k)}}{4^k(2k)!}\right) \times \prod_{i=s+1}^{s+m}\left(\sum_{k=0}^{t_i}\frac{v_{t_i-k}^{(2k+1)}}{4^k(2k+1)!}\right)\right\} + h^r\theta_7,$$

where

$$|\theta_7| \le c_6 \quad \forall\, x \in \omega_h.$$

Note that because $v_0 = u$ the term $-v_0'' + f(x, u, u')$ is zero. From (6.12) all terms in h^{2j}, $j = 1, \ldots, l$ cancel out. Therefore

$$-w_{\bar{x}\bar{x}} + f_{\bar{x}}(x, w_{\bar{x}}, w_{\bar{x}}) = h^r\theta \quad \text{on } \omega_h. \tag{6.18}$$

Here θ_8 denotes a net function with modulus uniformly bounded by a constant independent of h:

$$|\theta_8| \le c_7 \quad \text{on } \omega_h. \tag{6.19}$$

Note that from the homogenuous boundary conditions, (6.13) and the relation (6.14) for the function w we have $w(0) = u_0$ and $w(1) = u_1$. Therefore one may use the difference form of strong monotonicity (6.8), assuming that $u(x) = u^h(x)$ and $v(x) = w(x)$. From (6.6) and (6.18) we get

$$-h^r \sum_{\omega_h} \theta_8(u^h - w)h \ge c_1 \sum_{\bar{\omega}_h}(u_{\bar{x}}^h - w_{\bar{x}})^2 h. \tag{6.20}$$

Since $u^h(0) - w(0) = 0$ and $u^h(1) - w(1) = 0$ one may apply the difference analog of the embedding estimate from $\mathring{W}^1_2(0, 1)$ to $C[0, 1]$ (see [112]):

$$\max_{\bar{\omega}_h} |u^h - w| \le \frac{1}{2} \left(\sum_{\bar{\omega}^h} (u^h_{\bar{x}} + w_{\bar{x}})^2 h \right)^{1/2}. \tag{6.21}$$

From (6.19) it follows that

$$\sum_{\omega_h} \theta_8 |u^h - w| h \le c_7 \max_{\bar{\omega}_h} |u^h - w|. \tag{6.22}$$

Combining the inequalities (6.20) and (6.21) we have

$$4c_1 \left(\max_{\bar{\omega}_h} |u^h - w| \right)^2 \le c_7 h^r \max_{\bar{\omega}_h} |u^h - w|.$$

Divide this inequality by $\max |u^h - w|$. We get

$$\max_{\bar{\omega}_h} |u^h - w| \le \frac{c_7}{4c_1} h^r.$$

Thus if we write

$$\eta^h = (u^h - w)h^{-r} \quad \text{on } \bar{\omega}_h$$

then the inequality

$$\max_{\bar{\omega}_h} |\eta^h| \le c_7/(4c_1)$$

is valid for this function, and coincides with (6.11) if we assume that

$$c_3 = c_7/(4c_1).$$

Theorem 6.2 is proved. $\square$

Let us now demonstrate how we might apply Theorem 6.2. Let $l = [(r - 1)/2]$. For the $l + 1$ integers $N_k = kN$, $k = 1, \ldots, l + 1$, we will construct nets $\bar{\omega}_{h_k}$ with mesh-sizes $h_k = h/k$, where $h = 1/N$. Then let us then solve the difference equation (6.6), (6.7). Since the solutions of these problems are unique it is irrelevant which algorithms we use. Several iterative methods for solving nonlinear systems of the form (6.6), (6.7) are studied in [103].

All solutions u^{h_k} we obtain are defined on the net $\bar{\omega}_h$ with mesh-size h. Let us add them up, with weights

$$\gamma_k = \frac{2(-1)^{l-k+1} k^{2l+2}}{(l - k + 1)! \, (l + k + 1)!}. \tag{6.23}$$

We obtain the corrected solution

$$U = \sum_{k=1}^{l+1} \gamma_k u^{h_k} \quad \text{on } \bar{\omega}_h. \tag{6.24}$$

Theorem 6.3. *Suppose the hypotheses of Theorem 6.2 are satisfied. Then the corrected solution (6.24) obeys the estimate*

$$\max_{\bar{\omega}_h} |U - u| \leq c_8 h^r.$$

The proof duplicates that of Theorem 4.6, and gives

$$c_8 = 2c_3 \sum_{k=1}^{l+1} \frac{k^{2l+2-r}}{(l-k+1)! \, (l+k+1)!}.$$

To illustrate the effect of this correction numerically let us consider

$$u'' = \cos(u' + 4u) + 16e^{-4x} - \cos(4x + 1) \quad \text{on } (0, 1),$$
$$u(0) = 1, \qquad u(1) = 1 + e^{-4}. \tag{6.25}$$

The exact (unique) solution of this problem is the function

$$u(x) = e^{-4x} + x.$$

Difference nets $\bar{\omega}_{h_i}$ were constructed for a certain set of integers N_i and the solutions u^{h_i} of the difference equations (6.6), (6.7) were found. Since the solutions u^{h_i} were found iteratively we actually get some v^{h_i} whichs differs from u^{h_i}. So that the error $v^{h_i} - u^{h_i}$ can be ignored we iterate until the error is less than 10^{-12}, i.e.,

$$|v^{h_i}(x) - u^{h_i}(x)| \leq 10^{-12} \quad \forall \, x \in \bar{\omega}_{h_i}.$$

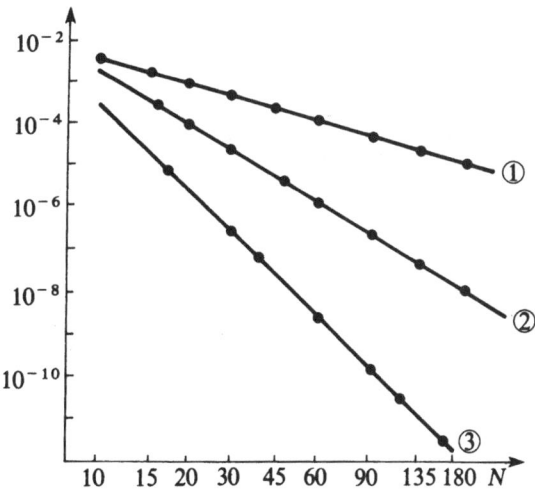

Figure 3.4. Maximal errors of approximate solutions for (6.25).

(1) The error of the difference solution of (6.6), (6.7); (2) the error of the solution extrapolated on two nets with mesh-sizes in a ratio of $1:2$; (3) the error of the solution extrapolated on three nets with mesh-sizes in a ratio of $1:2:3$.

We then compute the maximal error

$$\xi(N_i) = \max_{\bar{\omega}_{h_i}} |u^{h_i} - u|,$$

the graph of which is given in Figure 3.4. We have thus constructed corrected solutions (6.24), extrapolating two or three solutions with mesh-sizes in the rations $1:2$ and $1:2:3$ respectively. The dependence of the errors of (6.24) on the number of knots used in the extrapolation, is given in Figure 3.4.

CHAPTER 4
Elliptic Equations

Boundary-value problems for elliptic equations are the most common problems of mathematical physics. They have many practical applications. They also appear in the reduced forms of parabolic and hyperbolic equations. Therefore it is quite natural to pay particular attention to problems related to elliptic operators. Here we will not focus on well-known facts connected with the statement of the boundary-value problem of elliptic type, and the dependence of the solutions on the properties of the input data. This information can be found in the literature. We will give only the necessary facts from the theory, and focus on the application of the various approaches to improving the difference solutions to such problems. In addition to linear problems a solvable nonlinear problem will also be considered. This nonlinear problem, as well as a simple diffraction problem, have been chosen because we wanted to demonstrate the improvement in the accuracy of the solutions for relatively simple problems. Our aim was also to acquaint the reader with basic algorithmic techniques used.

4.1. The Statement of the Problem

Let R^2 be the two-dimensional Euclidean space of points $x = (x_1, x_2) = (x, y)$ with the metric

$$|x - x'| = ((x_1 - x_1')^2 + (x_2 - x_2')^2)^{1/2}.$$

Let Ω be a bounded, connected, strictly Lipschitz region (see [73]) with boundary Γ. In this chapter we will study the equation

$$Lu \equiv -\frac{\partial}{\partial x_1} p \frac{\partial u}{\partial x_1} - \frac{\partial}{\partial x_2} p \frac{\partial u}{\partial x_2} + qu = f \quad \text{on } \Omega, \tag{1.1}$$

where the coefficients satisfy the conditions

$$p \geq c_1 > 0, \qquad q \geq 0 \quad \text{on } \Omega. \tag{1.2}$$

Our boundary conditions will be the Dirichlet problem

$$u(x) = \varphi(x) \quad \forall\, x \in \Gamma. \tag{1.3}$$

Let $\alpha \in (0, 1)$. Define

$$\langle u \rangle_\Omega^\alpha = \sup_{x, x' \in \Omega} |u(x) - u(x')| / |x - x'|^\alpha.$$

Define the Banach space $C^{l+\alpha}(\overline{\Omega})$ as in [73]. The elements of this space are continuous functions on $\overline{\Omega}$. They have derivatives up to order l, which are continuous on Ω. For these functions introduce the norm ($k_i \geq 0$ are integers)

$$\|u\|_{C^{l+\alpha}(\overline{\Omega})} = \sum_{0 \leq k_1 + k_2 \leq l} \sup_\Omega \left| \frac{\partial^{k_1 + k_2} u}{\partial x_1^{k_1}\, \partial x_2^{k_2}} \right| + \sum_{k_1 + k_2 = l} \left\langle \frac{\partial^{k_1 + k_2} u}{\partial x_1^{k_1}\, \partial x_2^{k_2}} \right\rangle_\Omega^\alpha,$$

which will make this space a Banach space.

Denote by $C^{l+\alpha}(\Omega)$ the set of functions belonging to $C^{l+\alpha}(\overline{\Omega}')$ for any strictly interior subdomain $\Omega' \subset \Omega$ (i.e., a domain such that the distance from Ω' to Γ is nonzero).

Let $L_2(\Omega)$ be the Banach space of measurable functions which are square integrable on Ω with respect to Lebesgue measure. Denote $\|u\|_{L_2(\Omega)} = (\int_\Omega u^2\, dx)^{1/2}$.

$W_2^l(\Omega)$ is the Banach space which consists of all elements $u \in L_2(\Omega)$ having generalized derivatives $\partial^{k_1 + k_2} u / \partial x_1^{k_1}\, \partial x_2^{k_2}$, $0 \leq k_1 + k_2 \leq l$, which are square integrable on Ω. Generalized derivatives are interpreted in a usual sense (for example, as in [133, 94]). The norm is given by

$$\|u\|_{W_2^l(\Omega)} = \int_\Omega \left(\sum_{0 \leq k_1 + k_1 \leq l} \frac{\partial^{k_1 + k_2} u}{\partial x_1^{k_1}\, \partial x_2^{k_2}}\, dx \right)^{1/2}.$$

$\mathring{W}_2^l(\Omega)$ is the subspace of $W_2^l(\Omega)$ which consists of all infinitely differentiable functions with supports on Ω, with the closure taken in the norm $\|\cdot\|_{W_2^l(\Omega)}$.

For a convenience we will sometimes use another norm for the space $\mathring{W}_2^1(\Omega)$

$$|u| = \left(\int_\Omega \left\{ \left(\frac{\partial u}{\partial x_1} \right)^2 + \left(\frac{\partial u}{\partial x_2} \right)^2 \right\} dx \right)^{1/2}.$$

The norms $\|u\|_{W_2^1(\Omega)}$ and $|u|$ are equivalent (see [94]).

A curve K will be said to be a curve of class $C^{k+\alpha}$ (or C^k) if there is a number $a_0 > 0$, such that in a neighborhood of each point $x_0 \in K$ one can introduce orthogonal coordinates (y_1, y_2) centered at the point x_0, so that in these coordinates the boundary points will be described by the equation $y_2 = g(y_1)$, and the function $g(y_1)$ belongs to the class $C^{k+\alpha}[a, b]$ (or $C^k[a, b]$), where the interval $[a, b]$ is obtained by projecting the set $\{x \in K; |x - x_0| \leq a_0\}$ onto the line $y_2 = 0$.

Let us state the following result on the solvability of (1.1)–(1.3) without proof.

Theorem 1.1 (see [73]). *Assume that for* (1.1)–(1.3) *the coefficients satisfy*:

$$p \in C^{l+1+\alpha}(\overline{\Omega}), \qquad q, f \in C^{l+\alpha}(\overline{\Omega}) \qquad (1.4)$$

with integer $l \geq 0$, *and* $\alpha \in (0, 1)$. *Then there is a unique solution* u *and* $u \in C^{l+2+\alpha}(\Omega)$.

If one also assumes in Theorem 1.1 that the boundary and boundary values are smooth, then one can guarantee the smoothness of the solution "up to the boundary."

Theorem 1.2 (see [73]). *Assume that* (1.4) *and*

$$\Gamma \in C^{l+2+\alpha}, \qquad \varphi \in C^{l+2+\alpha}(\Gamma)$$

are satisfied for (1.1)–(1.3). *Then* $u \in C^{l+2+\alpha}(\overline{\Omega})$.

In addition to the operator formulation of problem (1.1)–(1.3) a weak (integral) formulation is possible. A function $u \in \mathring{W}_2^1(\Omega)$, which satisfies the integral identity

$$\int_\Omega \left(p \frac{\partial u}{\partial x_1} \frac{\partial v}{\partial x_1} + p \frac{\partial u}{\partial x_2} \frac{\partial v}{\partial x_2} + quv \right) dx = \int_\Omega fv \, dx \; \forall \, v \in \mathring{W}_2^1(\Omega) \quad (1.5)$$

will be called a generalized solution (see [73]) of (1.1) with homogeneous boundary condition (1.3) (i.e., $\varphi \equiv 0$).

If the coefficient p has generalized derivatives of the first order and $u \in W_2^2(\Omega)$, then identity (1.5) follows from equation (1.1), and conversely. Let us give a criterion for the solvability of (1.5).

Theorem 1.3 (see [73]). *Assume that the coefficients* p, q *in equation* (1.5) *are measurable functions, bounded on* $\overline{\Omega}$ *satisfying condition* (1.2). *Then for any function* f *with the bounded norm* $\| f \|_{L_2(\Omega)}$ *there is a unique generalized solution of* (1.5).

In some cases it is possible to prove additional smoothness.

Theorem 1.4. *Let the coefficients of* (1.5) *satisfy the conditions of Theorem 1.3 and*

$$\left| \frac{\partial p}{\partial x_i} \right| \leq c_2 < \infty, \qquad i = 1, 2,$$

where the Ω *is either a circle, an annulus, a rectangle, a triangle, or a region that can be transformed into one of these domains by a regular transformation of* $C^2(\overline{\Omega})$. *Then* (1.5) *has a unique solution in* $W_2^2(\Omega)$.

PROOF. The theorem, in the cases where Ω is a circle, an annulus, a rectangle, or a domain which can be transformed into one of these is proved in [73]. Let us now assume that Ω is an open triangle. Consider the problem

$$-\Delta v = f \quad \text{in } \Omega,$$
$$v = 0 \quad \text{on } \Gamma. \tag{1.6}$$

Augment the triangle Ω with another triangle obtained by reflecting Ω about its longest side. Add this side to the set the points thus obtained, to obtain a convex quadrangle $\tilde{\Omega}$ with boundary $\tilde{\Gamma}$. Extend the function f to the reflected triangle, so that the extended function is antisymmetric with respect to the longest side, and is zero on this side. We have

$$-\Delta \tilde{v} = f \quad \text{in } \tilde{\Omega},$$
$$\tilde{v} = 0 \quad \text{on } \tilde{\Gamma}. \tag{1.7}$$

Since the convex quadrangle $\tilde{\Omega}$ can be transformed into a rectangle by a bilinear regular transformation, and the extended function f belongs to $L_2(\tilde{\Omega})$, then (1.7) has a unique solution $\tilde{v} \in W_2^2(\tilde{\Omega})$. Let us prove that $\tilde{v}$ satisfies (1.6) on Ω. Indeed, $\tilde{v}$ satisfies the equation $-\Delta \tilde{v} = f$ almost everywhere on Ω, and $\tilde{v} \in W_2^2(\Omega)$. The values of the function $\tilde{v}$ are antisymmetric relative to the symmetry axis, and therefore the function $\tilde{v}$ is zero on the longest side of the triangle Ω. The function $\tilde{v}$ is zero on the other two sides because of the homogeneous boundary condition (1.7). Thus, we have found the unique solution of (1.6). $\qquad\Box$

Further conclusions related to the change of variables will be found in [73].

4.2. Difference Methods for the Dirichlet Problem on a Domain with a Smooth Boundary

The difference methods which will be introduced in the present section can be applied to many types of multidimensional and quasilinear equations. But proving the basic theorems and obtaining results on multidimensional quasilinear problems would require a great deal of effort. We will investigate simple boundary-value problems for elliptic equations, so that we can concentrate on the ideas behind the constructions which lead to the improvement of approximate solutions.

In this section we will consider the Dirichlet problem for the two-dimensional Poisson equation

$$-\Delta u = f \quad \text{in } \Omega, \tag{2.1}$$
$$u = \varphi \quad \text{on } \Gamma. \tag{2.2}$$

Let us assume that the conditions of Theorem 1.2 are satisfied by an integer $l \geq 2$ and an α from the interval $(0, 1)$.

4.2.1. Approximation of the Boundary Condition

This section is devoted to developing an idea introduced in [138].

To construct a difference problem let us assume that the domain Ω is enclosed in a square $\{-b < x < b, -b < y < b\}$. Cover it with a square grid with a mesh-size $h = b/N$ formed by the lines $x_i = ih$ and $y_j = jh$, where $i, j = -N, \ldots, N$. The points of intersection of these lines we will call knots. The knot $x = (x_i, y_j)$ will be called an interior knot if $x \in \Omega$. Let us denote the set of all interior knots by Ω_h. Each line of grid which crosses Ω, also crosses the boundary Γ; the crossing points divide the boundary into some intervals, because of the smoothness of the boundary ($\Gamma \in C^{4+\alpha}$). We call the endpoints of these intervals boundary knots in the x- (or y-) direction, if the grid line passing through this knot is parallel to the coordinate axis Ox (Oy, respectively). Note that the point A (Figure 4.1) is a boundary knot in the x-direction but is not a boundary knot in the y-direction even though it lies on a grid line parallel to the axis Oy. We denote the sets of all boundary knots in the x-direction and in the y-direction by $\Gamma_{h,x}$ and $\Gamma_{h,y}$ correspondingly. The union of these sets (the set of boundary knots) we denote by Γ_h. An interior knot is said to be regular if $\overline{\Omega}$ contains this knot together with four closed intervals of the grid, connecting this knot with its four nearest neighboring grid knots of Ω_h. The set of all regular knots is denoted by Ω_h^r; and the rest of the interior knots are called irregular; the set of these is denoted by Ω_h^{ir}.

Let us write the usual fivepoint difference equation at each regular knot

$$-u_{\bar{x}x}^h - u_{\bar{y}y}^h = f \quad \text{on } \Omega_h^r. \tag{2.3}$$

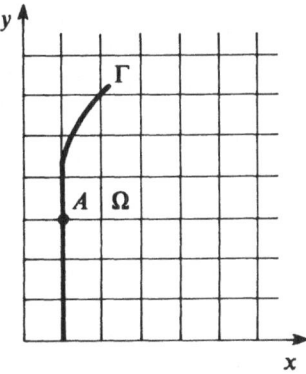

Figure 4.1. A boundary knot in the x-direction.

which for $u \in C^{l+\alpha+2}(\overline{\Omega})$ has an approximation error in the following form

$$-u_{\xi\xi} - u_{\check{y}\check{y}} = -\Delta u - \sum_{k=1}^{[l/2]} h^{2k} \frac{2}{(2k+2)!} \left(\frac{\partial^{2k+2}u}{\partial x^{2k+2}} + \frac{\partial^{2k+2}u}{\partial y^{2k+2}} \right) + \frac{2h^{l+\alpha}}{(2l+2)!} \theta_1,$$

(2.4)

where

$$|\theta_1(x)| \leq \|u\|_{C^{l+2+\alpha(\overline{\Omega})}} \quad \forall x \in \Omega_h^r.$$

Note that the coefficients of the even powers of h do not depend on h, and that they are smooth functions on $\overline{\Omega}$. Therefore from §1.2 one hopes that the terms of the approximation error yield similar contributions to the solution error. However, in constructing the equations for irregular knots the following problem arises. The usual fivepoint equations as applied to these knots, which are not uniformly arranged, do not give the required form for the approximation error. Instead of these equations other grid equations, based on the Lagrange interpolation formula, are used.

On the axis Ot let us put a knot $\delta\overline{h}$ to the right of the origin and put n uniformly spaced knots $(-\overline{h}, -2\overline{h}, \ldots, -n\overline{h})$ to the left of the origin. Using the Lagrange formula to interpolate $\psi(t)$ at these points we have

$$\psi(0) = \sum_{k=1}^{n} (-1)^{k-1} \frac{n!}{k!(n-k)!} \frac{\delta}{\delta+k} \psi(-k\overline{h}) + \prod_{k=1}^{n} \frac{k}{\delta+k} \psi(\delta\overline{h}) + R(0),$$

(2.5)

where the remainder $R(0)$ satisfies the inequality

$$|R(0)| \leq \overline{h}^{n+1} \frac{\delta}{n+1} \max_{[-n\overline{h}, \delta h]} \left| \frac{d^{n+1}\psi}{dt^{n+1}} \right|.$$

(2.6)

Let us define the operator

$$I_\delta^n \psi(0) = \sum_{k=1}^{n} (-1)^{k-1} \frac{n!}{k!(n-k)!} \frac{\delta}{\delta+k} \psi(-k\overline{h}),$$

(2.7)

and the number

$$\lambda_\delta^n = \prod_{k=1}^{n} \frac{k}{\delta+k}.$$

When $\delta > 0$ goes monotonically to zero the quantity

$$B(\delta) = \sum_{k=1}^{n} \frac{n!}{k!(n-k)!} \frac{\delta}{\delta+k}$$

also goes monotonically to zero, therefore there is a δ_n, such that

$$B(\delta) \leq \tfrac{1}{2} \quad \forall \delta \leq \delta_n.$$

(2.8)

Let us apply the interpolation formula at each irregular knot $x \in \Omega_h^{ir}$, choosing the axis Ot parallel to one of the coordinate axes with the origin at the point x. The irregularity of the knot x is expressed in the fact that there is a boundary knot ξ_x at a distance less than h in at least in one of four directions parallel to the coordinate axes. Now we choose the direction of the axis Ot so that it connects the point x to the point ξ_x, and assume that the value $\delta\bar{h}$ is equal to the distance from x to ξ_x. Let us assume $\bar{h}$ is equal to

$$h([1/\delta_n] + 1). \tag{2.9}$$

The equation for the irregular knot will have the form

$$u^h(x) = I_\delta^n u^h(x) + \lambda_\delta^n \varphi(\xi_x), \qquad x \in \Omega_h^{ir}, \tag{2.10}$$

where (2.8) is satisfied for each knot x.

Note that the line segment $[-n\bar{h}, \delta\bar{h}]$ on the axis Ot is assumed to belong to the closed domain $\bar{\Omega}$, this being true if h is sufficiently small. Indeed, let only the line segment which connects the knots $x = (x, y)$ and $(x \pm h, y)$ is not in $\bar{\Omega}$. If the boundary Γ crosses the axis Ot twice at a distance less than $(n + 1)h$, then the boundary would have curvature of order $1/h$. This fact contradicts the condition $\Gamma \in C^2$ when $h \to 0$. This is illustrated by the example shown in Figure 4.2. Here the direction of the axis Ot coincides with the direction of the axis Ox, and the second crossing point of the axis Ot with the boundary Γ is denoted by ρ. Let the boundary arc between the points ρ and ξ_x be described by the curve $\sigma = g(t)$ in the orthogonal frame with axes Ot and $O\sigma$; the axis $O\sigma$ is parallel to Oy and originates from the point x. Denote the crossing point of the axis $O\sigma$ with the boundary Γ by θ. If we keep in mind that $|\theta - x| \geq h$, then according to Lagrange's theorem there is a point t_1 on the axis Ot between the points x and ξ_x where $g'(t_1) < -1$. According to Rolle's theorem there is a point t_2 on the axis Ot between the points ρ and ξ_x, where $g'(t_2) = 0$. Using again Lagrange's theorem we conclude that there is the point t_3 between the points t_1 and t_2 where

$$g''(t_3) = \{g'(t_2) - g'(t_1)\}/(t_2 - t_1) \geq 1/\{(n + 1)h\}.$$

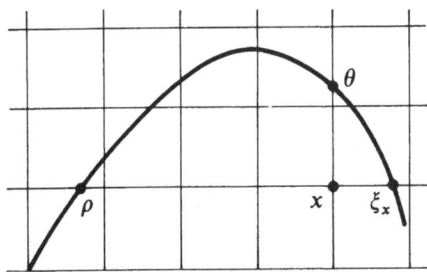

Figure 4.2. A knot irregular in the x-direction.

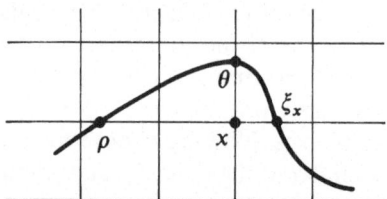

Figure 4.3. A knot which is irregular in two directions

If the boundary crosses two intervals connecting the knots $(x + h, y)$, $(x, y + h)$, say, with $x = (x, y)$ then we choose the direction of the axis Ot so that it goes to the nearest boundary knot. Suppose this is the knot ξ_x (Figure 4.3); the demonstration is then exactly as above. Note that condition (2.8) makes the system of algebraic equations *diagonally dominant* with strong diagonal dominance at the irregular knots. This fact allows to prove the stability of the difference problem (2.3), (2.10), and this in turn gives us the following result.

Theorem 2.1 (see [138]). *Let*

$$\Gamma \in C^{2m+2+\lambda}, \qquad \varphi \in C^{2m+2+\lambda}(\Gamma), \qquad f \in C^{2m+\lambda}(\overline{\Omega})$$

in (2.1), (2.2), for a natural number m and $\lambda \in (0, 1)$. Then the solution u^h of the system of difference equations (2.3), (2.10) (where $n = 2m$) can be represented in the form

$$u^h = u + \sum_{k=1}^{m} h^{2k}w_k + h^{2m+\lambda}r_h \quad on \ \Omega_h. \tag{2.11}$$

Here u is the solution of (2.1), (2.2). The functions w_k belong to $C^{2m+2-2k+\lambda}(\overline{\Omega})$ and do not depend on h, the grid function r_h is bounded:

$$|r_h(x)| \le c_2 \quad \forall \ x \in \Omega_h.$$

The expansion (2.11) can, in principal, be used in the multidimensional case, and will not differ from the one-dimensional case. We only note that interpolation for a multidimensional region is more complex than in the one-dimensional case, and it becomes preferable to correct the mesh-sizes h, $h/2$, $h/3, \ldots$, since this does not require interpolation, at least for knots of the grid $\overline{\Omega}_h$.

It should be noted that there are two inconveniences associated with using an approximation of the form (2.10). To guarantee (2.8) for $n > 2$ we have to take a mesh-size $\bar{h}$ several times greater than h. For small h this condition substantially degrades the approximating properties of the Lagrange polynomials (the constant in (2.6) increases). For large h the interpolation knots may leave the domain Ω, and the construction (2.10) will be ill-defined. To attain a reasonable degree of accuracy one can apply other approximations to the boundary conditions which use information about equation (2.1) itself instead of (2.10). For example, to attain accuracy of order h^3 it is enough to use the approximation suggested in [95, 117], which has been discussed in [119]. To attain accuracy of order h^4 one can use the approximations suggested in [136, 30, 2, 19]. An approximation providing h^5 accuracy is given in [140]; it has a compact knot arrangement. At the end of the present section we will approximate the Poisson equation using a nonuniform difference net for the irregular knots, rather than approximate the boundary condition. The difference equations obtained have a more compact arrangement of the knots than in (2.10). These equations are convenient because they can be used in a splitting scheme for the heat equation. At the same time an approximation of the form (2.10) can be applied directly to equations with variable coefficients, while other constructions become significantly more complicated.

In solving (2.3), (2.10) another disadvantage in comparison with the usual fivepoint schemes is encountered. It is connected with the fact that the matrix of the algebraic system is asymmetric which excludes some efficient approaches. To avoid this problem we consider the iterative process

$$-w^k_{\bar{x}x} - w^k_{\bar{y}y} = f \quad \text{on } \Omega^r_h, \qquad w^k(x) = I^{2m}_\delta w^{k-1}(x) + \lambda^{2m}_\delta \varphi(\xi_x), \qquad x \in \Omega^{ir}_h,$$

$$(2.12)$$

with zero initial condition $w^0 = 0$, say. At each step of the process we solve a system with a symmetric positive-definite matrix. We will show that the error decreases by a factor of $\frac{1}{2}$ at each iteration. Assume

$$\psi^k = w^k - u^h \quad \text{on } \Omega_h,$$

where ψ^k is the error of the solution w^k. Then

$$-\psi^k_{\bar{x}x} - \psi^k_{\bar{y}y} = 0 \quad \text{on } \Omega^r_h,$$

$$\psi^k = I^{2m}_\delta \psi^{k-1} \quad \text{on } \Omega^{ir}_h. \qquad (2.13)$$

Note that the equalities (2.13) satisfy the maximum principle. Therefore the maximum $|\psi^k|$ is attained on the set Ω^{ir}_h. Let it be at the point $x_0 \in \Omega^{ir}_h$. Then

$$\max_{\Omega_h} |\psi^k| = |\psi^k(x_0)| = |I^{2m}_\delta \psi^{k-1}(x_0)|.$$

From the definition (2.7) of the operator I^{2m}_δ the relation

$$\max_{\Omega_h} |\psi^k| \leq B(\delta) \max_{\Omega_h} |\psi^{k-1}|$$

holds, and from inequality (2.8) one has

$$\max_{\Omega_h} |\psi^k| \leq \tfrac{1}{2} \max_{\Omega_h} |\psi^{k-1}|.$$

Note also that the solution of (2.12) need not be constructed with maximal accuracy for each k. If an iterative process is used to solve the system, then it is enough to have enough iterations so that the remainder ξ^k for the approximate solution $\tilde{w}^k$ of the problem

$$-\tilde{w}^k_{\hat{x}\hat{x}} - \tilde{w}^k_{\hat{y}\hat{y}} = f + \xi^k \quad \text{on } \Omega^r_h,$$

$$\tilde{w}^k(x) = I^{2m}_\delta \tilde{w}^{k-1}(x) + \lambda^{2m}_\delta \varphi(\xi_x), \qquad x \in \Omega^{ir}_h$$

satisfies the inequality

$$\max_{\Omega^r_h} |\xi^k| \leq 2^{-k} \max_{\Omega^r_h} |\xi^0| = 2^{-k} \max_{\Omega^r_h} |f|.$$

If $\tilde{w}^{k-1}$ is sufficiently close one does not need many iterations to find $\tilde{w}^k$.

4.2.2. Refinements Using Higher-Order Differences

Now we present an approach for improving approximate solutions directly on the grid Ω_h that was suggested by Fox [40] and Volkov [135], who were the first to lay the foundation for approximate solutions of partial differential equations.

As in [138] we denote by $D_x^{(p,\,n)}$ the difference operator which approximates the pth derivative with respect to x, leaving constant mesh-size h on the axis Ox. The operator $D_x^{(p,\,n)}$ can be obtained by p-fold differentiation of the Newton interpolation formula with differences of up to the n-order included (see [14]). Note that the operators

$$D_x^{(m)} = \sum_{k=2}^{m+1} \frac{2h^{2k-2}}{(2k)!} D_x^{(2k,\,2m+2)}, \qquad D_y^{(m)} = \sum_{k=2}^{m+1} \frac{2h^{2k-2}}{(2k)!} D_y^{(2k,\,2m+2)} \quad (2.14)$$

allow one to approximate the second derivatives with an accuracy of order $h^{2m+\lambda}$ for any function $\psi \in C^{2m+2+\lambda}(\overline{\Omega})$, for example,

$$-\Delta\psi = -\psi_{\hat{x}\hat{x}} - \psi_{\hat{y}\hat{y}} + D_x^{(m)}\psi + D_y^{(m)}\psi + h^{2m+\lambda}\theta_3, \qquad (2.15)$$

where

$$|\theta_3(x)| \leq c_3 \|\psi\|_{C^{2m+2+\lambda}(\overline{\Omega})}$$

with the condition that all knots used in the operators $D_x^{(m)}$ and $D_y^{(m)}$ belong to $\overline{\Omega}$. This can be achieved by using different definitions for the operators $D_x^{(m)}$ and $D_y^{(m)}$ near the boundary Γ. One should use only central differences where possible, while in the rest of the cases other modifications are to be used. If we introduce the difference operators

$$\nabla f(x) = f(x) - f(x-h),$$

$$\triangle f(x) = f(x+h) - f(x),$$

$$\square f(x) = f(x+h/2) - f(x-h/2)$$

and recursively define

$$\nabla^k f = \nabla(\nabla^{k-1} f)$$

(similarly, Δ^k and $\square^k$), then the appliable operators $D_x^{(m)}$ are

$$D_x^{(m)} = \frac{\Delta^4}{12} - \frac{\Delta^5}{12} + \frac{13}{180}\Delta^6 - \frac{11}{180}\Delta^7 + \frac{87}{1680}\Delta^8 + \cdots,$$

$$D_x^{(m)} = \nabla\left(\frac{\Delta^3}{12} - \frac{\Delta^5}{90} + \frac{\Delta^6}{90} - \frac{47}{5040}\Delta^7 + \cdots\right),$$

$$D_x^{(m)} = \nabla^2\left(\frac{\Delta^2}{12} + \frac{\Delta^3}{12} - \frac{\Delta^4}{90} + \frac{1}{560}\Delta^6 + \cdots\right),$$

$$\cdots\cdots\cdots\cdots\cdots\cdots\cdots\cdots$$

$$D_x^{(m)} = \frac{\square^4}{12} - \frac{\square^6}{90} + \frac{\square^8}{560} - \cdots,$$

$$\cdots\cdots\cdots\cdots\cdots\cdots\cdots\cdots$$

$$D_x^{(m)} = \Delta^2\left(\frac{\nabla^2}{12} - \frac{\nabla^2}{12} - \frac{\nabla^4}{90} + \frac{1}{560}\nabla^6 + \cdots\right),$$

$$D_x^{(m)} = \Delta\left(\frac{\nabla^3}{12} - \frac{\nabla^5}{90} - \frac{\nabla^6}{90} - \frac{47}{5040}\nabla^7 + \cdots\right),$$

$$D_x^{(m)} = \frac{\nabla^4}{12} + \frac{\nabla^5}{12} + \frac{13}{180}\nabla^6 + \frac{11}{180}\nabla^7 + \frac{87}{1680}\nabla^8 + \cdots$$

Note that if h is sufficiently small and m is finite then for regular knots one can always use one of the variants $D_x^{(m)}$ and $D^{(m)}$ so that all the knots used belong to $\bar{\Omega}$.

Assume, for instance, that we have the situation shown in Figure 4.4. The point (x, y) is regular, and consequently the distance from this point to the boundary point (x, η) is larger than h. If the distance between the boundary points (θ_1, y) and (θ_2, y) is less than nh, then according to Lagrange's theorem there is the point on the arc $(\theta_1, y), (x, \eta)$ at which the derivative is more than $1/n$, and there is the point on the arc $(\theta_2, y), (x, \eta)$ at which the derivative is less

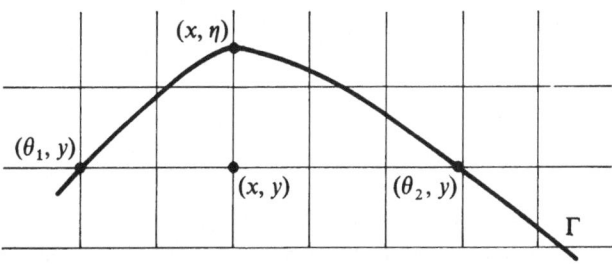

Figure 4.4. A possible arrangement for a regular knot (x, y).

than $-1/n$. According to the same theorem there is the point at which the curvature is of order $O(h^{-1})$, but this is impossible if h is sufficiently small.

Let us formulate the difference problem in the form

$$-u_{\bar{x}x}^h - u_{\bar{y}y}^h = f - D_x^{(m)}u^h - D_y^{(m)}u^h \quad \text{on } \Omega_h^r,$$

$$u^h(x) = I_\delta^{2m}u^h(x) + \lambda_\delta^{2m}\varphi(\xi_x), \qquad x \in \Omega_h^{ir}.$$

The approximation errors are of order $h^{2m+\lambda}$ for functions $u \in C^{2m+2+\lambda}(\bar{\Omega})$. However the question of whether this problem is stable is open and in a number of cases this scheme results in an unsatisfactory order of accuracy.

Let us construct instead the set of the difference problems

$$-v_{\bar{x}x}^q - v_{\bar{y}y}^q = f - D_x^{(m)}v^{q-1} - D_y^{(m)}v^{q-1} \quad \text{on } \Omega_h^r, \qquad (2.16)$$

$$v^q(x) = I_\delta^{2m}v^q(x) + \lambda_\delta^{2m}\varphi(\xi_x), \qquad x \in \Omega_h^{ir}, \quad q = 1, 2, \ldots, m+1. \qquad (2.17)$$

As an initial approximation assume $v^0 = 0$. Then obviously v^1 coincides with the solution of (2.3), (2.10) and differs from the solution of the differential equation by a term of order h^2. Later iterations v^q have increasing orders of accuracy, bounded by the degree of smoothness of the data of the differentional equation.

Theorem 2.2 (see [139]). *Assume that the following conditions*

$$\Gamma \in C^{2m+2+\lambda}, \qquad \varphi \in C^{2m+2+\lambda}(\Gamma), \qquad f \in C^{2m+\lambda}(\bar{\Omega})$$

are satisfied in (2.1), (2.2), *where* $m \geq 1$ *and* $\lambda \in (0, 1)$. *Then the solution* v^{m+1} *of problem* (2.16), (2.17) *is representable in the form*

$$v^{m+1} = u + h^{2m+\lambda} \ln^m(2b/h) \quad \text{on } \Omega_h. \qquad (2.18)$$

Here u *is the solution of* (2.1), (2.2) *and* r_h *is the bounded grid function*

$$|r_h| \leq c_3 \quad \text{on } \Omega_h. \qquad (2.19)$$

Note that the implementation of the method (2.16), (2.17) is significantly simplified if one replaces equation (2.17) by

$$v^q(x) = I_\delta^{2m}v^{q-1}(x) + \lambda_\delta^{2m}\varphi(\xi_x), \qquad x \in \Omega_h^{ir},$$

since now for each function v^q the problem becomes a selfadjoint problem. However, in this case v^q will not approximate the solution u with order h^{2q}. Nevertheless, we can use this method as an auxiliary one to carry out iterations at fixed q. This has already been mentioned in the previous discussion.

4.2.3. Multipoint Approximation of the Second Derivative

Let us consider some approaches to the approximation of the second derivatives. We desire to find an approximation which provide a particular kind of approximation error for the irregular knots.

Let $\psi(t)$ be $C^{m+\alpha}[-1, 1]$ function, where $m \geq 2$, $\alpha \in (0, 1)$. We know that when we approximate using three values at equidistant knots $-h$, 0, h the approximation error for the second derivative has the form

$$\psi_{\bar{t}t}(0) - \psi''(0) = \sum_{k=1}^{[(m-2)/2]} h^{2k} \frac{2\psi^{(2k+2)}(0)}{(2k+2)!} + h^{m+\alpha}\theta, \qquad (2.20)$$

where

$$|\theta| \leq \frac{2}{m!} \|\psi\|_{C^{m+\alpha}[-1, 1]}. \qquad (2.21)$$

Let us attempt to obtain the same form for the approximation error for a nonuniform arrangement of the knots. Suppose $n + 2$ points on the axis Ot are given: $\delta h, 0, -h, \ldots, -nh$, where $\delta \in (0, \frac{4}{3})$. To obtain formula (2.20) it is necessary to find the value $\psi(h)$. Calculate it using a Lagrange interpolation polynomial:

$$\psi(h) = \sum_{k=0}^{n} a_{n,k}\psi(-kh) + a_{n,\delta}\psi(\delta h) + r^h, \qquad (2.22)$$

where

$$a_{n,k} = \frac{(-1)^{k+1}(n+1)!\,(1-\delta)}{(k+1)!\,(k+\delta)(n-k)!}, \qquad a_{n,\delta} = \prod_{k=0}^{n} \frac{k+1}{k+\delta}.$$

If $m \geq n + 2$ then the remainder is estimated as follows

$$|r^h| \leq |h - \delta h|h(h + h)\cdots(h + nh)\max_{[-1, 1]} |\psi^{(n+2)}|/(n + 2)!$$

$$\leq h^{n+2}|1 - \delta|\max_{[-1, 1]} |\psi^{(n+2)}|/(n + 2). \qquad (2.23)$$

But if the function ψ is not sufficiently smooth ($m < n + 2$) then the remainder is of a different order. Let us prove this. Consider the polynomial

$$\varphi(t) = \sum_{i=0}^{m} \alpha_i t^i.$$

Since its $(n + 2)$th derivative is identically zero then from relation (2.22) for $\varphi(h)$ it follows that

$$\varphi(h) = \sum_{k=0}^{n} a_{n,k}\varphi(-kh) + a_{n,\delta}\varphi(\delta h). \qquad (2.24)$$

Let us return to the function ψ. From the Taylor expansion we have

$$\psi(-kh) = \sum_{i=0}^{m} h^i \frac{(-k)^i}{i!} \psi^{(i)}(0) + h^{m+\alpha}|k|^{m+\alpha}\theta_k/m!, \qquad k = -1, 0, \ldots, n,$$

$$(2.25)$$

and

$$\psi(\delta h) = \sum_{i=0}^{m} h^i \frac{\delta^i}{i!} \psi^{(i)}(0) + h^{m+\alpha}\delta^{m+\alpha}\theta_\delta/m!,$$

where

$$\max_{-1 \leq k \leq n} \{|\theta_k|, |\theta_\delta|\} \leq \|\psi\|_{C^{m+\alpha}[-1, 1]}.$$

Assume that the coefficients α_i of the polynomial φ are equal to $\psi^{(i)}(0)/i!$. Then the expansions (2.25) are transformed to

$$\psi(-kh) = \varphi(-kh) + h^{m+\alpha}|k|^{m+\alpha}\theta_k/m!,$$

$$\psi(\delta h) = \varphi(\delta h) + h^{m+\alpha}\delta^{m+\alpha}\theta_\delta/m!.$$

Substituting these expansions into (2.22), and keeping in mind (2.24), we have

$$\frac{h^{m+\alpha}}{m!} \theta_1 = \frac{h^{m+\alpha}}{m!} \sum_{k=1}^{n} a_{n,k} k^{m+\alpha}\theta_k + a_{n,\delta}\frac{h^{m+\alpha}\delta^{m+\alpha}}{m!} \theta_\delta + r^h.$$

Hence

$$|r^h| \leq \frac{h^{m+\alpha}}{m!} \|\psi\|_{C^{m+\alpha}[-1, 1]}\left(1 + \sum_{k=1}^{n} |a_{n,k}| k^{m+\alpha} + a_{n,\delta}\delta^{m+\alpha}\right).$$

Since

$$|a_{n,k}| \leq (n + 1)! \quad \text{for } k = 1, \ldots, n, \quad \delta \in (0, \tfrac{4}{3})$$

and

$$\delta a_{n,\delta} \leq n + 1,$$

then

$$|r^h| \leq \frac{h^{m+\alpha}}{m!} \|\psi\|_{C^{m+\alpha}[-1, 1]}(1 + n(n + 1)! \, n^{m+\alpha} + (n + 1)2^{m+\alpha-1}).$$

Assuming

$$c_4 = \frac{1}{m!} \|\psi\|_{C^{m+\alpha}[-1, 1]}(1 + (n + 1)\{(n + 1)! \, n^{m+\alpha} + 2^m\}),$$

we get the estimate

$$|r^h| \leq c_4 h^{m+\alpha}, \qquad m < n + 2. \tag{2.26}$$

Setting aside the remainder r^h in (2.22), we have the approximate value

$$\tilde{\psi}(h) = \sum_{k=0}^{n} a_{n,k}\psi(-kh) + a_{n,\delta}\psi(\delta h). \tag{2.27}$$

Apply it to approximate the second derivative:

$$\tilde{\psi}_{\bar{t}t}(0) \equiv \frac{\tilde{\psi}(h) - 2\psi(0) + \psi(-h)}{h^2} = \sum_{k=0}^{n} b_{n,k}\psi(-kh) + b_{n,\delta}\psi(\delta h), \tag{2.28}$$

where

$$b_{n,0} = (a_{n,0} - 2)/h^2, \qquad b_{n,1} = (a_{n,1} + 1)/h^2,$$
$$b_{n,k} = a_{n,k}/h^2, \qquad\qquad b_{n,\delta} = a_{n,\delta}/h^2, \qquad k = 2, \ldots, n.$$

For this derivative the equalities

$$\tilde{\psi}_{\mathit{ff}}(0) - \psi_{\mathit{ff}}(0) = \frac{\tilde{\psi}(h) - \psi(h)}{h^2} = r^h/h^2$$

hold true. Therefore, applying expansion (2.20) and the estimates (2.23), (2.26) we get the expansion

$$\tilde{\psi}_{\mathit{ff}}(0) - \psi''(0) = \sum_{k=1}^{s} \frac{2h^{2k}}{(2k+2)!} \psi^{(2k+2)}(0) + r_1^h, \qquad (2.29)$$

where

$$s = [\min\{n-1, m-2\}/2],$$

and for the remainder we have the estimate

$$|r_1^h| \le \begin{cases} c_5 h^n & \text{if } n+2 \le m, \\ c_6 h^{m-2+\alpha} & \text{if } 1 \le m \le n+1 \end{cases} \qquad (2.30)$$

holds, the constants c_5, c_6 being independent of h and δ.

For $n = 1, 2, 3$ we have the following important property of the coefficients $b_{n,k}$:

$$\left(-b_{n,0} - \sum_{k=1}^{n} |b_{n,k}|\right) \Big/ \left(2/h^2 + \sum_{k=1}^{n} |b_{n,k}|\right) \ge c_7 > 0, \qquad (2.31)$$

where the constant c_7 does not depend on h and δ, but it does depend on n. Let us prove this inequality.

Let $n = 1$; then

$$b_{1,\delta} = \frac{2}{\delta(\delta+1)h^2}, \qquad b_{1,0} = -\frac{2}{\delta h^2}, \qquad b_{1,1} = \frac{2}{(1+\delta)h^2}.$$

The left-hand side of (2.31) is equal to

$$\left(\frac{2}{\delta h^2} - \frac{2}{(1+\delta)h^2}\right) \Big/ \left(\frac{2}{h^2} + \frac{2}{(1+\delta)h^2}\right) = \frac{1}{2\delta + \delta^2}.$$

Since $\delta \in (0, \frac{4}{3})$, then

$$\frac{1}{2\delta + \delta^2} \ge \frac{9}{40}$$

and one can assume that the constant c_7 for $n = 1$ is equal to $\frac{9}{40}$.

Let now $n = 2$; then

$$b_{2,\delta} = \frac{6}{\delta(\delta + 1)(\delta + 2)h^2}, \qquad b_{2,0} = -\frac{3 - \delta}{\delta h^2},$$

$$b_{2,1} = \frac{4 - 2\delta}{(1 + \delta)h^2}, \qquad b_{2,2} = -\frac{1 - \delta}{(2 + \delta)h^2}.$$

Let us investigate the case $\delta \in (0, 1]$ first. On the left-hand side of (2.31) the denominator is positive, and therefore

$$\left(\frac{3 - \delta}{\delta h^2} - \frac{4 - 2\delta}{(1 + \delta)h^2} - \frac{1 - \delta}{(2 + \delta)h^2}\right)\bigg/\left(\frac{2}{h^2} + \frac{4 - 2\delta}{(1 + \delta)h^2} + \frac{1 - \delta}{(2 + \delta)h^2}\right)$$

$$\geq \left(\frac{3 - \delta}{\delta} - \frac{4 - 2\delta}{1 + \delta} - \frac{1 - \delta}{2}\right)\bigg/\left(2 + \frac{4 - 2\delta}{1 + \delta} + \frac{1 - \delta}{2}\right)$$

$$= \frac{6 - 5\delta + 2\delta^2 + \delta^3}{\delta(13 - \delta^2)} \geq (6 - 5\delta + 2\delta^2)/13.$$

The minimum of $6 - 5\delta + 2\delta^2$ is attained in $(0, 1]$ at $\delta = 1$, and therefore

$$(6 - 5\delta + 2\delta^2)/13 \geq \tfrac{3}{13}.$$

Now consider the case where $\delta \in (1, \tfrac{4}{3}]$. We will calculate the left-hand side of expression (2.31) for this case and use the positivity of the denominator. We have

$$\left(\frac{3 - \delta}{\delta h^2} - \frac{4 - 2\delta}{(1 + \delta)h^2} + \frac{1 - \delta}{(2 + \delta)h^2}\right)\bigg/\left(\frac{2}{h^2} + \frac{4 - 2\delta}{(1 + \delta)h^2} - \frac{1 - \delta}{(2 + \delta)h^2}\right)$$

$$\geq \left(\frac{3 - \delta}{\delta} - \frac{4 - 2\delta}{1 + \delta} + \frac{1 - \delta}{3}\right)\bigg/\left(2 + \frac{4 - 2\delta}{1 + \delta} - \frac{1 - \delta}{3}\right)$$

$$= \frac{9 - 5\delta + 3\delta^2 - \delta^3}{17\delta + \delta^3}.$$

The numerator of the last fraction monotonically decreases when $\delta \to 0$ and the denominator monotonically increases. Therefore, assuming $\delta = \tfrac{4}{3}$, we have

$$(9 - 5\delta + 3\delta^2 - \delta^3)/(17\delta + \delta^3) \geq \tfrac{143}{676} \geq \tfrac{1}{5}.$$

Thus, taking both cases together, we assume that $c_7 = \tfrac{1}{5}$ for $n = 2$.

Let us determine the constant c_7 for $n = 3$. Write out the coefficients:

$$b_{3,\delta} = \frac{24}{\delta(1 + \delta)(2 + \delta)(3 + \delta)h^2}, \qquad b_{3,0} = -\frac{2(2 - \delta)}{\delta h^2},$$

$$b_{3,1} = \frac{7 - 5\delta}{(1 + \delta)h^2}, \qquad b_{3,2} = \frac{-4(1 - \delta)}{(2 + \delta)h^2}, \qquad b_{3,3} = \frac{1 - \delta}{(3 + \delta)h^2}.$$

The first step is to consider the case $\delta \in (0, 1]$. Calculate the left-hand side of (2.31). We have

$$\left(\frac{2(2-\delta)}{\delta h^2} - \frac{7-5\delta}{(1+\delta)h^2} - \frac{4(1-\delta)}{(2+\delta)h^2} - \frac{1-\delta}{(3+\delta)h^2}\right) \Big/$$

$$\left(\frac{2}{h^2} + \frac{7-5\delta}{(1+\delta)h^2} + \frac{4(1-\delta)}{(2+\delta)h^2} + \frac{1-\delta}{(3+\delta)h^2}\right)$$

$$\geq \left(\frac{2(2-\delta)}{\delta} - \frac{7-5\delta}{1+\delta} - \frac{4(1-\delta)}{2+\delta} - \frac{1}{3}\right) \Big/ \left(2 + \frac{7-5\delta}{1+\delta} + \frac{4(1-\delta)}{2+\delta} + \frac{1}{3}\right)$$

$$= (24 - 32\delta + 20\delta^3)/(68\delta + 12\delta^2 - 20\delta^3).$$

Discarding the positive addend in the numerator, and the negative one in the denominator, we arrive at the inequality

$$(24 - 32\delta + 20\delta^3)/(68\delta + 12\delta^2 - 20\delta^3) \geq (6 - 8\delta + 8\delta^3/3)/(17\delta + 3\delta^2).$$

The numerator of the last fraction decreases monotonically in the interval $(0, 1]$ and the denominator increases. Taking $\delta = 1$ we have

$$(6 - 8\delta + 8\delta^3/3)/(17\delta + 3\delta^2) = \tfrac{1}{30}.$$

It remains only to consider the case $\delta \in (1, \tfrac{4}{3}]$. Let us calculate the left-hand side in (2.31) for this case. We have

$$\left(\frac{2(2-\delta)}{\delta h^2} - \frac{7-5\delta}{(1+\delta)h^2} + \frac{4(1-\delta)}{(2+\delta)h^2} + \frac{1-\delta}{(3+\delta)h^2}\right) \Big/$$

$$\left(\frac{2}{h^2} + \frac{7-5\delta}{(1+\delta)h^2} - \frac{4(1-\delta)}{(2+\delta)h^2} - \frac{1-\delta}{(3+\delta)h^2}\right)$$

$$\geq \left(\frac{2(2-\delta)}{\delta} - \frac{7-5\delta}{1+\delta} + \frac{4(1-\frac{4}{3})}{3} + \frac{1-\frac{4}{3}}{4}\right) \Big/$$

$$\left(2 + \frac{7-5\delta}{1+\delta} - \frac{4(1-\frac{4}{3})}{3} - \frac{1-\frac{4}{3}}{4}\right)$$

$$= \left(\frac{4 - 5\delta + 3\delta^2}{\delta(1+\delta)} - \frac{19}{36}\right) \Big/ \left(2 + \frac{7-5\delta}{1+\delta} + \frac{19}{36}\right).$$

The polynomial $4 - 5\delta + 3\delta^2$ increases on the interval $(1, \tfrac{4}{3}]$ and therefore for the numerator the estimate

$$\frac{4 - 5\delta + 3\delta^2}{\delta(1+\delta)} - \frac{19}{36} \geq \frac{2}{(1 + \frac{4}{3})\frac{4}{3}} - \frac{19}{36} \geq \frac{1}{2}$$

holds. The denominator is estimated as follows:

$$2 + (7 - 5\delta)/(1 + \delta) + \tfrac{19}{36} \leq 2 + \tfrac{2}{2} + \tfrac{19}{36} = \tfrac{127}{36}.$$

Thus, the whole fraction is bounded below by $\frac{18}{127}$. Taking both cases together, we get $c_7 = \frac{1}{30}$ for $n = 3$. Note that we have made rather crude calculations of the constant c_7 so as to prove that it exists.

Later we will see that property (2.31) will guarantee that the difference scheme is stable. Now our calculation of c_7 in inequality (2.31) fails for $n \geq 4$, because the expression on the left-hand side of the inequality becomes negative for certain $\delta \in (0, \frac{4}{3}]$.

To obtain an estimate of the form (2.31) and to guarantee the stability of the difference scheme we proceed as follows. Let us apply the interpolation formula with the mesh-size $\bar{h}$, some multiple of h, as has been done at the beginning of this section. Namely, assume $\bar{h} = ph$, p a natural number, and repeat the manipulations (2.22)–(2.27). We come to the formula

$$\tilde{\psi}_h(h) = \sum_{k=0}^{n} \alpha_{n,k} \psi(-kph) + \alpha_{n,\delta} \psi(\delta h), \tag{2.32}$$

where

$$\alpha_{n,k} = (-1)^{k+1} \frac{(1 - \delta)(p + 1)(2p + 1) \cdots (np + 1)}{(kp + 1)(kp + \delta)k!\,(n - k)!\,p^n},$$

$$\alpha_{n,\delta} = \frac{(p + 1)(2p + 1) \cdots (np + 1)}{\delta(p + \delta)(2p + \delta) \cdots (np + \delta)}. \tag{2.33}$$

This approximate value differs from the exact one by

$$r_2^h = \tilde{\psi}(h) - \psi(h),$$

which depends on the relation of m to n. If $m \geq n + 2$ then

$$|r_2^h| \leq h^{n+2}|1 - \delta|(p + 1)(2p + 1) \cdots (np + 1) \max_{[-1,1]} |\psi^{(n+2)}|/(n + 2)!$$

$$\leq h^{n+2}p^n|1 - \delta| \max_{[-1,1]} |\psi^{(n+2)}|/(n + 2).$$

But if $m \leq n + 1$, then

$$|r_2^h| \leq \left(h^{m+\alpha}/m! + \sum_{k=1}^{n} |\alpha_{n,k}| h^{m+\alpha}(kp)^{m+\alpha}/m! \right.$$

$$\left. + |\alpha_{n,\delta}| h^{m+\alpha}\delta^{m+\alpha}/m! \right) \|\psi\|_{C^{m+\alpha}[-1,1]}. \tag{2.34}$$

From (2.33) the estimates

$$|\alpha_{n,k}| \leq \frac{|1 - \delta|(n + 1)!}{k^2 p^2 k!\,(n - k)!} \leq \frac{|1 - \delta|(n + 1)!}{p^2}, \qquad k = 1, \ldots, n,$$

$$\delta|\alpha_{n,k}| \leq n + 1 \tag{2.35}$$

follow. Therefore

$$|r_2^h| \leq h^{m+\alpha} \|\psi\|_{C^{m+\alpha}[-1,1]}$$

$$\times (1 + n(np)^{m+\alpha}|1 - \delta|(n + 1)!/p^2 + (n + 1)\delta^{m+\alpha-1})/m!.$$

Since $\delta \in (0, \frac{4}{3}]$, then arrive at the estimate

$$|r_2^h| \leq c_8 h^{m+\alpha}, \qquad m \leq n + 1, \tag{2.36}$$

with the constant c_8 depending on p.

Let us apply the equality (2.32) to the approximation of the second derivative

$$\tilde{\psi}_{\overline{t}t}(0) \equiv \{\tilde{\psi}(h) - 2\psi(0) + \psi(-h)\}/h^2$$

$$= \sum_{i=0}^{np} \beta_{n,i} \psi(-ih) + \beta_{n,\delta} \psi(\delta h), \tag{2.37}$$

where

$$\beta_{n,\delta} = \alpha_{n,\delta}/h^2, \qquad \beta_{n,0} = (\alpha_{n,0} - 2)/h^2,$$

$$\beta_{n,1} = \begin{cases} 1/h^2 & \text{if } p > 1, \\ (\alpha_{n,1} + 1)/h^2 & \text{if } p = 1, \end{cases} \tag{2.38}$$

$$\beta_{n,i} = \begin{cases} \alpha_{n,k}/h^2 & \text{if } i = pk, \\ 0 & \text{for } i = 2, \ldots, np; \quad i \neq pk. \end{cases}$$

The relation (2.37) is the extension of formula (2.28) for p a natural number. Let us choose the parameter p so as to satisfy the inequality

$$\left(-\beta_{n,0} - \sum_{k=1}^{np} |\beta_{n,k}|\right) \bigg/ \left(2/h^2 + \sum_{k=1}^{np} |\beta_{n,k}|\right) \geq c_7 \tag{2.39}$$

with the positive constant c_7 independent of $\delta \in (0, \frac{4}{3}]$. One can always satisfy (2.39) for sufficiently large p and sufficiently small h, since as $p \to \infty$ from (2.35) and equalities (2.38) it follows that

$$\beta_{n,0} \to -2/h^2, \qquad \beta_{n,1} \to 1/h^2, \qquad \beta_{n,i} \to 0, \qquad i = 2, \ldots, np.$$

Note that for $n = 4$ or $n = 5$ (eventually providing an accuracy of the approximate solution of order h^{n+2}) one can choose comparatively small values p from the condition (2.37): $p = 2$ or $p = 3$, respectively.

Assuming p to be fixed, apply the estimates (2.34) and (2.36) to the difference $\tilde{\psi}_{\overline{t}t}(0) - \psi_{\overline{t}t}(0)$. It becomes clear that the expansion (2.29), with the remainder estimate (2.30), is also valid for the approximation (2.37).

Let us now start to construct the difference problem. Assume that the domain Ω is contained in the square $\{-b < x < b, -b < y < b\}$. Cover it with an orthogonal grid with the mesh-size $h = b/N$ formed by the lines $x_i = ih$ and $y_j = jh$ where $i, j = -N, \ldots, N$. The knot $x \in \Omega_h$ will be called regular in the x-direction if the closed interval connecting the points $(x - h, y)$ and $(x + h, y)$ belongs to $\overline{\Omega}$. A knot regular in the y-direction is defined similarly. Let us denote the set of knots regular in the x-direction by $\Omega_{h,x}^r$ and the set of knots regular in the y-direction by $\Omega_{h,y}^r$. It is natural to call $\Omega_{h,x}^{ir} = \Omega_h \backslash \Omega_{h,x}^r$ the set of knots irregular in the x-direction and $\Omega_{h,y}^{ir} = \Omega_h \backslash \Omega_{h,y}^r$ the set of knots irregular in the y-direction. It is obvious that these sets are related to the sets Ω_h^r and Ω_h^{ir}, introduced previously, in the following way:

$$\Omega_h^r = \Omega_{h,x}^r \cap \Omega_{h,y}^r, \qquad \Omega_h^{ir} = \Omega_{h,x}^{ir} \cup \Omega_{h,y}^{ir}.$$

In each knot $(x, y) \in \Omega_{h,x}^r$ we approximate the derivative $\partial^2 u/\partial x^2$ by the second difference $u_{\bar{x}x}^h(x, y)$ and at the irregular knots of $\Omega_{h,x}^{ir}$ we use the approximations constructed above. We approximate the derivative $\partial^2 u/\partial y^2$.

First, let us choose the number n, which characterizes the order of approximation at the irregular knots. Assume that the assumptions of Theorem 1.2 are satisfied for the differential equation for some integer $l \geq 2$ and some $\alpha \in (0, 1)$. Assume $n = l - 1$.

Let $x \in \Omega_{h,x}^{ir}$; then there is a point $\xi_x \in \Gamma_{h,x}$ closest to x in the direction parallel to the axis Ox. In this case let us put the origin of the positive Ot axis at the point x, have it pass through the point ξ_x, and use the approximation (2.37) for the second derivative $\partial^2 u/\partial x^2(x)$:

$$\tilde{u}_{\bar{t}\bar{t}}(x) = J_{l-1}^x u(x) + \rho_{l-1}^x \varphi(\xi_x), \tag{2.40}$$

where

$$J_n^x u(x) = \sum_{i=0}^{np} \beta_{n,i}^x u(x \pm ih, y), \qquad \rho_n^x = \beta_{n,\delta}^x,$$

keeping in mind that $\delta^x = |\xi_x - x|$ and the plus or minus sign is chosen depending on whether the direction from x to ξ_x coincides with the direction of Ox or is opposite to it.

Remark. There can be a situation when all the points covered by the operator J_{l-1}^x do not belong to $\overline{\Omega}$ (as in Figure 4.5) and the approximation (2.40) makes no sense.

In this case it is necessary to eliminate the points marked by $*$ from Ω_h. The points marked by $\circ$ then become irregular in the y-direction. The distance between these points and the boundary Γ exceeds h but differs from h by a quantity of order h^2 due to the smoothness of the boundary Γ. It is because of this that we verified the properties (2.31), (2.39) not only for $\delta \in (0, 1]$ but for δ greater than one: $\delta \in (1, \frac{4}{3}]$.

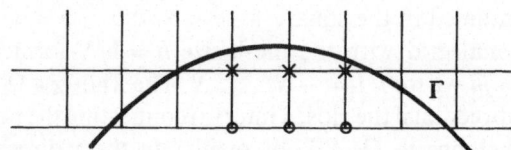

Figure 4.5. The change in grid region with an unsuccessful arrangement of the knot x. The Symbol $*$ denotes a point to be excluded from Ω_h, and the symbol $\circ$ denotes a point which has become irregular in the y-direction.

Similarly, if $x \in \Omega^r_{h,y}$ then the derivative $\partial^2 u / \partial y^2$ is approximated by the difference-derivative $u_{\bar{y}\bar{y}}(x)$. But if $x \in \Omega^{ir}_{h,y}$, then there is a point $\eta_x \in \Gamma_{h,y}$ closest to x in the direction parallel to the axis Oy. Let us put the origin of the positive Ot axis at the point x and have it pass through η_x. Then the approximation (2.37) of the second derivative $\partial^2 u / \partial y^2$ has the form

$$\tilde{u}_{\overline{tt}}(x) = J^y_{l-1} u(x) + \rho^y_{l-1} \varphi(\eta_x), \qquad (2.41)$$

where

$$J^y_n u(x) = \sum_{i=0}^{np} \beta^y_{n,i} u(x, y \pm ih), \qquad \rho^y_{l-1} = \beta^y_{n,\delta},$$

with $\delta^y = |\eta_x - x|$ and we choose the plus or minus sign depending on whether the direction from x to η_x coincides with the direction of the axis Oy or not. The possibility that some knots in (2.41) will be outside the domain still exists and the method of eliminating such a situation given above is still effective for this approximation.

Thus, four kinds of approximation of (2.1) are used at the knots of the grid Ω_h:

$$-u^h_{\bar{x}\bar{x}}(x) - u^h_{\bar{y}\bar{y}}(x) = f(x), \qquad x \in \Omega^r_h; \qquad (2.42)$$

$$-J^x_{l-1} u^h(x) - u^h_{\bar{y}\bar{y}}(x) = f(x) + \rho^x_{l-1} \varphi(\xi_x), \qquad x \in \Omega^{ir}_{h,x} \cap \Omega^r_{h,y}; \quad (2.43)$$

$$-u^h_{\bar{x}\bar{x}}(x) - J^y_{l-1} u^h(x) = f(x) + \rho^y_{l-1} \varphi(\eta_x), \qquad x \in \Omega^r_{h,x} \cap \Omega^{ir}_{h,y}; \quad (2.44)$$

$$-J^x_{l-1} u^h(x) - J^y_{l-1} u^h(x) = f(x) + \rho^x_{l-1} \varphi(\xi_x) + \rho^y_{l-1} \varphi(\eta_x),$$
$$x \in \Omega^{ir}_{h,x} \cap \Omega^{ir}_{h,y}. \quad (2.45)$$

Note that the number of unknowns is equals to the number of points of the set Ω_h, and the system (2.42)–(2.45) is a linear algebraic system with a square matrix. We will derive the *a priori* estimates to justify the claim that this system is solvable.

Lemma 2.3. *Let u^h be the solution of system* (2.42)–(2.45) *where $\varphi = 0$ on Γ_h. Then the estimate*

$$\max_{\Omega_h} |u^h| \leq \frac{b^2(1 + c_7)}{2c_7} \max_{\Omega^r_h} |f| + \frac{h^2}{2c_7} \max_{\Omega^{ir}_h} |f| \qquad (2.46)$$

holds.

PROOF. The first step is to consider the system

$$-v^h_{\bar{x}\bar{x}} - v^h_{\bar{y}\bar{y}} = f \quad \text{on } \Omega^r_h,$$

$$v^h = 0 \quad \text{on } \Omega^{ir}_h \cup \Gamma_h.$$

Since it satisfies the maximum principle, then there is a unique solution, and the comparison theorem is valid (see [112]). It says that if

$$-v_{\hat{x}\hat{x}} - v_{\hat{y}\hat{y}} = g \quad \text{on } \Omega_h^r,$$

$$v \geq 0 \quad \text{on } \Omega_h^{ir} \cup \Gamma_h$$

then from the inequality

$$|f| \leq g \quad \text{on } \Omega_h^r$$

it follows that

$$|v^h| \leq v \quad \text{on } \Omega_h.$$

Assume

$$v(x) = (2b^2 - x^2 - y^2) \max_{\Omega_h^r} |f|/4.$$

It is obvious that $v \geq 0$ on the entire set Ω_h including $\Omega_h^{ir} \cup \Gamma_h$. Besides

$$-v_{\hat{x}\hat{x}}(x) - v_{\hat{y}\hat{y}}(x) = -\Delta v(x) = \max_{\Omega_h^r} |f|$$

due to the absence of an approximation error. Hence

$$|f| \leq g \quad \text{on } \Omega_h^r,$$

$$|v^h| \leq v \leq b^2 \max_{\Omega_h^r} |f|/2 \quad \text{on } \Omega_h.$$

The solution of the system

$$-w_{\hat{x}\hat{x}}^h - w_{\hat{y}\hat{y}}^h = 0 \quad \text{on } \Omega_h^r,$$

$$w^h = u^h \quad \text{on } \Omega_h^{ir} \cup \Gamma_h$$

is also unique. By comparing w^h with the constant function

$$w(x) = \max_{\Omega_h^{ir}} |u^h|$$

we have

$$|w^h| \leq \max_{\Omega_h^{ir}} |u^h| \quad \text{on } \Omega_h.$$

For any fixed u^h the solution of the system of equations

$$-z_{\hat{x}\hat{x}}^h - z_{\hat{y}\hat{y}}^h = f \quad \text{on } \Omega_h^r,$$

$$z^h = u^h \quad \text{on } \Omega_h^{ir} \cup \Gamma_h$$

is unique and therefore on Ω_h^r z^h coinsides with u^h, since it also satisfies this system. Taking into account that $z^h = v^h + w^h$ we get the estimates

$$|u^h| = |z^h| \leq |v^h| + |w^h|$$

$$\leq \max_{\Omega_h^{ir}} |u^h| + b^2 \max_{\Omega_h^r} |f|/2 \quad \text{on } \Omega_h^r. \tag{2.47}$$

Consider the point $\bar{x}$ at which $\max_{\Omega_h^{ir}} |u^h|$ is attained.

Suppose, for example, that

$$\bar{x} = (\bar{x}, \bar{y}) \in \Omega_{h,x}^{ir} \cap \Omega_{h,y}^{r}.$$

Rewrite equation (2.43) using the definition (2.40):

$$(2/h^2 - \beta_{n,0}^x)u^h(\bar{x}, \bar{y}) - u^h(\bar{x}, \bar{y} + h)/h^2$$

$$- u^h(x, \bar{y} - h)/h^2 - \sum_{i=1}^{np} \beta_{n,i}^x u^h(\bar{x} \pm ih, \bar{y}) = f(\bar{x}, \bar{y}). \quad (2.48)$$

Here $n = l - 1$, and the sign in $\bar{x} \pm ih$ is irrelevant for further calculations. For the knots $x \in \Omega_h^{ir}$ the inequality

$$|u^h(x)| \le |u^h(\bar{x})|$$

is valid, and for the knots $x \in \Omega_h^r$ from inequality (2.47) it follows that

$$|u^h(x)| \le |u^h(\bar{x})| + b^2 \max_{\Omega_h^r} |f|/2. \quad (2.49)$$

Therefore one may assume that inequality (2.49) holds true for all $x \in \Omega_h$. Use it to estimate the summands in relation (2.48):

$$|2/h^2 - \beta_{n,0}^x||u^h(\bar{x})| \le \left(2/h^2 + \sum_{i=1}^{np} |\beta_{n,i}^x|\right)$$

$$\times \left(|u^h(\bar{x})| + b^2 \max_{\Omega_h^r} |f|/2\right) + \max_{\Omega_h^{ir}} |f|. \quad (2.50)$$

From inequality (2.39) it follows that $\beta_{n,0}^x < 0$, and consequently, the factor multiplying $|u^h(\bar{x})|$ on the left-hand side of (2.50) is equal to

$$2/h^2 + |\beta_{n,0}^x|.$$

The factor multiplying $|u^h(x)|$ on the right-hand side of (2.50) is positive. Dividing both sides by this factor and carrying out some simplification we have

$$\left\{(2/h^2 + |\beta_{n,0}^x|)\Big/\left(2/h^2 + \sum_{i=1}^{np} |\beta_{n,i}^x|\right)\right\}|u^h(\bar{x})|$$

$$\le |u^h(\bar{x})| + b^2 \max_{\Omega_h^r} |f|/2 + h^2 \max_{\Omega_h^{ir}} |f|/2.$$

Subtracting $|u^h(\bar{x})|$ from both sides of the inequality and using (2.39) we have

$$c_7|u^h(\bar{x})| \le b^2 \max_{\Omega_h^r} |f|/2 + h^2 \max_{\Omega_h^{ir}} |f|/2. \quad (2.51)$$

But if $\bar{x} \in \Omega_{h,x}^r \cap \Omega_{h,y}^{ir}$ then, as above, we get (2.51) again. Dividing both sides of it by c_7, we have

$$|u^h(\bar{x})| \le b^2 \max_{\Omega_h^r} |f|/(2c_7) + h^2 \max_{\Omega_h^{ir}} |f|/(2c_7). \quad (2.52)$$

It remains to consider the case when

$$\bar{x} \in \Omega^{ir}_{h,x} \cap \Omega^{ir}_{h,y}.$$

Then instead of relation (2.48) we obtain

$$(-\beta^x_{n,0} - \beta^y_{n,0})u^h(\bar{x}, \bar{y}) - \sum_{i=1}^{np} \beta^x_{n,i} u^h(\bar{x} \pm ih, \bar{y})$$

$$- \sum_{j=1}^{np} \beta^y_{n,j} u^h(\bar{x}, \bar{y} \pm jh) = f(\bar{x}, \bar{y}).$$

Here $\beta^x_{n,i}$ are the coefficients of the operator J^x_n and $\beta^y_{n,j}$ are the coefficients of the operator J^y_n. The signs in the expressions $\bar{x} \pm ih$ and $\bar{y} \pm jh$ are irrelevent for later calculation. Using (2.49) we have

$$|-\beta^x_{n,0} - \beta^y_{n,0}||u^h(\bar{x})| \le \left(\sum_{i=1}^{np} |\beta^x_{n,i}| + \sum_{j=1}^{np} |\beta^y_{n,j}| \right)$$

$$\times \left(|u^h(\bar{x})| + b^2 \max_{\Omega^r_h} |f|/2 \right) + \max_{\Omega^{ir}_h} |f|. \quad (2.53)$$

From (2.39) it follows that for $\beta^x_{n,0}$ and $\beta^y_{n,0}$ the inequalities

$$\beta^x_{n,0} < 0, \qquad \beta^y_{n,0} < 0$$

are valid, and therefore the factor multiplying $|u^h(\bar{x})|$ on the left-hand side of (2.53) is equal to $|\beta^x_{n,0}| + |\beta^y_{n,0}|$. Let us add

$$4|u^h(\bar{x})|/h^2,$$

to the left-hand side of the inequality and

$$4\left(|u^h(\bar{x})| + b^2 \max_{\Omega^r_h} |f|/2 \right) \Big/ h^2$$

to the right-hand side and divide the inequality obtained by

$$\sum_{i=1}^{np} |\beta^x_{n,i}| + \sum_{j=1}^{np} |\beta^y_{n,j}| + 4/h^2.$$

As a result we have

$$\left\{ 1 + \left(|\beta^x_{n,0}| - \sum_{i=1}^{np} |\beta^x_{n,i}| + |\beta^y_{n,0}| - \sum_{j=1}^{np} |\beta^y_{n,j}| \right) \right/$$

$$\left(2/h^2 + \sum_{i=1}^{np} |\beta^x_{n,i}| + 2/h^2 + \sum_{j=1}^{np} |\beta^y_{n,j}| \right) \right\} |u^h(\bar{x})|$$

$$\le |u^h(\bar{x})| + b^2 \max_{\Omega^r_h} |f|/2 + h^2 \max_{\Omega^{ir}_h} |f|/4. \quad (2.54)$$

The quantities

$$\alpha_1 = |\beta_{n,0}^x| - \sum_{i=1}^{np} |\beta_{n,i}^x|, \qquad \beta_1 = 2/h^2 + \sum_{i=1}^{np} |\beta_{n,i}^x|,$$

$$\alpha_2 = |\beta_{n,0}^y| - \sum_{j=1}^{np} |\beta_{n,j}^y|, \qquad \beta_2 = 2/h^2 + \sum_{j=1}^{np} |\beta_{n,j}^y|$$

because of (2.39), satisfy the inequalities

$$\alpha_1/\beta_1 \geq c_7, \qquad \alpha_2/\beta_2 \geq c_7$$

from which it follows that

$$(\alpha_1 + \alpha_2)/(\beta_1 + \beta_2) \geq c_7. \tag{2.55}$$

Using this relation, we obtain from (2.54) the inequality

$$c_7|u^h(\bar{x})| \leq b^2 \max_{\Omega_h^r} |f|/2 + h^2 \max_{\Omega_h^{ir}} |f|/4.$$

Replacing the last term by the larger one and dividing the inequality obtained by c_7 we get (2.52) again.

Combining (2.52) and (2.49) we have inequality (2.46). Lemma 2.3 is proved. □

Applying the estimate (2.46) to the homogeneous system corresponding to the difference problem (2.42)–(2.45) we see that the system (2.42)–(2.45) is uniquely solvable for any continuous functions f and φ.

Let us construct the corrector from several solutions with different mesh-sizes h according to the following rule. Assume $s = [l/2]$. Construct and solve the difference problem (2.42)–(2.45) for $k = 1, \ldots, s + 1$ on each of the grids $\Omega_{h/k}$ with mesh-size h/k. All solutions $u^{h/k}$ obtained are defined on the grid Ω_h. Form the linear combination

$$U^H(x) = \sum_{k=1}^{s+1} \frac{2(-1)^{s-k+1} k^{2s+2}}{(s-k+1)!\,(s+k+1)!} u^{h/k}(x) \quad \text{on } \Omega_h \tag{2.56}$$

and determine its order of accuracy.

Theorem 2.4. *Assume that the conditions of Theorem 1.2 with $l \geq 2$ and $\alpha \in (0, 1)$ are satisfied for (2.1), (2.2). Then for the corrected solution (2.56) the estimate*

$$\max_{\Omega_h} |U^H - u| \leq c_9 h^{l+\alpha} \tag{2.57}$$

holds.

PROOF. Let us verify that the conditions of Theorem 2.2 (Chapter 1) hold, taking into account the remarks at the end of §1.2. Assume

$$M_k(\Omega) = C^{k+\alpha}(\overline{\Omega}), \qquad P_k(\overline{\Omega}) = C^{k+2+\alpha}(\overline{\Omega}), \qquad N_k(D) = C^{k+2+\alpha}(\Gamma).$$

Then Condition A of §1.2 is a consequence of Theorem 1.2 of the present chapter. For the difference problem (2.3) of §1.2 we assume $\bar{\Omega}_h = \Omega_h \cup \Gamma_h$, $\mathring{\Omega}_h = \Omega_h$, $D_h = \Gamma_h$ and choose the norms

$$\|u\|_{\bar{\Omega}_h} = \max_{\bar{\Omega}_h} |u|, \qquad \|u\|_{\mathring{\Omega}_h} = \max_{\Omega_h^c} |u| + h^2 \max_{\Omega_h^{|\Gamma}} |u|.$$

Then Condition B' coincides with the *a priori* estimate of Lemma 2.3. The expansion of the approximation error in even powers of h follows from the construction of the difference scheme. Thus Condition D is satisfied with constant $\beta = \alpha$, the boundary condition being approximated exactly, i.e., $l_h u = lu$ on Γ_h. Now we use Theorem 3.2 of §1.3. Condition (3.15) of the theorem is satisfied with the constant $d_3 = 1/s$. Application of Theorem 3.2 yields the estimate (2.57). The theorem is proved. $\square$

Note that the system (2.42)–(2.45) has a nonsymmetric matrix. This fact restricts the type of methods we can use to solve it. But one can use the method presented at the end of §4.2.1, which reduces the solution of system (2.42)–(2.45) to a sequence of algebraic problems with symmetric matrices. Inequality (2.46) is of great importance in determining when to stop the iterative processes, since it allows one to estimate the error of approximation to the solution of system (2.42)–(2.45) using the residual value.

4.3. The Dirichlet Problem in a Rectangle

It is well known that in the case of domains with corners the smoothness of a solution in the vicinity of a corner usually decreases, and in the general case this is an obstruction to the methods used to increase the accuracy of approximate solutions. But when the initial data satisfies certain coherence conditions such a loss of a smoothness does not occur. This case will be considered in the present section.

Consider the equation

$$-\Delta u + au = f \qquad (3.1)$$

with a nonnegative function a in the region $\Omega \subset R^2$ which is the rectangle:

$$\Omega = \{(x_1, x_2); 0 < x_1 < b_1, 0 < x_2 < b_2\}.$$

Let the coefficients of the equation satisfy

$$a, f \in C^{3+\alpha}(\bar{\Omega}), \qquad \alpha \in (0, 1). \qquad (3.2)$$

We will seek a solution satisfying the homogeneous boundary condition

$$u = 0 \quad \text{on } \Gamma, \qquad (3.3)$$

where Γ is the boundary of the rectangle Ω.

Note that there is a solution of the given problem in $C^{5+\alpha}(\Omega)$, but we can not conclude that the solution belongs to $C^{5+\alpha}(\overline{\Omega})$, or even to $C^{2+\alpha}(\overline{\Omega})$ because the boundary is not sufficiently smooth (it only belongs to C^γ, $\gamma \in (0, 1]$). However, it is enough to require that the right-hand side satisfies some coherence conditions to guarantee that the solutions belong to the indicated class.

Lemma 3.1. *Let*

$$a, f \in C^{1+\alpha}(\Omega), \qquad f(x) = 0 \tag{3.4}$$

in (3.1)–(3.3) *at each of four corners of the rectangle. Then* $u \in C^{3+\alpha}(\overline{\Omega})$.

PROOF. Since $\Gamma \subset C^\alpha$, then from [73] we can conclude that the solution belongs to $C^\alpha(\overline{\Omega})$. Subtract au from both sides, and consider the problem

$$-\Delta v = f - au \quad \text{in } \Omega,$$

$$v = 0 \qquad \text{on } \Gamma. \tag{3.5}$$

The function $f - au$ belongs to $C^\alpha(\overline{\Omega})$, and the solution v exists and is unique. Therefore $v = u$. But in (3.5) the coherence conditions are satisfied, and consequently from [137, 43]) the solution belongs to $C^{2+\alpha}(\overline{\Omega})$: the difference $f(x) - a(x)u(x)$ is equal to zero at each of the corners of the rectangle, since $f(x)$ and $u(x)$ become zero due to condition (3.4) and the boundary condition (3.3), respectively. Thus $u \in C^{2+\alpha}(\overline{\Omega})$. Now one can see that in (3.5) the right-hand side belongs to $C^{1+\alpha}(\overline{\Omega})$. But since the coherence condition is satisfied, then the same condition (from the papers mentioned above) guarantees that the solution belongs to $C^{3+\alpha}(\overline{\Omega})$ with this right-hand side. The lemma is proved. □

One can guarantee greater smoothness if one requires an additional coherence condition.

Lemma 3.2. *Let the right-hand side of* (3.1)–(3.3) *satisfy the coherence conditions*

$$f(x) = 0, \qquad \frac{\partial^2 f}{\partial x_1^2}(x) - \frac{\partial^2 f}{\partial x_2^2}(x) = 0 \tag{3.6}$$

at each corner of the rectangle. Then $u \in C^{5+\alpha}(\overline{\Omega})$.

PROOF. On the basis of the previous lemma $u \in C^{3+\alpha}(\overline{\Omega})$. Let us return to problem (3.5). Under the assumptions of this lemma the right-hand side belongs to $C^{3+\alpha}(\overline{\Omega})$. In order that the fact that the solution v belongs to $C^{5+\alpha}(\overline{\Omega})$ follows from the above it is enough to require (see [137, 43]) that the coherence condition is satisfied:

$$\frac{\partial^2}{\partial x_1^2}(f - au) - \frac{\partial^2}{\partial x_2^2}(f - au) = 0$$

at each of the corners of the rectangle. Due to the condition of the lemma this is equivalent to

$$\frac{\partial^2}{\partial x_1^2}(au) - \frac{\partial^2}{\partial x_2^2}(au) = 0.$$

Since the second derivatives of au are continuous on $\bar{\Omega}$, let us calculate them along the sides of the rectangle. On the boundary $au = 0$, because of the boundary condition. Thus the coherence condition for problem (3.5) coincides with the condition of the lemma and $v \in C^{5+\alpha}(\bar{\Omega})$. Since $v = u$, the lemma is proved.   □

Let us construct the difference scheme. Consider the uniform rectangular grid

$$\bar{\Omega}_h = \{(x_1, x_2); x_1 = ih_1, x_2 = jh_2, 0 \le i, j \le N\}$$

with mesh-sizes $h_1 = b_1/N$ and $h_2 = b_2/N$. Take

$$\Gamma_h = \bar{\Omega}_h \cap \Gamma, \qquad \Omega_h = \bar{\Omega}_h \backslash \Gamma_h.$$

To solve (3.1)–(3.3) numerically we use the usual fivepoint difference scheme

$$-u^h_{\bar{x}_1 x_1}(x) - u^h_{\bar{x}_2 x_2}(x) + a(x)u^h(x) = f(x), \qquad x \in \Omega_h, \qquad (3.7)$$

with the boundary conditions

$$u^h = 0 \quad \text{on } \Gamma_h. \qquad (3.8)$$

Lemma 3.3. *The inequality*

$$\max_{\bar{\Omega}_h}|u^h| \le (b_1^2 + b_2^2) \max_{\bar{\Omega}}|f|/16 \qquad (3.9)$$

holds for the difference problem (3.7), (3.8).

PROOF. The comparison theorem proved in [113] allows us to assert that the solution of the problem

$$-W_{\bar{x}_1 x_1} - W_{\bar{x}_2 x_2} + aW = g_1 \quad \text{on } \Omega_h,$$

$$W = g_2 \quad \text{on } \Gamma_h$$

under the conditions

$$|f| \le g_1 \quad \text{on } \Omega_h, \quad 0 \le g_2 \quad \text{on } \Gamma_h$$

majorizes the solution of (3.7), (3.8):

$$|u^h| \le W \quad \text{on } \bar{\Omega}_h.$$

Let us take the Gerschgorin majorant function as our function W;

$$W(x) = \{(b_1^2 + b_2^2)/4 - (x_1 - b_1/2)^2 - (x_2 - b_2/2)^2\} \max_{\Omega_h}|f|/4.$$

It is easy to see that $W \geq 0$ on the whole rectangle $\overline{\Omega}_h$. All derivatives of the function W beginning with the third derivatives are zero. Therefore the second difference-derivatives are exactly equal to corresponding partial derivatives:

$$W_{\bar{x}_i x_i} = \frac{\partial^2 W}{\partial x_i^2} \quad \text{on } \Omega_h.$$

Taking this into account we have

$$g_1 = -\Delta W + aW \geq -\Delta W = \max_{\Omega_h}|f| \geq |f|.$$

Thus the conditions of the comparison theorem are satisfied and

$$|u^h| \leq W \quad \text{on } \overline{\Omega}_h$$

Hence

$$\max_{\overline{\Omega}_h}|u^h| \leq \max_{\overline{\Omega}_h}|W| \leq (b_1^2 + b_2^2)\max_{\Omega_h}|f|/16.$$

The lemma is proved. □

Assume that $h = 1/N$ and define the expansion in terms of h for the approximate solution.

Theorem 3.4. *Assume that the assumptions of Lemma 3.2 are satisfied for (3.1)–(3.3), then we have the following expansion for the solution u^h of (3.7), (3.8):*

$$u^h = u + h^2 v + h^{3+\alpha}\eta^h \quad \text{on } \overline{\Omega}_h \tag{3.10}$$

with v a continuous function independent of h, and with bounded grid function η^h:

$$\max_{\overline{\Omega}_h}|\eta^h| \leq c_1. \tag{3.11}$$

PROOF. The function v is defined as the solution of the differential equation

$$-\Delta v + av = \frac{b_1^2}{12}\frac{\partial^4 u}{\partial x_1^4} + \frac{b_2^2}{12}\frac{\partial^4 u}{\partial x_2^4} \quad \text{on } \Omega, \tag{3.12}$$

$$v = 0 \qquad \text{on } \Gamma.$$

The first step is to make sure that the function v belongs to $C^{3+\alpha}(\overline{\Omega})$. Indeed, if the conditions of Lemma 3.2 are satisfied, then $u \in C^{5+\alpha}(\overline{\Omega})$ and therefore the right-hand side of equation (3.12) belongs to $C^{1+\alpha}(\overline{\Omega})$. Consequently the assumptions of Lemma 3.1 on the smoothness of the right-hand side are satisfied for (3.12). The coherence condition is satisfied, since at each corner of the rectangle the derivatives $\partial^4 u/\partial x_1^4$, $\partial^4 u/\partial x_2^4$ are zero (this can be easily verified since because they are continuous on $\overline{\Omega}$ they can be calculated along the sides of the rectangle where $u = 0$).

Thus the functions u^h, u, v are determined uniquely. Assume

$$\eta^h = (u^h - u - h^2 v)/h^{3+\alpha} \quad \text{on } \overline{\Omega}_h \tag{3.13}$$

and substitute this function into the difference operator of problem (3.7):

$$-\eta^h_{\hat{x}_1 \hat{x}_1} - \eta^h_{\hat{x}_2 \hat{x}_2} + a\eta^h = \{-u^h_{\hat{x}_1 \hat{x}_1} - u^h_{\hat{x}_2 \hat{x}_2} + au^h$$
$$+ u_{\hat{x}_1 \hat{x}_1} + u_{\hat{x}_2 \hat{x}_2} - au + h^2(v_{\hat{x}_1 \hat{x}_1} + v_{\hat{x}_2 \hat{x}_2} - av)\}/h^{3+\alpha}.$$

Transform this expression using equation (3.7) and the expansions (2.4) of the difference derivatives:

$$-\eta^h_{\hat{x}_1 \hat{x}_1} - \eta^h_{\hat{x}_2 \hat{x}_2} + a\eta^h = \left(f + \frac{\partial^2 u}{\partial x_1^2} + \frac{h_1^2}{12} \frac{\partial^4 u}{\partial x_1^4} + \frac{\partial^2 u}{\partial x_2^2} + \frac{h_2^2}{12} \frac{\partial^4 u}{\partial x_2^4} - au - h^2 av \right.$$

$$\left. + h^2 \frac{\partial^2 v}{\partial x_1^2} + h^2 \frac{\partial^2 v}{\partial x_2^2} + h_1^{3+\alpha}\xi_1 + h_2^{3+\alpha}\xi_2 \right) \Big/ h^{3+\alpha}.$$

Here $|\xi_1| \le c_2, |\xi_2| \le c_3$ uniformly in x and h. Keeping in mind that $h_1 = b_1 h$, $h_2 = b_2 h$ and using equation (3.1) we have

$$-\eta^h_{\hat{x}_1 \hat{x}_1} - \eta^h_{\hat{x}_2 \hat{x}_2} + a\eta^h = \left(\frac{b_1^2}{12} \frac{\partial^4 u}{\partial x_1^4} + \frac{b_2^2}{12} \frac{\partial^4 u}{\partial x_2^4} + \Delta v - av \right) \Big/ h^{1+\alpha} + \xi_3,$$

where

$$|\xi_3| \le c_4 = c_2 b_1^{3+\alpha} + c_3 b_2^{3+\alpha}. \tag{3.14}$$

The cancellation of the term of order $h^{-1-\alpha}$ follows from the definition of the function v and leads to the equality

$$-\eta^h_{\hat{x}_1 \hat{x}_1} - \eta^h_{\hat{x}_2 \hat{x}_2} + a\eta^h = \xi_3 \quad \text{on } \Omega_h. \tag{3.15}$$

From (3.13) it follows that

$$\eta^h = 0 \quad \text{on } \Gamma_h \tag{3.16}$$

since all three functions u^h, u, v are zero on Γ_h. For (3.15), (3.16) the estimate of Lemma 3.3 is satisfied. This estimate leads to the inequality

$$\max_{\overline{\Omega}_h} |\eta^h| \le c_4 (b_1^2 + b_2^2)/16,$$

where c_4 is determined in (3.14). Thus the inequality (3.11) is established. The theorem is proved. $\square$

Consider one example of this theorem. Suppose that two problems (3.7), (3.8) with mesh-sizes h and $h/2$ have been solved. Then on the net Ω_h there are two approximate solutions u^h and $u^{h/2}$. Adding them with weights $-\frac{1}{3}$ and $\frac{4}{3}$ we have an approximate solution with accuracy $O(h^{3+\alpha})$:

$$\max_{\Omega_h} |u - (\tfrac{4}{3}u^{h/2} - \tfrac{1}{3}u^h)| \le c_5 h^{3+\alpha}, \tag{3.17}$$

where the constant c_5 depends neither on h, nor on x. The proof is based on the expansion (3.10). As in previous sections we choose weights $-\frac{1}{3}$ and $\frac{4}{3}$ to eliminate the terms of order h^2 in the sum.

Let us make some remarks concerning the results we have obtained.

Remark 1. It is clear that when the amount of smoothness of the right-hand side decreases, for example, when $f \in C^{2+\alpha}(\overline{\Omega})$ the magnitude of remainder increases and the expansion takes the form

$$u^h = u + h^2 v + h^{2+\alpha} \eta^h \quad \text{on } \overline{\Omega}_h.$$

However, when the amount of smoothness increases ($f \in C^{k+\alpha}(\overline{\Omega}), k \geq 4$) the quality of the expansion does not increase substantially (the magnitude of the irregular remainder does not decrease) even if the solution is assumed to be as smooth as possible. This is because the auxiliary equations used to define the functions v_i in the expansion

$$u^h = u + h^2 v_1 + h^4 v_2 + \cdots + h^{r+\alpha} \eta^h \tag{3.18}$$

do not necessarily satisfy the coherence conditions, which guarantees the smoothness of the solution. At the same time from [137, 43] it follows that these conditions are not only sufficient; they are also necessary for the existence of a smooth solution.

In [138] it is suggested that one replace the solution u by the sum $W + Q$, where W is an unknown function for which the same equation (with a different right-hand side) is valid, and Q is a known function. It is proved that as a result of this replacement one can arrive at an equation for the sufficiently smooth function W, so that the coherence conditions are satisfied not only for the function W itself, but also for all auxiliary problems equations, which will guarantee that the solutions are sufficiently smooth. Thus in this paper expansion (3.18) is proved on the assumption $f \in C^{r+\alpha}(\overline{\Omega})$.

Remark 2. Expansion (3.10) allows us to justify computing the first and second derivatives of the solution using the difference solution. For example,

$$\{u^h(x_1 + h_1, x_2) - u^h(x_1 - h_1, x_2)\}/(2h_1) = \frac{\partial u}{\partial x}(x_1, x_2) + h^2 \xi_3(x_1, x_2),$$

where $|\xi_3| \leq c_7$ and c_7 does not depend on x or h. Indeed using the expansion (3.10) we have

$$\{u(x_1 + h_1, x_2) - u(x_1 - h_1, x_2)\}/(2h_1)$$
$$+ h^2\{v(x_1 + h_1, x_2) - v(x_1 - h_1, x_2)\}/(2h_1)$$
$$+ h^{3+\alpha}\{\eta^h(x_1 + h_1, x_2) - \eta^h(x_1 - h_1, x_2)\}/(2h_1)$$
$$= \frac{\partial u}{\partial x_1}(x_1, x_2) + \frac{h_1^2}{3}\frac{\partial^3 u}{\partial x_1^3}(\theta_1, x_2) + h_1^2 \frac{\partial v}{\partial x_1}(\theta_2, x_2)$$
$$+ \frac{h^{2+\alpha}}{2b_1}\{\eta^h(x_1 + h_1, x_2) - \eta^h(x_1 - h_1, x_2)\},$$
$$\theta_i \in (x_1 - h_1, x_1 + h_1).$$

Estimating the last terms we can estimate the remainder

$$|\xi_3| \le \frac{b_1^2}{3} \max_{\bar{\Omega}} \left| \frac{\partial^3 u}{\partial x_1^3} \right| + b_1^2 \max_{\bar{\Omega}} \left| \frac{\partial v}{\partial x_1} \right| + c_1/b_1.$$

The computation of the derivative $\partial u/\partial y$ is similar. The second derivatives can be determined with accuracy $h^{1+\alpha}$. For example,

$$\frac{\partial^2 u}{\partial x_1^2} = u_{\bar{x}_1 \bar{x}_1}^h + h^{1+\alpha}\xi_4.$$

Indeed, using (3.17) we have

$$u_{\bar{x}_1 \bar{x}_1}^h(x) = u_{\bar{x}_1 \bar{x}_1}(x) + h^2 v_{\bar{x}_1 \bar{x}_1}(x) + h^{3+\alpha}\eta_{\bar{x}_1 \bar{x}_1}^h(x)$$

$$= \frac{\partial^2 u}{\partial x_1^2}(x) + \frac{h_1^2}{12} \frac{\partial^4 u}{\partial x_1^4}(\theta_3, x_2) + h^2 \frac{\partial^2 v}{\partial x_1^2}(\theta_4, x_2)$$

$$+ \frac{h^{1+\alpha}}{b_1^2} \{\eta^h(x_1 + h_1, x_2) - 2\eta^h(x_1, x_2) + \eta^h(x_1 - h_1, x_2)\},$$

where $\theta_i \in (x_1 - h_1, x_1 + h_1)$. Estimating the terms on the right-hand side of the equality we get

$$|\xi_4| \le \frac{b_1^2}{12} \max_{\bar{\Omega}} \left| \frac{\partial^4 u}{\partial x_1^4} \right| + \max_{\bar{\Omega}} \left| \frac{\partial^2 v}{\partial x_1^2} \right| + 4c_2/b_1^2.$$

A similar formula holds for $\partial^2 u/\partial x_2^2$. The mixed derivative is determined using central differences, and has the same accuracy:

$$\frac{\partial^2 u}{\partial x_1 \partial x_2}(x_1, x_2) = \{u^h(x_1 + h_1, x_2 + h_2) - u^h(x_1 - h_1, x_2 + h_2)$$

$$+ u^h(x_1 - h_1, x_2 - h_2) - u^h(x_1 + h_1, x_2 - h_2)\}/(h_1 h_2) + O(h^{1+\alpha}).$$

4.4. A Quasilinear Equation in a Triangular Region

In solving quasilinear equations of elliptic type for two-dimensional problems with a boundary with corners one has to put up with a loss of smoothness of a solution due to the appearance of singularities near the corners. Typical singularities appear at the corners of a triangular region, since any region with corners can be represented as a set of triangular ones. We will try to construct an algorithm, which will give a solution with an order of accuracy close to h^2 for noncoherent initial data. This is the first step toward enhancing the solution of the problem.

In the present section we consider the effect of corners and investigate them within the context of refining the grid for a quasilinear equation in a triangular region, using the usual fivepoint difference scheme.

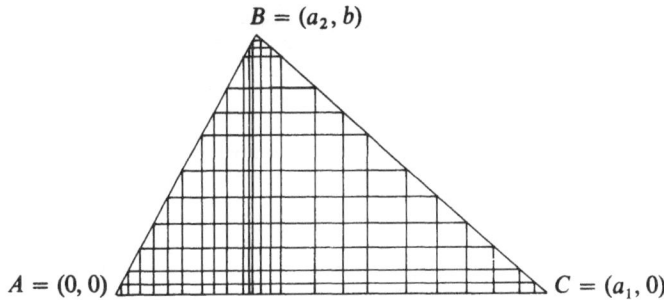

Figure 4.6. A triangle with a nonuniform grid.

Let Ω be the open triangle with vertices $(0, 0)$, $(a_1, 0)$, (a_2, b), where $b > 0$, $a_1 > a_2 > 0$ (Figure 4.6).

Consider the equation

$$-\Delta u(x) = f(u(x), x), \qquad x \in \Omega, \tag{4.1}$$

where

$$f(p, x) \in C^2((-\infty, \infty) \times \overline{\Omega}). \tag{4.2}$$

We will seek a solution satisfying the homogeneous condition

$$u = 0 \quad \text{on } \Gamma, \tag{4.3}$$

where Γ is the boundary of the region Ω. It is assumed that there is at least one solution of this problem in $L_2(\Omega)$ with bounded modulus, i.e.,

$$\textit{vrai} \sup_{\Omega} |u| \le c_0. \tag{4.4}$$

We also assume that the condition

$$\frac{\partial f}{\partial p}(p, x) \le 0, \qquad p \in (-\infty, \infty), \quad x \in \overline{\Omega}, \tag{4.5}$$

is satisfied.

Lemma 4.1. *Assume condition (4.4) is satisfied for the solution $u \in L_2(\Omega)$, and (4.2), (4.5) hold for the right-hand side. Then this solution is unique and belongs to $W_2^2(\Omega) \cap C^\gamma(\overline{\Omega}) \ \forall \ \gamma \in (0, 1)$.*

PROOF. Due to condition (4.4) $f(u(x), x) \in L_2(\Omega)$ as a function of x. Thus the solution of problem

$$-\Delta v = f(u(x), x) \quad \text{on } \Omega,$$

$$v = 0 \qquad\qquad \text{on } \Gamma,$$

from Theorem 1.4, belongs to $W_2^2(\Omega)$. But the function u also satisfies this equation. Thus $v = u$ and consequently $u \in W_2^2(\Omega)$. From the embedding theorem [73] it follows that $u \in C^\gamma(\overline{\Omega}) \ \forall \ \gamma \in (0, 1)$. Now we show that there are no other solutions of problem (4.1), (4.3) satisfying condition (4.4). Indeed, suppose there is another solution u_1. Then $u_1 \in W_2^2(\Omega)$ and for the difference $w = u_1 - u$ we have the following equality as a result of equation (4.1):

$$\int_\Omega \{-\Delta w + f(u, x) - f(u_1, x)\} w \, dx = 0.$$

Using the homogeneous boundary conditions we apply the first Green's theorem. We get

$$\int_\Omega \left\{ \left(\frac{\partial w}{\partial x_1}\right)^2 + \left(\frac{\partial w}{\partial x_2}\right)^2 + (f(u, x) - f(u_1, x))w \right\} dx = 0.$$

Application of Lagrange's formula for the last group of terms yields

$$f(u, x) - f(u_1, x) = \frac{\partial f}{\partial p}(\xi, x)(u - u_1).$$

Therefore using inequality (4.5) we have

$$\int_\Omega \left\{ \left(\frac{\partial w}{\partial x_1}\right)^2 + \left(\frac{\partial w}{\partial x_2}\right)^2 \right\} dx \le 0.$$

Hence $w = 0$ in the space $\mathring{W}_2^1(\Omega)$. By virtue of the embedding theorem cited above $u = u_1$ in the space $L_2(\Omega)$. The lemma is proved. $\qquad\square$

As for the solvability of problem (4.1), (4.3) in $L_2(\Omega)$ we refer the reader to [73], where a series of sufficient conditions for a wide variety of quasilinear problems are described.

Note that one can weaken condition (4.5) (this would somewhat complicate the procedure), replacing zero on the right-hand side by a positive number $d_1 < 1/d_2^2$, where d_2 is the constant in Friedrichs' inequality (see [94]):

$$\|u\|_{L_2(\Omega)} \le d_2 |u| \quad \forall u \in \mathring{W}_2^1(\Omega).$$

The classes $C^{l+\alpha}(\overline{\Omega})$ do not give a complete representation of the solution of (4.1), (4.3), since even when the right-hand sides are analytic the solution may not belong to $C^2(\overline{\Omega})$ if the equation does not satisfy some additional conditions. We have already encountered this phenomenon in the previous section, where we managed to formulate coherence conditions in an explicit form. Here, if the angles of the triangle are not equal to π/k (where k is an integer) the conditions which guarantee the smoothness of the solution are not quite so simple (they are integral relations which are difficult to verify in practice).

Nevertheless, the corners do not prevent the solutions from being sufficiently smooth in the strict interior of the region. Consider the problem

$$-\Delta v(x) = f(u(x), x), \qquad x \in \Omega,$$
$$v(x) = 0 \qquad\qquad x \in \Gamma. \tag{4.6}$$

The function $f(u(x), x)$ belongs to $C^\gamma(\overline{\Omega})$ and, according to Theorem 1.1, $v \in C^{2+\gamma}(\Omega)$. But $v = u$ due to the uniqueness of the solution of (4.6) and (4.1), (4.3). Thus $u \in C^{2+\gamma}(\Omega)$.

But the fact of belonging to $C^{l+\gamma}(\Omega)$ does not impose any bounds on the derivatives or on the function itself; they may be unbounded on Ω. Therefore before the difference scheme is constructed it is necessary to determine how these functions behave.

Consider the behavior of the solution near the corners, as in [137, 64, 101, 43]. Let us fix a corner and introduce polar coordinates (r, φ) in a neighborhood of it. Let us investigate, for example, the behavior of the solution u near the point $(0, 0)$. We consider the circular sector $S = \{(r, \varphi); 0 < r < \varepsilon, 0 < \varphi < \beta\}$ as a neighborhood of this point. Here the Cartesian coordinates are related to the polar coordinates by $x_1 = r \sin \varphi$, $x_2 = z \cos \varphi$; here β is the angle BAC, and ε is equal to half the length of the smallest side of the triangle. In this case $S \subset \Omega$ and two radii of S lie along the sides of the triangle. Then from [29] the solution v of (4.6) is representable in the form

$$v = w + \sum_{i=1}^{M} r^{\mu_i} \ln^{q_i} r \theta_i(\varphi). \tag{4.7}$$

Here the smooth component† w belongs to $C^{2+\gamma}(\overline{\Omega})$, $\{\mu_i\}_{i=1}^{M}$ is a certain nonincreasing sequence of positive numbers, $\{q_i\}_{i=1}^{M}$ is a set of nonnegative integers, and $\{\theta_i(\varphi)\}_{i=1}^{M}$ is a set of infinitely differentiable trigonometric polynomials. We are interested in the main term of asymptotic decomposition (4.7) whose behavior is responsible for the degree of the grid refinement. A technique exists in the theory of partial differential equations to elucidate this behavior; we will only touch on it.

The equation

$$-\Delta v = q \quad \text{on } S,$$

where $g(x) = f(u(x), x)$, will, in polar coordinates, have the form

$$-\frac{\partial^2 v}{\partial r^2} - \frac{1}{r}\frac{\partial v}{\partial r} - \frac{1}{r^2}\frac{\partial^2 v}{\partial \varphi^2} = g(r, \varphi) \quad \text{on } S, \tag{4.8}$$

and

$$
\begin{aligned}
v(r, 0) &= 0, & r &\in (0, \varepsilon), \\
v(r, \beta) &= 0, & r &\in (0, \varepsilon), \\
v(\varepsilon, \varphi) &= u(\varepsilon, \varphi), & \varphi &\in [0, \beta].
\end{aligned}
\tag{4.9}
$$

In order that the solution exist and be unique at $r = 0$ the requirement $|v(\delta, \varphi)| < \infty$ is imposed for any positive δ, with δ going to zero. Note that

---

† To be more exact, it follows from [64] that $w \in W_2^4(\Omega)$ but according to the embedding theorem (see [73]) $w \in C^{2+\gamma}(\overline{\Omega}) \ \forall \ \gamma \in (0, 1)$.

the arc (ε, φ) has distance > 0 from the vertices of the triangle, and therefore $u(\varepsilon, \varphi) \in C^{2+\gamma}[0, \beta]$ as a function of the argument φ.

Let us expand the functions v and g in Fourier series

$$v(r, \varphi) = \sum_{k=1}^{\infty} v_k(r) \sin k\lambda\varphi,$$

$$g(r, \varphi) = \sum_{k=1}^{\infty} g_k(r) \sin k\lambda\varphi,$$

(4.10)

where $\lambda = \pi/\beta$. The Fourier coefficients are computed by

$$v_k(r) = \frac{2}{\beta} \int_0^\beta v(r, \varphi) \sin k\lambda\varphi \, d\varphi,$$

$$g_k(r) = \frac{2}{\beta} \int_0^\beta g(r, \varphi) \sin k\lambda\varphi \, d\varphi.$$

(4.11)

Substitute the expansion (4.10) into (4.8) and equate the coefficients of the functions $\sin k\lambda\varphi$. We have

$$-v_k'' - \frac{1}{r} v_k' + \frac{k^2\lambda^2}{r^2} v_k = g_k, \qquad k = 1, 2, \ldots .$$

From (4.9), (4.11) we get the conditions

$$v_k(\varepsilon) = z_k \equiv \frac{2}{\beta} \int_0^\beta u(\varepsilon, \varphi) \sin k\lambda\varphi \, d\varphi,$$

$$|v_k(\delta)| < \infty \quad \text{at } \delta \to 0.$$

From [101] it follows that the given problem has the unique solution

$$v_k(r) = -\frac{r^{k\lambda}}{2k\lambda\varepsilon^{2k\lambda}} \int_0^r g_k(\rho)\rho^{k\lambda+1} \, d\rho$$

$$+ \frac{r^{k\lambda}}{2k\lambda} \int_r^\varepsilon g_k(\rho)\rho^{1-k\lambda} \, d\rho + \frac{r^{-k\lambda}}{2k\lambda} \int_0^r g_k(\rho)\rho^{k\lambda+1} \, d\rho + \frac{r^{k\lambda}}{\varepsilon^{k\lambda}} z_k.$$

This is established by a simple but tedious calculation, assuming that the coefficient $v_1(r)$ has the worst possible smoothness. Two cases arise. If $\beta = \pi/2$, then $\lambda = 2$ and the term of order $r^2 \ln r$ becomes the main term of the asymptotic expansion (4.7). But if $\beta \neq \pi/2$ then the terms of order z^λ become the main terms of the asymptotic expansion of $v_1(r)$. Therefore $v_1(r) \sin \lambda\varphi \in C^1(\bar{S})$. The other terms in the Fourier series of the function v belong to $C^1(\bar{S})$, and one can prove that the Fourier series for $\partial v/\partial x$ and $\partial v/\partial y$ converge absolutely. Thus $v \in C^1(\bar{S})$. Moreover,

$$r^\gamma \frac{dv_k}{dx}(r) \sin k\lambda\varphi \in C^\gamma(\bar{S}), \qquad \gamma \in (0, 1), \quad k = 1, 2, \ldots$$

whence it follows that $r^\gamma(\partial v/\partial x) \in C^\gamma(\bar{S})$. Similarly $r^\gamma(\partial v/\partial y) \in C^\gamma(\bar{S})$.

Introduce the function $d(x)$, defined to be the distance between the point x and the nearest vertex of the triangle. It is clear that within S the equality $d(x) = r$ holds. Therefore $d^\gamma(\partial v/\partial x), d^\gamma(\partial v/\partial y) \in C^\gamma(\bar{S})$. But this is valid for each vertex of the triangle. This means that

$$u \in C^1(\bar{\Omega}), \qquad d^\gamma \frac{\partial u}{\partial x}, d^\gamma \frac{\partial u}{\partial y} \in C^\gamma(\bar{\Omega}). \qquad (4.12)$$

These properties give additional information about the right-hand side g of equation (3.8). From (4.12) and the properties of the function f it follows that

$$g \in C^1(\bar{\Omega}), \qquad d^\gamma \frac{\partial g}{\partial x}, d^\gamma \frac{\partial g}{\partial y} \in C^\gamma(\bar{\Omega}).$$

Using similar considerations for the second and third derivatives of the function u we get the following estimates:

$$\max_{\bar{\Omega}} \left(\left| d \frac{\partial^2 u}{\partial x^2} \right|, \left| d \frac{\partial^2 u}{\partial y^2} \right|, \left| d^2 \frac{\partial^3 u}{\partial x^3} \right|, \left| d^2 \frac{\partial^3 u}{\partial y^3} \right| \right) \le c_1,$$

$$\max_{x, x' \in \Omega} \left(\min\{d^{2+\alpha}(x), d^{2+\alpha}(x')\} \left| \frac{\partial^3 u}{\partial x^3}(x) - \frac{\partial^3 u}{\partial x^3}(x') \right| \middle/ |x - x'|^\alpha \right) \le c_1, \quad (4.13)$$

$$\max_{x, x' \in \Omega} \left(\min\{d^{2+\alpha}(x), d^{2+\alpha}(x')\} \left| \frac{\partial^3 u}{\partial y^3}(x) - \frac{\partial^3 u}{\partial y^3}(x') \right| \middle/ |x - x'|^\alpha \right) \le c_1$$

for any $\alpha \in (0, 1)$, and the constant c_1 depends on α. We therefore fix α in what follows. In addition to (4.13) for u, from Theorem 1.1 it follows that $u \in C^{3+\alpha}(\Omega)$.

For the constitution of our difference scheme we introduce the function

$$\varphi(t) = bt^2(3 - 2t).$$

Note that $\varphi(0) = 0$, and $\varphi(1) = b$. We fix an integer $N \ge 2$ and assume $h = 1/N$. Let us draw a family of parallel lines

$$x_2 = \varphi(ih), \qquad i = 1, \ldots, N - 1.$$

Through the points at which these lines cross AB and BC we draw another family of parallel lines orthogonal to the first one (see Figure 4.6). Thus we have a difference grid set up within the triangle. Denote by Ω_h the set of all points in Ω, which are crossing points of these lines, and by Γ_h the set of crossing points of these lines with the boundary of the triangle; assume $\bar{\Omega}_h = \Omega_h \cup \Gamma_h$.

To solve (4.1), (4.3) numerically we use the usual fivepoint scheme on the nonuniform rectangular grid:

$$L^h u^h(x) = -u^h_{\bar{x}_1 \hat{x}_1}(x) - u^h_{\bar{x}_2 \hat{x}_2}(x) = f(u^h(x), x), \qquad x \in \Omega_h, \qquad (4.14)$$

with the boundary conditions

$$u^h(x) = 0, \qquad x \in \Gamma_h. \qquad (4.15)$$

Here the symbolic notation for the difference-derivatives on the nonuniform grid has the following meaning. Assume, say, that $(x_1, \varphi(ih)) \in \Omega_h$; then

$$-v_{\hat{x}_2 \hat{x}_2}(x_1, \varphi(ih)) = \{\{v(x_1, \varphi(ih)) - v(x_1, \varphi((i-1)h))\}/\{\varphi(ih) - \varphi((i-1)h)\}$$
$$- \{v(x_1, \varphi((i+1)h)) - v(x_1, \varphi(ih))\}/\{\varphi((i+1)h) - \varphi(ih)\}\}$$
$$\times 2/\{\varphi((i+1)h) - \varphi((i-1)h)\}.$$

To prove that the solution of the nonlinear algebraic problem (4.14), (4.15) is unique we used the following *a priori* estimate.

Theorem 4.2. *Let u^h be the solution of* (4.14), (4.15); *then for any function v, defined on $\bar{\Omega}_h$ and equal to zero on Γ_h the inequality*

$$\max_{\Omega_h} |u^h - v| \le c_2 \delta$$

holds, where

$$\delta = \max_{x \in \Omega_h}(|L^h v(x) - f(v(x), x)|(\min\{x_2, b - x_2\})^{2-\gamma})$$

and the constant c_2 does not depend on x or h (but it does depend on $\gamma \in (0, 2]$).

PROOF. Let w be the solution of

$$L^h w = (\min\{x_2, b - x_2\})^{-2+\gamma} \quad \text{on } \Omega_h,$$
$$w = 0 \quad \text{on } \Gamma_h. \tag{4.16}$$

Introduce the grid function $p = \delta w$. Problem (4.16) satisfies the discrete maximum principle (see [114]), and, according to the comparison theorem (which is the consequence of it), we have

$$w \ge 0 \quad \text{on } \bar{\Omega}_h.$$

But since $\delta \ge 0$ the function p satisfies the same property

$$p \ge 0 \quad \text{on } \bar{\Omega}_h.$$

Whenever the function $f(p, x)$ is continuously differentiable with respect to the first argument then the Lagrange theorem holds true:

$$f(v(x), x) - f(u^h(x), x) = \frac{\partial f}{\partial p}(\xi(x), x)\{v(x) - u^h(x)\}$$

for a certain discrete function $\xi(x)$, each value of which is between $u^h(x)$ and $v(x) \; \forall \; x \in \bar{\Omega}_h$.

Let us use these facts to transform the expression

$$\left(L^h - \frac{\partial f}{\partial p}(\xi(x), x)\right)\{v(x) + p(x) - u^h(x)\}$$

$$= L^h v(x) - f(v(x), x) + \left(L^h - \frac{\partial f}{\partial p}(\xi(x), x)\right)p(x).$$

Taking into account the fact that $\partial f/\partial p$ is nonpositive, that ρ is nonnegative, and using formula (4.16) we obtain

$$\left(L^h - \frac{\partial f}{\partial p}(\xi(x), x)\right)\{v(x) + \rho(x) - u^h(x)\}$$

$$\geq L^h v(x) - f(v(x), x) + \delta(\min\{x_2, b - x_2\})^{-2+\gamma} \geq 0.$$

The last inequality follows from

$$\delta \geq -\{L^h v(x) - f(v(x), x)\}(\min\{x_2, b - x_2\})^{2-\gamma}$$

due to the definition of δ.

On the grid boundary Γ_h the relation

$$v - u^h + \rho = 0$$

is satisfied. Note that the difference operator satisfies the maximum principle since $\partial f/\partial p$ is nonpositive. On this basis we have

$$v - u^h + \rho \geq 0 \quad \text{on } \bar{\Omega}_h.$$

The inequality

$$u^h - v + \rho \geq 0 \quad \text{on } \bar{\Omega}_h$$

is proved by repeating the foregoing manipulations. From these two inequalities it is evident that

$$|u^h - v| \leq \rho = \delta w$$

and therefore

$$\max_{\bar{\Omega}_h} |u^h - v| \leq \delta \max_{\bar{\Omega}_h} |w|. \tag{4.17}$$

Estimate $\max |w|$ by using the Lagrange theorem for the operator L^h. First of all let $\gamma \in (0, 1)$. Introduce the function

$$z(x_1, x_2) = 4^{2-\gamma}(x_2^\gamma + (b - x_2)^\gamma)/\{\gamma(1 - \gamma)\}$$

and consider the grid function

$$Z = z \quad \text{on } \bar{\Omega}_h.$$

Let us calculate $L^h Z$. Recall that for the function g, which is continuous on the interval $[t - h_1, t + h_2]$ and twice continuously differentiable on the interval $(t - h_1, t + h_2)$, the Taylor expansion and Lagrange theorem hold true for the second derivative. This yields

$$\{(g(t + h_2) - g(t))/h_2 - (g(t) - g(t - h_1))/h_1\} \times 2/(h_1 + h_2) = g''(\eta),$$

where $\eta \in (t - h_1, t + h_2)$. An application of this formula results in

$$Z_{\hat{x}_1 \hat{x}_1} = 0 \quad \text{on } \Omega_h,$$

$$Z_{\hat{x}_2 \hat{x}_2}(x_1, \varphi(ih)) = \frac{\partial^2 z}{\partial x_2^2}(x_1, \eta_i),$$

where $\eta_i \in (\varphi((i-1)h), \varphi((i+1)h))$. Therefore

$$L^h Z(x_1, \varphi(ih)) = -\frac{\partial^2 z}{\partial x_2^2}(x_1, \eta_i) = 4^{2-\gamma}\{\eta_i^{-2+\gamma} + (b - \eta_i)^{-2+\gamma}\}.$$

In virtue of the fact that the functions $x_2^{-2+\gamma}$ and $(b - x_2)^{-2+\gamma}$ are monotone we obtain

$$L^h Z(x_1, \varphi(ih)) \geq 4^{2-\gamma}\{\varphi^{-2+\gamma}((i+1)h) + (b - \varphi((i-1)h))^{-2+\gamma}\}.$$

Consider the extremum of

$$\varphi(t+h)/\varphi(t) = \{3(t+h)^2 - 2(t+h)^3\}/\{3t^2 - 2t^3\},$$

where $t \in [h, 1 - h]$. The maximum and the minimum are attained at the ends of the interval $[h, 1 - h]$ and are equal to, respectively,

$$(12h^2 - 16h^3)/(3h^2 - 2h^3) \leq 4, \qquad (3h^2 - 2h^3)/(12h^2 - 16h^3) \geq \tfrac{1}{4}.$$

Thus

$$L^h Z(x_1, x_2) \geq x_2^{-2+\gamma} + (b - x_2)^{-2+\gamma} \geq (\min\{x_2, b - x_2\})^{-2+\gamma}.$$

Also

$$Z \geq w = 0 \quad \text{on } \Gamma_h.$$

On the basis of the comparison theorem we have

$$Z = z \geq w \quad \text{on } \overline{\Omega}_h.$$

Since

$$w(x) \leq \max_{\overline{\Omega}_h} z \leq b^\gamma 2^{5-3\gamma}/\{\gamma(1 - \gamma)\},$$

then inequality (4.17) leads to the statement of the theorem, if we assume

$$c_2 = b^\gamma 2^{5-3\gamma}/\{\gamma(1 - \gamma)\}.$$

If $\gamma \in [1, 2]$, then the result of the theorem follows from the above using the the inequality

$$(\min\{x_2, b - x_2\})^{\mu_1} \leq (b/2)^{\mu_1 - \mu_2}(\min\{x_2, b - x_2\})^{\mu_2}, \qquad \mu_1 \geq \mu_2 \geq 0.$$

From this theorem it follows that the solution of (4.14), (4.15) is unique. Let u^h and v^h be two different solutions of the difference problem (4.14), (4.15). Then the inequality of Theorem 4.2 in which $\delta = 0$ holds. But since the difference $u^h - v^h$ is different from zero at least at one point then

$$\max_{\Omega_h}|u^h - v^h| > 0$$

and we have a contradiction.

To show that the solution of (4.14), (4.15) exists, consider the mapping of a linear space of the vectors ξ with components $\xi(x)$, $x \in \Omega_h$ into itself defined by the rule

$$P: \xi(x) \to \{L^h\xi(x) - f(\xi(x), x)\}\hbar_1(x)\hbar_2(x), \qquad x \in \Omega_h.$$

Here for notational purposes we assume that $\xi = 0$ on Γ_h. Let us explain the symbols $\hbar_i(x)$. Let $x \in \Omega_h$; then there are four neighboring points in the grids $\overline{\Omega}_h$: on the left is $(x_1 - \mu_1, x_2)$, on the right is $(x_1 + \mu_2, x_2)$, above is $(x_1, x_2 + \mu_3)$, and below is $(x_1, x_2 - \mu_4)$. Put

$$\hbar_1(x) = (\mu_1 + \mu_2)/2, \qquad \hbar_2(x) = (\mu_3 + \mu_4)/2,$$

$$\hbar_1^+(x) = \mu_2, \qquad \hbar_1^-(x) = \mu_1, \qquad \hbar_2^+(x) = \mu_3, \qquad \hbar_2^-(x) = \mu_4.$$

Note that the mapping P is continuous. Now we use Lemma 6.1 of §3.6. For this purpose we compute the sum

$$\sum_{x \in \Omega_h} \xi(x)P_\xi(x) = \sum_{x \in \Omega_h} \{-\xi_{\hat{x}_1 \hat{x}_1}(x) - \xi_{\hat{x}_2 \hat{x}_2}(x) - f(\xi(x), x)\}\xi(x)\hbar_1(x)\hbar_2(x).$$

$$(4.18)$$

We will consider all terms in this sum. For fixed x_1, apply the first difference Green's formula (see [113]) to get

$$-\sum_{\substack{\Omega_h \\ x_1 = \text{const}}} \xi_{\hat{x}_2 \hat{x}_2}\xi\hbar_1\hbar_2$$

$$= -\sum_{i=1}^{k} \xi_{\hat{x}_2 \hat{x}_2}(x_1, \varphi(ih))\xi(x_1, \varphi(ih))\hbar_1(x_1, \varphi(ih))\hbar_2(x_1, \varphi(ih))$$

$$= \sum_{i=0}^{k} (\{\xi(x_1, \varphi((i+1)h)) - \xi(x_1, \varphi(ih))\}/\hbar_2(x_1, \varphi(ih)))^2$$

$$\times \hbar_1(x_1, \varphi(ih)) \times \hbar_2(x_1, \varphi(ih)).$$

Here k is chosen from the requirement that $(x_1, \varphi((k+1)h)) \in \Gamma_h$; therefore the equalities

$$\xi(x_1, 0) = \xi(x_1, \varphi((k+1)h)) = 0$$

hold. Now we apply the Friedrichs difference inequality for the lower bound [114]:

$$-\sum_{\substack{\Omega_h \\ x_1 = \text{const}}} \xi_{\hat{x}_2 \hat{x}_2}\xi\hbar_1\hbar_2 \geq \{4/\varphi^2((k+1)h)\}$$

$$\times \sum_{i=1}^{k} \xi^2(x_1, \varphi(ih))\hbar_1(x_1, \varphi(ih))\hbar_2(x_1, \varphi(ih))$$

$$\geq \{4/b^2\} \times \sum_{i=1}^{k} \xi^2\hbar_1\hbar_2.$$

Summing up these inequalities with respect to the x_1, which form the difference grid, we get the inequality

$$- \sum_{\Omega_h} \xi_{\bar{x}_2 x_2} \xi \hbar_1 \hbar_2 \geq \frac{4}{b^2} \sum_{\Omega_h} \xi^2 \hbar_1 \hbar_2. \tag{4.19}$$

The inequality

$$- \sum_{\Omega_h} \xi_{\bar{x}_1 x_1} \xi \hbar_1 \hbar_2 \geq \frac{4}{a_1^2} \sum_{\Omega_h} \xi^2 \hbar_1 \hbar_2 \tag{4.20}$$

is proved similarly. Rewrite the third term in (4.18):

$$- \sum_{x \in \Omega_h} f(\xi(x), x) \xi(x) \hbar_1(x) \hbar_2(x)$$

$$= - \sum_{x \in \Omega_h} f(0, x) \xi(x) \hbar_1(x) \hbar_2(x)$$

$$- \sum_{x \in \Omega_h} \{f(\xi(x), x) - f(0, x)\} \xi(x) \hbar_1(x) \hbar_2(x)$$

$$\geq - \max_{\bar{\Omega}} |f(0, x)| \sum_{\Omega_h} |\xi| \hbar_1 \hbar_2 - \sum_{x \in \Omega_h} \frac{\partial f}{\partial p} (\eta(x), x) \xi^2(x) \hbar_1(x) \hbar_2(x).$$

Here $\eta(x)$ lies between 0 and $\xi(x)$, $\forall x \in \Omega_h$. Let us discard the second term on the right-hand side of the inequality, and apply the Cauchy–Schwartz–Bunyakovsky inequality to the first. Then we get

$$\sum_{x \in \Omega_h} f(\xi(x), x) \xi(x) \hbar_1(x) \hbar_2(x) \geq - \max_{\bar{\Omega}} |f(0, x)| \left(\sum_{\Omega_h} \xi^2 \hbar_1 \hbar_2 \right)^{1/2} \left(\sum_{\Omega_h} \hbar_1 \hbar_2 \right)^{1/2}$$

$$\geq -c_3 \left(\sum_{\Omega_h} \xi^2 \hbar_1 \hbar_2 \right)^{1/2}. \tag{4.21}$$

Here

$$c_3 = \max_{\bar{\Omega}} |f(0, x)| (a_2 b/2)^{1/2}.$$

Combining inequalities (4.19)–(4.21) we obtain

$$\sum_{\Omega_h} \xi P_\xi \geq (4/b^2 + 4/a_1^2) \sum_{\Omega_h} \xi^2 \hbar_1 \hbar_2 - c_3 \left(\sum_{\Omega_h} \xi^2 \hbar_1 \hbar_2 \right)^{1/2}.$$

For all

$$\sigma \geq \sigma_0 = c_3/(4/b^2 + 4/a_1^2)$$

the inequality

$$(4/b^2 + 4/a_1^2)\sigma^2 - c_3 \sigma \geq 0$$

holds. Therefore for an arbitrary net function $\xi(x)$ defined on Ω_h and such that

$$\left(\sum_{\Omega_h} \xi^2\right)^{1/2} = \sigma^0 \left(\min_{\Omega_h} \hbar_1 \hbar_2\right)^{-1/2},$$

we have

$$\left(\sum_{\Omega_h} \xi^2 \hbar_1 \hbar_2\right)^{1/2} \geq \sigma_0$$

and thus

$$\sum_{\Omega_h} \xi P_\xi \geq 0.$$

In virtue of these inequalities according to Lemma 6.1 of §3.6 there is a solution ξ of

$$P_\xi(x) = 0, \qquad x \in \Omega_h,$$

satisfying the homogeneous boundary condition

$$\xi = 0 \quad \text{on } \Gamma_h.$$

Therefore the solution problem (4.14), (4.15) exists. Moreover, from Theorem 4.2 an estimate of the modulus of the solution can be derived. If we assume $v \equiv 0$ then for any fixed $\gamma \in (0, 2]$ we have

$$\max_{\bar{\Omega}_h} |u^h| \leq c_2 \max_{\bar{\Omega}} |d^{2-\gamma} f(0, x)|. \tag{4.22}$$

We again exploit Theorem 4.2, but now we take the function v as a solution of (4.1)–(4.4). We compute the difference

$$L^h u(x) - f(u(x), x).$$

Suppose first that the point $x = (x_1, \varphi(ih)) \in \Omega_h$ has index $i \neq 1$. Then

$$u(x_1, \varphi((i + 1)h)) = u(x) + h_2^+(x) \frac{\partial u}{\partial x_2}(x) + \tfrac{1}{2}(h_2^+(x))^2 \frac{\partial^2 u}{\partial x_2^2}(x)$$

$$+ \tfrac{1}{6}(h_2^+(x))^3 \frac{\partial^3 u}{\partial x_2^3}(x) + \tfrac{1}{6}(h_2^+(x))^{3+\alpha} \xi^+(x).$$

The inequality

$$|\xi^+(x)| \leq \left|\frac{\partial^3 u}{\partial x_2^3}(x) - \frac{\partial^3 u}{\partial x_2^3}(x^+)\right| \bigg/ |x - x^+|^\alpha$$

$$\leq c_1 (\min\{d(x), d(x^+)\})^{-2-\alpha}$$

follows from (4.13). The point x^+ lies on the line segment between the knots $(x_1, \varphi(ih))$ and $(x_1, \varphi((i + 1)h))$. As a result

$$|\xi^+(x)| \leq c_1(\min\{\varphi((i + 1)h), \varphi((i - 1)h),$$
$$b - \varphi((i + 1)h), b - \varphi((i - 1)h)\})^{-2-\alpha}. \quad \square$$

In proving Theorem 4.2 it was established that

$$|\varphi((i \pm 1)h)| \leq 4\varphi(ih), \qquad |b - \varphi((i \pm 1)h)| \leq 4(b - \varphi(ih)), \qquad (4.23)$$

therefore

$$|\xi^+(x)| \leq 4^{2+\alpha}c_1(\min(x_2, b - x_2))^{-2-\alpha}. \qquad (4.24)$$

The relation

$$u(x_1, \varphi((i-1)h)) = u(x) - \hbar_2^-(x)\frac{\partial u}{\partial x_2}(x) + \tfrac{1}{2}(\hbar_2^-(x))^2\frac{\partial^2 u}{\partial x_2^2}(x)$$

$$- \tfrac{1}{6}(\hbar_2^-(x))^3\frac{\partial^3 u}{\partial x_2^3}(x) + \tfrac{1}{6}(\hbar_2^-(x))^{3+\alpha}\xi^-(x) \quad (4.25)$$

and the estimate

$$|\xi^-(x)| \leq 4^{2+\alpha}c_1(\min\{x_2, b - x_2\})^{-2-\alpha}$$

are derived similarly. Both expansions yield

$$-u_{\hat{x}_2\hat{x}_2} = -\frac{\partial^2 u}{\partial x_2^2} - \frac{\hbar_2^+ - \hbar_2^-}{3}\frac{\partial^3 u}{\partial x_2^3}$$

$$- \tfrac{1}{3}\{(\hbar_2^+)^{2+\alpha}\xi^+ + (\hbar_2^-)^{2+\alpha}\xi^-\}/(\hbar_2^+ + \hbar_2^-). \qquad (4.26)$$

The equality

$$\hbar_2^+(x) - \hbar_2^-(x) = \varphi((i+1)h) - 2\varphi(ih) + \varphi((i-1)h) = bh^2(6 - 12\eta_i)$$

follows from the definition of the function φ, where $\eta_i \in ((i-1)h, (i+1)h)$, and therefore

$$|\hbar_2^+(x) - \hbar_2^-(x)|/3 \leq 2bh^2,$$

but $\varphi(h) = b - \varphi(1 - h)$ and, consequently,

$$|\hbar_2^+(x) - \hbar_2^-(x)|/3 \leq h^{1+\alpha}(\min\{x_2, b - x_2\})^{(1-\alpha)/2}(2b)^{(1+\alpha)/2}. \qquad (4.27)$$

From (4.13) it follows that

$$\left|\frac{\partial^3 u}{\partial x_2^3}(x)\right| \leq c_1 d^{-2}(x). \qquad (4.28)$$

Let us estimate the other expression on the right-hand side of (4.26). We have

$$(\hbar_2^+(x))^{2+\alpha}/\{\hbar_2^+(x) + \hbar_2^-(x)\} \leq (\hbar_2^+(x))^{1+\alpha} = h^{1+\alpha}|\varphi'(\xi_i)|^{1+\alpha},$$

where $\xi_i \in (ih, (i+1)h)$. Let $t \leq \tfrac{1}{2}$. Since

$$\varphi'(t) = 6b(t - t^2),$$

then

$$|\varphi'(t)| \leq 6bt.$$

Further

$$\varphi(t) \geq 2t^2 b.$$

Therefore

$$|\varphi'(t)| \leq 3\sqrt{2b}(\varphi(t))^{1/2}.$$

If $t \geq \frac{1}{2}$, then similarly

$$|\varphi'(t)| \leq 3\sqrt{2b}(b - \varphi(t))^{1/2}.$$

Taking into account (4.23) we have

$$|\varphi'(\xi_i)| \leq 3\sqrt{2b}(\min\{\varphi(\xi_i), b - \varphi(\xi_i)\})^{1/2}$$
$$\leq 6\sqrt{2b}(\min\{x_2, b - x_2\})^{1/2}. \qquad (4.29)$$

Combining relations (4.24), (4.25), (4.27)–(4.29) we have

$$\left| -u_{\hat{x}_2 \hat{x}_2}(x) + \frac{\partial^2 u}{\partial x_2^2}(x) \right|$$

$$\leq h^{1+\alpha} c_1 d^{-2}(x)(\min\{x_2, b - x_2\})^{(1-\alpha)/2}(2b)^{(1+\alpha)/2}$$
$$+ h^{1+\alpha}(6\sqrt{2b}(\min\{x_2, b - x_2\})^{1/2})^{1+\alpha} 4^{2+\alpha} c_1(\min\{x_2, b - x_2\})^{-2-\alpha}$$
$$\leq c_4 h^{1+\alpha}(\min\{x_2, b - x_2\})^{(-3-\alpha)/2}. \qquad (4.30)$$

These inequalities were established under the assumption that $x = (x_1, \varphi(ih)) \in \Omega_h$, $i \neq 1$. But if $i = 1$, then using the fact that u belongs to $C^1(\bar{\Omega})$ we have

$$u(x_1, 0) = u(x) - \varphi(h)\zeta^-(x)$$

and

$$u(x_1, \varphi(2h)) = u(x) + (\varphi(2h) - \varphi(h))\zeta^+(x),$$

where

$$|\zeta^{\pm}(x)| \leq c_5.$$

Substitution of these expansions into the difference relation yields

$$|u_{\hat{x}_2 \hat{x}_2}(x)| = 2|\zeta^+(x) - \zeta^-(x)|/\varphi(2h) \leq 4c_5/\varphi(2h).$$

But for the function φ the inequalities $\varphi(2h) \geq \varphi(h)$ and $\varphi(h) \leq 3bh^2$ hold. From these inequalities there follows

$$|u_{\hat{x}_2 \hat{x}_2}(x)| \leq 4c_5/\varphi(h) \leq c_6 h^{1+\alpha}(\varphi(h))^{(-3-\alpha)/2}.$$

Recall that $x_2 = \varphi(h)$; therefore from (4.13) it follows that

$$\left| \frac{\partial^2 u}{\partial x_2^2}(x) \right| \leq c_1 d^{-1}(x) \leq c_1/\varphi(h) \leq c_7 h^{1+\alpha}(\varphi(h))^{(-3-\alpha)/2}$$

Combining the two estimates obtained we get the inequality

$$\left| -u_{\hat{x}_2\hat{x}_2}(x) + \frac{\partial^2 u}{\partial x_2^2}(x) \right| \le c_8 h^{1+\alpha}(\varphi(h))^{(-3-\alpha)/2} \qquad (4.31)$$

if $x_2 = \varphi(h)$.

Let us choose $c_9 = \max\{c_4, c_8\}$. Inequalities (4.30), (4.31) allow us to write

$$\left| -u_{\hat{x}_2\hat{x}_2}(x) + \frac{\partial^2 u}{\partial x_2^2}(x) \right| \le c_9 h^{1+\alpha}(\min\{x_2, b - x_2\})^{(-3-\alpha)/2} \quad \forall\, x \in \Omega_h.$$

(4.32)

Similar manipulations lead to an inequality with respect to another variable:

$$\left| -u_{\hat{x}_1\hat{x}_1}(x) + \frac{\partial^2 u}{\partial x_1^2}(x) \right| \le c_{10} h^{1+\alpha}(\min\{x_2, b - x_2\})^{(-3-\alpha)/2} \quad \forall\, x \in \Omega_h.$$

(4.33)

Combining inequalities (4.32) and (4.33) we have

$$|L^h u(x) + \Delta u(x)| \le c_{11} h^{1+\alpha}(\min\{x_2, b - x_2\})^{(-3-\alpha)/2} \quad \forall\, x \in \Omega_h.$$

From equation (4.1) it follows that $\Delta u = -f(u(x), x)$, and therefore

$$|L^h u(x) - f(u(x), x)| \le c_{11} h^{1+\alpha}(\min\{x_2, b - x_2\})^{(-3-\alpha)/2} \quad \forall\, x \in \Omega_h.$$

Hence on the basis of the estimate of Theorem 4.2 we have:

Theorem 4.3. *Suppose conditions* (4.2), (4.5) *are satisfied for* (4.1), (4.3). *Then we have the estimate*

$$\max_{\bar{\Omega}_h} |u^h - u| \le c_{12} h^{1+\alpha}$$

for the difference solution u^h of (4.14), (4.15) *where α is a fixed number in* (0, 1).

Note that it is impossible to pass to the limit $\alpha \to 1$ in the estimate obtained, since it follows from the above that the constant c_{12} will increase when $\alpha \to 1$.

4.5. On the Diffraction Problem

The diffraction problem leads to a boundary problem for an equation whose coefficients have discontinuities of the first kind. On the lines of discontinuity conservation laws come into play, and lead to certain compatibility conditions. The discontinuity of coefficients is an expression of the fact that the medium consists of two or more heterogeneous materials.

When the lines of discontinuity, are smooth and do not have selfintersection points, and do not intersect the boundary the solution will not have the expected degree of smoothness on the entire domain. But the solution remains smooth in each subregion banded by the lines of discontinuity and the smoothness is preserved up to these lines. But if the lines of discontinuity intersect the boundary of the region then the smoothness of the solutions will change as in the case of corners. This will require special methods when constructing variational-difference schemes. However, we will investigate a case in which one can attain sufficient accuracy by applying the usual variational-difference scheme. To increase the accuracy we use a simple approach involving passing from one norm to a weaker one.

Consider the simplest diffraction problem, when the medium consists of two heterogeneous materials. We will seek the solution of the equation

$$-\frac{\partial}{\partial x_1} a \frac{\partial u}{\partial x_1} - \frac{\partial}{\partial x_2} a \frac{\partial u}{\partial x_2} = f \tag{5.1}$$

in the rectangle $\Omega = \{(x_1, x_2); 0 < x_1 < b_1, 0 < x_2 < b_2\}$ with the boundary condition

$$u = 0 \quad \text{on } \Gamma, \tag{5.2}$$

where Γ is the boundary. The straight line $x_1 = b_1/2$ divides Ω into two rectangles:

$$\Omega_1 = \{(x_1, x_2); 0 < x_1 < b_1/2, 0 < x_2 < b_2\},$$
$$\Omega_2 = \{(x_1, x_2); b_1/2 < x_1 < b_1, 0 < x_2 < b_2\}.$$

Let S be the common part of their boundary. Let the coefficient $a(x)$ be piecewise constant:

$$a(x) = a_i > 0$$
$$\qquad\qquad\qquad\qquad \text{at } x \in \Omega_i.$$
$$f(x) = f_i(x)$$

Two equations in two regions are thus given. To insure unique solvability additional conditions on S are necessary. Assume that the solution satisfies the two compatibility conditions

$$[u(x)]_S = 0, \qquad x \in S, \tag{5.3}$$

$$\left[a(x) \frac{\partial u}{\partial x_1}(x) \right]_S = 0, \qquad x \in S. \tag{5.4}$$

The symbol $[\varphi(x)]_S$ denotes the difference between the limiting values of the function φ computed when we approach the point $x \in S$ from the regions Ω_1 and Ω_2. Assume that

$$f \in C^\alpha(\bar{\Omega}_1) \cap C^\alpha(\bar{\Omega}_2), \qquad \alpha \in (0, 1). \tag{5.5}$$

Under these conditions (see [72]) the solution u exists, is unique and in each subregion belongs to $C^{2+\alpha}(\Omega_i)$ though, generally speaking, it does not belong to $C^{2+\alpha}(\overline{\Omega}_i)$. The problem is thus equivalent to finding a function $u \in \mathring{W}\,^1_2(\Omega)$ satisfying the integral identity

$$\int_\Omega a \sum_{i=1}^2 \frac{\partial u}{\partial x_i} \frac{\partial v}{\partial x_i}\, dx = \int_\Omega fv\, dx \quad \forall\, v \in \mathring{W}\,^1_2(\Omega). \tag{5.6}$$

On the basis of Theorem 1.3 the above-mentioned conditions provide the unique solvability of (5.6) in $\mathring{W}\,^1_2(\Omega)$. Additional smoothness for u is established in [72]. From general properties of the generalized solution it follows that $u \in W^2_2(\Omega'_i)$ for any subregion $\Omega'_i \subset \Omega_i$ which has distance >0 from S. We then have the estimate

$$\|u\|_{W^2_2(\Omega'_i)} \le c_1 \|f\|_{L_2(\Omega_i)}, \tag{5.7}$$

where the constant is independent of f and u, but is dependent on Ω'_i. An application of the compatibility conditions (5.3), (5.4) gives $u \in W^2_2(\Omega_i \cap \Omega')$, $i = 1, 2$ for any subregion $\Omega' \subset \Omega$ which has distance >0 from Γ, and

$$\|u\|_{W^2_2(\Omega' \cap \Omega_1)} + \|u\|_{W^2_2(\Omega' \cap \Omega_2)} \le c_2 \|f\|_{L_2(\Omega)}. \tag{5.8}$$

To decide if u belongs to W^2_2 we need only investigate the solution only the crossing points of S and Γ.

To investigate the smoothness of the solution in this region one can use the methods used for corners given in the previous section. Namely, polar coordinates are introduced near the singular points and a solution is sought in the form of a Fourier series. Note that the solution and the function on the right-hand side are expanded near the singularity in series, using the special system of piecewise smooth functions in [101] using a scalar product with a piecewise constant weight. When one investigates convergence of the series in detail one observes that the angle $\pi/2$, at which Γ and S intersect is special. In fact, the solution u of the diffraction problem will belong to the space $W^2_2(\Omega_i)$ in each subregion Ω_i. This smoothness is enough so that the method of finite elements gives a satisfactory degree of accuracy.

We will give a shorter proof of this statement. Consider the rectangle $\Omega^S = \{(x_1, x_2);\ 0 < x_1 < b_1,\ -b_2 < x_2 < b_2\}$ containing the region Ω. Extend the function a from Ω to Ω^S so that it will be even, and extend the function f so that it will be odd relative to the axis Ox_1:

$$\tilde{a}(x) = \begin{cases} a_1 & \text{if } x_1 < b_1/2, \\ a_2 & \text{if } x_1 > b_1/2, \end{cases}$$

$$\tilde{f}(x) = \begin{cases} f(x_1, x_2) & \text{if } x_2 \ge 0, \\ -f(x_1, -x_2) & \text{if } x_2 < 0. \end{cases} \tag{5.9}$$

Note that $\tilde{f} \in L_2(\Omega^S)$, $\|\tilde{f}\|_{L_2(\Omega^S)} \le \sqrt{2}\,\|f\|_{L_2(\Omega)}$.

Consider the problem of finding a function, $\tilde{u} \in \mathring{W}_2^1(\Omega^S)$ satisfying the integral identity

$$\int_{\Omega^S} \tilde{a} \sum_{i=1}^2 \frac{\partial \tilde{u}}{\partial x_i} \frac{\partial \tilde{v}}{\partial x_i}\, dx = \int_{\Omega^S} \tilde{f}\tilde{v}\, dx \quad \forall\, \tilde{v} \in \mathring{W}_2^1(\Omega^S). \tag{5.10}$$

On the basis of Theorem 1.3 this problem has a unique solution in this space. Moreover, because of the property mentioned above it belongs to $W_2^2(\Omega_i \cap \Omega')$, $i = 1, 2$, for any subregion $\Omega' \subset \Omega$ which has distance > 0 from the boundary of the region Ω^S (but not Ω!). This requirement on Ω' is satisfied if a sufficiently small neighborhood of the point $(b_1/2, 0)$ is used as Ω'. The estimate

$$\|\tilde{u}\|_{W_2^2(\Omega' \cap \Omega_1)} + \|\tilde{u}\|_{W_2^2(\Omega' \cap \Omega_2)} \le c_3 \|\tilde{f}\|_{L_2(\Omega^S)} \le \sqrt{2}c_3 \|f\|_{L_2(\Omega)}. \tag{5.11}$$

then holds.

Now we show that problem (5.10) has the solution

$$\tilde{u}(x) = \begin{cases} u(x_1, x_2) & \text{if } x_2 \ge 0, \\ -u(x_1, -x_2) & \text{if } x_2 < 0. \end{cases} \tag{5.12}$$

which is odd relative to the axis Ox_1. Indeed because $u \in \mathring{W}_2^1(\Omega)$ it follows that $\tilde{u} \in \mathring{W}_2^1(\Omega^S)$. This follows easily from the definition of the space $\mathring{W}_2^1$, given in [94]. Now we take an arbitrary function $\tilde{v} \in \mathring{W}_2^1(\Omega^S)$. For this function the following equalities

$$\int_{\Omega^S} a \sum_{i=1}^2 \frac{\partial \tilde{u}}{\partial x_i} \frac{\partial \tilde{v}}{\partial x_i}\, dx = \int_{\Omega^S; x_2 > 0} a \sum_{i=1}^2 \frac{\partial u}{\partial x_i} \frac{\partial \tilde{v}}{\partial x_i}\, dx$$

$$+ \int_{\Omega^S; x_2 < 0} \tilde{a}\left(-\frac{\partial u}{\partial x_1}(x_1, -x_2)\frac{\partial \tilde{v}}{\partial x_1} + \frac{\partial u}{\partial x_2}(x_1, -x_2)\frac{\partial \tilde{v}}{\partial x_2}\right) dx$$

$$= \int_\Omega a \sum_{i=1}^2 \frac{\partial u}{\partial x_i} \frac{\partial \tilde{v}}{\partial x_i}\, dx - \int_\Omega a\left\{-\frac{\partial u}{\partial x_1} \frac{\partial \tilde{v}}{\partial x_1}(x_1, -x_2)\right.$$

$$\left. + \frac{\partial u}{\partial x_2} \frac{\partial \tilde{v}}{\partial x_2}(x_1, -x_2)\right\} dx = \int_\Omega a \sum_{i=1}^2 \frac{\partial u}{\partial x_i} \frac{\partial w}{\partial x_i}\, dx$$

hold, where the function w is determined by the equality

$$w(x_1, x_2) = \tilde{v}(x_1, x_2) - \tilde{v}(x_1, -x_2). \tag{5.13}$$

From $v \in \mathring{W}_2^1(\Omega^S)$, using the definition given in [94], it follows that $w \in \mathring{W}_2^1(\Omega)$. Therefore the equality

$$\int_{\Omega^S} \tilde{a} \sum_{i=1}^2 \frac{\partial \tilde{u}}{\partial x_i} \frac{\partial \tilde{v}}{\partial x_i}\, dx = \int_\Omega f\tilde{v}\, dx - \int_\Omega f\tilde{v}(x_1, -x_2)\, dx$$

follows from (5.6). Because of the way we have constructed the function $\tilde{f}$ we have

$$\int_{\Omega^s} \tilde{a} \sum_{i=1}^{2} \frac{\partial \tilde{u}}{\partial x_i} \frac{\partial \tilde{v}}{\partial x_i} \, dx = \int_{\Omega^s} \tilde{f} \tilde{v} \, dx. \tag{5.14}$$

Since the function $\tilde{v}$ was arbitrary $\tilde{u}$ is a solution of (5.10). But (5.10) has a unique solution, and therefore it is determined by (5.12), and possesses the property (5.11).

Using estimates (5.7), (5.8), and (5.11), we see that the solution u of (5.6) belongs to $W_2^2(\Omega' \cap \Omega_i)$ for $i = 1, 2$ and for any subregion $\Omega' \subset \Omega$ which has distance >0 from the unique point $(b_1/2, b_2)$. But we can carry out the above procedure with an extension of the region to determine the solution in a neighborhood of this point also. Therefore the following statement

$$u \in W_2^2(\Omega_1) \cap W_2^2(\Omega_2)$$

holds, and we have the estimate

$$\|u\|_{W_2^2(\Omega_1)} + \|u\|_{W_2^2(\Omega_2)} \le c_4 \|f\|_{L_2(\Omega)}. \tag{5.15}$$

This completes our investigation of the smoothness of u. We now construct a varitional-difference scheme based on the Galerkin method.

Introduce a uniform rectangular grid

$$\bar{\Omega}_h = \{(x_1, x_2); x_1 = ih_1, x_2 = jh_2, i = 0, 1, \ldots, 2N_1, j = 0, 1, \ldots, N_2\}$$

with mesh-sizes $h_1 = b_1/(2N_1)$, $h_2 = b_2/N_2$, and with integer $N_i \ge 2$. We take an even number of subdivisions of the interval on the axis Ox_1 so that each net grid will be entirely in one of the regions Ω_i. Denote $\Omega_h = \bar{\Omega}_h \cap \Omega$, and at each point $y = (y_1, y_2) \in \Omega_h$ introduce the trial function $\varphi_y(x) \in \mathring{W}_2^1(\Omega)$. It is equal to 1 at the knot y, zero at all other knots of $\bar{\Omega}_h$ and linear on each open elementary triangle (Figure 4.7) obtained by dividing the rectangular

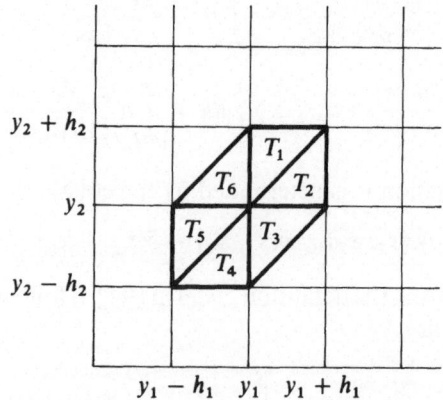

Figure 4.7. Triangulation of the region.

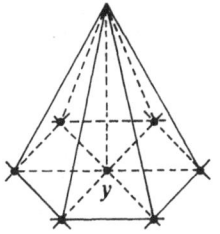

Figure 4.8. The trial function.

regions of the grid diagonally so as to form an acute angle with the axis Oy_1 (Figure 4.8). Direct computation of this function yields

$$\varphi_y(x) = \begin{cases} 1 + (y_2 - x_2)/h_2, & x \in \overline{T}_1, \\ 1 + (y_1 - x_1)/h_1, & x \in \overline{T}_2, \\ 1 + (y_1 - x_1)/h_1 - (y_2 - x_2)/h_2, & x \in \overline{T}_3, \\ 1 - (y_2 - x_2)/h_2, & x \in \overline{T}_4, \\ 1 - (y_1 - x_1)/h_1, & x \in \overline{T}_5, \\ 1 - (y_1 - x_1)/h_1 + (y_2 - x_2)/h_2, & x \in \overline{T}_6, \\ 0 & \text{in all other cases.} \end{cases} \tag{5.16}$$

Denote by H^h the linear subspace whose basis consists of these functions. It is clear that $H^h \subset \overset{\circ}{W}{}^1_2(\Omega)$.

The idea of the finite elements method (see [87, 125]) is to seek an approximate solution in the form of a sum

$$u^h(x) = \sum_{y \in \Omega_h} \alpha_y \varphi_y(x), \tag{5.17}$$

where $\{\alpha_y\}$ is a set of constants, defined from the equalities obtained when we substitute u for u^h and v for φ_y in (5.6):

$$L(u^h, \varphi_z) = (f, \varphi_z) \quad \forall z \in \Omega_h. \tag{5.18}$$

Here the scalar product and the bilinear form are determined by

$$(v, w) = \int_\Omega vw \, dx,$$

$$L(v, w) = \int_\Omega a \left(\frac{\partial v}{\partial x_1} \frac{\partial w}{\partial x_1} + \frac{\partial v}{\partial x_2} \frac{\partial w}{\partial x_2} \right) dx.$$

Note that the form L is linear in both arguments and therefore one can rewrite (5.18) as a system of linear algebraic equations for the coefficients α_y.

$$\sum_{y \in \Omega_h} \alpha_y L(\varphi_y, \varphi_z) = (f, \varphi_z), \quad z \in \Omega_h. \tag{5.19}$$

The number of unknowns in this system coincides with the number of equations, and is equal to $(2N_1 - 1)(N_2 - 1)$.

To implement this on a computer it is useful to write this system in matrix form. To do this we renumber the knots from Ω_h as follows: $(h_1, h_2), (h_1, 2h_2),$ $\ldots, (h_1, (N_2 - 1)h_2), (2h_1, h_2), \ldots, ((2N_1 - 1)h_1, (N_2 - 1)h_2)$. This ordering numbers the equations and unknowns in system (5.19). The matrix of this system is written in the following (block tridiagonal) form:

$$\begin{bmatrix} B_1 & A_2^T & & & 0 \\ A_2 & \ddots & \ddots & & \\ & \ddots & \ddots & \ddots & \\ & & \ddots & \ddots & A_{2N_1-1}^T \\ 0 & & & A_{2N_1-1} & B_{2N_1-1} \end{bmatrix}$$

Here A_i, B_i are $(N_2 - 1) \times (N_2 - 1)$ matrices, where the B_i are tridiagonal matrices, and the A_i are diagonal matrices:

$$B_i = \begin{cases} a_1\left(\dfrac{h_1}{h_2} D + \dfrac{2h_2}{h_1} I\right), & i = 1, 2, \ldots, N_1 - 1, \\[2mm] \dfrac{a_1 + a_2}{2}\left(\dfrac{h_1}{h_2} D + \dfrac{2h_2}{h_1} I\right), & i = N_1, \\[2mm] a_2\left(\dfrac{h_1}{h_2} D + \dfrac{2h_2}{h_1} I\right), & i = N_1 + 1, \ldots, 2N_1 - 1. \end{cases}$$

$$A_i = \begin{cases} -a_1 \dfrac{h_2}{h_1} I, & i = 2, 3, \ldots, N_1, \\[2mm] -a_2 \dfrac{h_2}{h_1} I, & i = N_1 + 1, \ldots, 2N_1 - 1. \end{cases}$$

Here I is the unit matrix and D is the tridiagonal $(N_2 - 1) \times (N_2 - 1)$ matrix:

$$I = \begin{bmatrix} 1 & & & 0 \\ & 1 & & \\ & & \ddots & \\ & & & \ddots 1 \\ 0 & & & 1 \end{bmatrix}, \qquad D = \begin{bmatrix} 2 & -1 & & & 0 \\ -1 & \ddots & \ddots & & \\ & \ddots & \ddots & \ddots & \\ & & \ddots & \ddots & -1 \\ 0 & & & -1 & 2 \end{bmatrix}.$$

From the matrix form of (5.19) it follows that the matrix is diagonal predominant and is indecomposable (see [11]), and therefore is nonsingular. Thus (5.19) has a unique solution $\{\alpha_y\}$ which uniquely defines the function u^h.

We will investigate the accuracy of this approximate solution. To do this we introduce $\tilde{u}^h$, a piecewise linear interpolant of the function u from H^h, which we define by the formula

$$\tilde{u}^h(x) = \sum_{y \in \Omega_h} u(y)\varphi_y(x).$$

Note that on the basis of the properties of the functions φ_y the equality $\tilde{u}^h(x) = u(x) \ \forall \ x \in \overline{\Omega}_h$ is valid, which justifies the name "interpolant" for the function $\tilde{u}^h$.

Lemma 5.1. *We have*

$$L(u^h - u, u^h - u) = L(u^h - u, \tilde{u}^h - u),$$

where u is the solution of (5.8).

PROOF. From the integral equality (5.6) the identities

$$L(u, \varphi_z) = (f, \varphi_z), \qquad z \in \Omega_h \tag{5.20}$$

follow. Summing up them with weights α_z we have

$$L(u, u^h) = (f, u^h) \tag{5.21}$$

because the scalar product and the form L are both linear in the second argument. If instead of (5.20) we take (5.18) then instead of (5.21) we have

$$L(u^h, u^h) = (f, u^h).$$

Subtracting this equality from (5.21) we come to

$$L(u^h - u, u^h) = 0.$$

The equality

$$L(u^h - u, \tilde{u}^h) = 0$$

is proved similarly. Whence it follows that

$$L(u^h - u, u^h) = L(u^h - u, \tilde{u}^h).$$

Subtracting $L(u^h - u, u)$ from both sides of this equality we arrive at the required result. □

We will obtain a simple inequality which relates the accuracy of u^h with the accuracy of the interpolant $\tilde{u}^h$.

Lemma 5.2. *We have*

$$|u^h - u| \leq |\tilde{u}^h - u| \max\{a_1, a_2\}/\min\{a_1, a_2\}.$$

PROOF. Let us first derive two auxiliary inequalities. We apply the Cauchy–Schwartz–Bunyakovsky inequality to the expression $L(v, w)$ and obtain

$$L(v, w) \leq \left(\int_\Omega a \left\{ \left(\frac{\partial v}{\partial x_1} \right)^2 + \left(\frac{\partial v}{\partial x_2} \right)^2 \right\} dx \right)^{1/2}$$

$$\times \left(\int_\Omega a \left\{ \left(\frac{\partial w}{\partial x_1} \right)^2 + \left(\frac{\partial w}{\partial x_2} \right)^2 \right\} dx \right)^{1/2}$$

In the integral we substitute the expression $a(x)$ for the larger $\max\{a_1, a_2\}$ and use the definition of the norms $|v|$ and $|w|$. We have

$$L(v, w) \leq \max\{a_1, a_2\}|v||w|. \tag{5.22}$$

Now we estimate $L(v, v)$. From the definition of the form L it follows that

$$L(v, v) = \int_\Omega a\left\{\left(\frac{\partial v}{\partial x_1}\right)^2 + \left(\frac{\partial v}{\partial x_2}\right)^2\right\} dx.$$

Substitute $a(x)$ for the smaller expression $\min\{a_1, a_2\}$ and use the definition of norm $|v|$. We have

$$L(v, v) \geq \min\{a_1, a_2\}|v|^2. \tag{5.23}$$

Apply the inequalities (5.22), (5.23) to both sides of the identity of Lemma 5.1, assuming $v = u^h - u$, $w = \tilde{u}^h - u$. We have

$$\min\{a_1, a_2\}|u^h - u|^2 \leq L(u^h - u, u^h - u) = L(u^h - u, \tilde{u}^h - u)$$
$$\leq \max\{a_1, a_2\}|u^h - u||\tilde{u}^h - u|. \tag{5.24}$$

If $|u^h - u| = 0$, then the lemma is proved. Let $|u^h - u| \neq 0$. Divide inequality (5.24) by $\min\{a_1, a_2\}|u^h - u|$. Then we arrive at the required estimate. Lemma 5.2 is proved. $\qquad\square$

Thus the error of u^h can be estimated from the error in $\tilde{u}^h$.

Theorem 5.3. *Suppose conditions (5.5) are satisfied for (5.6). Then for the approximate solution u^h of the Galërkin method (5.17), (5.18) we have the estimate*

$$|u^h - u| \leq c_5 h(\|u\|_{W_2^2(\Omega_1)} + \|u\|_{W_2^2(\Omega_2)}), \tag{5.25}$$

where $h = \max\{h_1, h_2\}$.

PROOF. First let us estimate $|\tilde{u}^h - u|$. We use the following inequality from [100]:

$$\int_{\Omega_i} \left(\frac{\partial \tilde{u}^h}{\partial x_j} - \frac{\partial u}{\partial x_j}\right)^2 dx \leq 4h^2 \|u\|_{W_2^2(\Omega_i)}^2, \qquad j = 1, 2,$$

where $i = 1, 2$. Summing over i we have

$$|\tilde{u}^h - u|^2 \leq 8h^2(\|u\|_{W_2^2(\Omega_1)} + \|u\|_{W_2^2(\Omega_2)}).$$

But from the inequality of Lemma 5.2 it follows that

$$|u^h - u| \leq 2\sqrt{2}h \frac{\max\{a_1, a_2\}}{\min\{a_1, a_2\}}(\|u\|_{W_2^2(\Omega_1)}^2 + \|u\|_{W_2^2(\Omega_2)}^2)^{1/2}.$$

Since the quantities $\|u\|_{W_2^2(\Omega_i)}$ are finite, we have (5.25). The theorem is proved. $\qquad\square$

Thus the accuracy of the solution u^h in the norm $|u^h - u|$ has been found. Applying the embedding theorem from $\mathring{W}^1_2(\Omega)$ to $L_2(\Omega)$ (see [73]) we can estimate the same error in the norm of $L_2(\Omega)$. Indeed, from the embedding theorem it follows that

$$\|u^h - u\|_{L_2(\Omega)} \leq c_6 |u^h - u|.$$

Applying (5.25) we have

$$\|u^h - u\|_{L_2(\Omega)} \leq c_7 h.$$

However, this estimate does not reflect the efficacy of the Galërkin method since the true error in the $L_2(\Omega)$ norm is of the second order in h. Let us prove this, using the approach described in [102, 7, 99].

Consider (5.6) with different right-hand side. We seek the solution w satisfying integral identity

$$L(w, v) = (u^h - u, v) \quad \forall v \in \mathring{W}^1_2(\Omega). \tag{5.26}$$

Since $(u^h - u) \in L_2(\Omega)$ then as in the beginning of this section if follows that $w \in W^2_2(\Omega_1) \cap W^2_2(\Omega_2)$, and we have the estimate

$$\|w\|_{W^2_2(\Omega_1)} + \|w\|_{W^2_2(\Omega_2)} \leq c_4 \|u^h - u\|_{L_2(\Omega)}. \tag{5.27}$$

By the Galërkin method (5.17), (5.18) we can find an approximate solution $w^h \in H^h$ of this problem. Theorem 5.3 is also valid for this solution and thus

$$|w^h - w| \leq c_5 h(\|w\|_{W^2_2(\Omega_1)} + \|w\|_{W^2_2(\Omega_2)}).$$

Applying (5.27) we have

$$|w^h - w| \leq c_4 c_5 h \|u^h - u\|_{L_2(\Omega)}. \tag{5.28}$$

In (5.26) we put $v = u^h - u$ and use the symmetry of the bilinear form L in its arguments. We have

$$\|u^h - u\|^2_{L_2(\Omega)} = L(w, u^h - u) = L(u^h - u, w). \tag{5.29}$$

Since $w^h \in H^h$ then we get

$$L(u^h - u, w^h) = 0.$$

Applying this to (5.29) we have

$$\|u^h - u\|^2_{L_2(\Omega)} = L(u^h - u, w - w^h).$$

Applying (5.22), (5.25), and (5.28) we have

$$\|u^h - u\|^2_{L_2(\Omega)} \leq \max\{a_1, a_2\} |u^h - u| |w - w^h|$$

$$\leq \max\{a_1, a_2\} c_4 c_5^2 h^2 \|u^h - u\|_{L_2(\Omega)} (\|u\|_{W^2_2(\Omega_1)} + \|u\|_{W^2_2(\Omega_2)}).$$

Thus if $\|u^h - u\|_{L_2(\Omega)} \neq 0$ then we divide the inequality by this quantity and apply (5.15). Thus

$$\|u^h - u\|_{L_2(\Omega)} \leq c_8 h^2 \|f\|_{L_2(\Omega)}. \tag{5.30}$$

If $\|u^h - u\|_{L_2(\Omega)} = 0$ then (5.30) is valid.

Thus, the error of the approximate solution u^h of (5.1)–(5.5), found by the Galërkin method (5.17), (5.18), is of order h^2 in the $L_2(\Omega)$ norm.

4.6. On the Separation of Singularities

In the present section we will discuss one of the approaches for separating the singular parts of the solution. These singularities occur in the neighborhood of the corners of the domain. In addition to refining the difference net near a singular point one can apply the "additive method" for two-dimensional problems. The main idea is that in the Ritz and Bubnov–Galërkin methods certain special functions which describe in detail the behavior of the irregular part of the solution near a corner (see [101, 125, 8, 13]) are added to the usual local trial functions (for example, to the Courant piecewise linear functions on a triangle (see [22]). Most often these functions are the main terms of an asymptotic expansion (4.7), adjusted so as to satisfy the boundary conditions of the problem. In some cases in order to attain an accuracy of order h^2 we have to use the terms of an asymptotic expansion following the main terms. We should note that in order to apply this approach one must know the required terms of the asymptotic expansion to within a constant multiple. These multiples become additional unknowns in the finite-dimensional algebraic system in the Ritz and Bubnov–Galërkin methods. In the two previous sections we have used one approach for finding the terms of the asymptotic expansion. This approach can be used to construct singular trial functions.

We will now give another approach for dealing with singularities at corners. Let us decompose the initial problem into several problems. One of these problems will have consistent data and a smooth solution, while the others will have noncoherent data, but simpler domains of definition will be chosen so that polar coordinates can be conveniently introduced, thereby simplifying numerical and analytical calculations. A circular sector with a given opening angle would be an example of such a domain. The problem thus obtained will be used to construct an overall solution by the Schwarz alternating process. This construction has been discussed in [141]. This work solves Laplace's equation on a polygon using the Schwarz alternating process with both polar and rectangular difference nets.

For simplicity we will consider a domain with one corner.

Let a bounded Lipschitz domain Ω have a boundary Γ which is everywhere smooth except at the point $(0, 0) \in \Gamma$. In the neighborhood of this point the

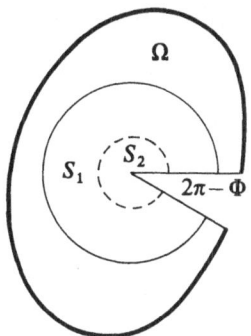

Figure 4.9. A domain with a corner.

boundary consists of two straight lines intersecting at an angle $\varphi \in (0, 2\pi]$ (Figure 4.9). Let us consider the equation

$$-\Delta u = f \qquad (6.1)$$

in the domain Ω with boundary condition

$$u = g \quad \text{on } \Gamma. \qquad (6.2)$$

Assume that

$$f \in C^{2+\alpha}(\overline{\Omega}) \qquad (6.3)$$

with some constant $\alpha \in (0, 1)$. Denote by Γ_1, the part of Γ, which results when we delete the straight lines adjacent to $(0, 0)$. This portion of the boundary we assume is sufficiently smooth:

$$\Gamma_1 \in C^{4+\alpha}. \qquad (6.4)$$

We require that

$$g \in C^{4+\alpha}(\Gamma \backslash (0, 0)). \qquad (6.5)$$

Let us also require that the function g itself, together with its derivatives along Γ up to the fourth order, have finite limits at the point $(0, 0)$.

A simple criterian for the data to ensure the smoothness of the solution of the Dirichlet problem is known when the angle is one of $\pi/j, j = 2, 3, \ldots$ (see [43]). If the angle is not one of these values then a more complicated integral condition arises. Therefore in the present section we do not impose any coherency conditions between f and g. Note that the solution of the given problem can even be noncontinuous on $\overline{\Omega}$. Nevertheless the solution of (6.1), (6.2) has the property

$$u \in C^{4+\alpha}(\overline{\Omega}_1) \qquad (6.6)$$

for any subdomain $\Omega_1 \subset \Omega$ which is at a distance > 0 from $(0, 0)$ (see [73]).

Introduce polar coordinates (r, φ) with center at $(0, 0)$. We will choose the number $a < 0$ so that the intersection of the domain Ω with the circle of radius a and center at $(0, 0)$ is a sector:

$$S_1 = \{(r, \varphi); 0 < r < a, 0 < \varphi < \phi\}.$$

We denote the circular sector of radius $a/6$ by S_2:

$$S_2 = \{(r, \varphi); 0 < r < a/6, 0 < \varphi < \phi\}.$$

It will become clear that it is desirable that we choose a as large as possible.

Assume that the domain Ω is contained in the square $\{(x, y); -b < x, y < b\}$. Cover it with an orthogonal grid with mesh-size $h = b/N$, formed by the lines $x_i = ih$ and $y_j = jh$, where $i, j = -N, \ldots, N$. We denote the set of all inner knots of the grid by Ω_h and the set of all boundary knots (see §4.2.1) by Γ_h. Let

$$D_h = \Omega_h \backslash S_2.$$

We will define a set of boundary knots ∂D_h for the grid domain D_h in the following way. Given a point of D_h the four neighboring knots will belong either to Γ_h or to Ω_h (neighbors being situated on lines parallel to the coordinate axes). Denote by $\bar{D}_h$ the union of the set of knots D_h and all knots adjoining them. Then it is natural to assume

$$\partial D_h = \bar{D}_h \backslash D_h.$$

Note that all knots $\bar{D}_h$ are not less than $a/6 - h$ distant from the point $(0, 0)$. Since we are interested in the behavior of the approximate solution as $h \to 0$ we will further assume that h is sufficiently small, for example:

$$a/6 - h \geq a/8.$$

At each knot $(x, y) \in D_h$ we define a fivepoint difference operator

$$L_h v(x, y) \equiv \frac{1}{h} \left[\frac{1}{\delta_1} v(x + \delta_1, y) + \frac{1}{\delta_2} v(x - \delta_2, y) + \frac{1}{\theta_1} v(x, y + \theta_1) \right.$$
$$\left. + \frac{1}{\theta_2} v(x, y - \theta_2) - \left(\frac{1}{\delta_1} + \frac{1}{\delta_2} + \frac{1}{\theta_1} + \frac{1}{\theta_2} \right) v(x, y) \right], \quad (6.7)$$

where $(x + \delta_1, y)$, $(x - \delta_2, y)$, $(x, y + \theta_1)$, $(x, y - \theta_2)$ are the points of $\bar{D}_h$, which are the neighbors of the knot (x, y). If all four distances $\delta_1, \delta_2, \theta_1, \theta_2$ are equal to h, then the knot (x, y) will be called regular, as before. The set of all regular knots on D_h will be denoted by D_h^r. All other knots of D_h will be called irregular, and we take $D_h^{ir} = D_h \backslash D_h^r$.

Since all knots of $\bar{D}_h$ are at a distance not less than $a/8$ from the point $(0, 0)$, then the difference operator L_h acting on the solution u provides a second-order approximation at the regular knots and a zeroth-order approximation at the irregular ones. In fact, it is easy to verify that

$$L_h u - \Delta u = \alpha^h \quad \text{on } D_h,$$

where

$$|\alpha^h| \leq h^2 c_1 \quad \text{on } D_h^r \quad (6.8)$$

and

$$|\alpha^h| \leq c_2 \quad \text{on } D_h^{ir} \tag{6.9}$$

with

$$c_1 = \frac{1}{12} \left(\max_{\bar{\Omega}_2} \left| \frac{\partial^4 u}{\partial x^4} \right| + \max_{\bar{\Omega}_2} \left| \frac{\partial^4 u}{\partial y^4} \right| \right), \tag{6.10}$$

$$c_2 = \frac{1}{2} \left(\max_{\bar{\Omega}_2} \left| \frac{\partial^2 u}{\partial x^2} \right| + \max_{\bar{\Omega}_2} \left| \frac{\partial^2 u}{\partial y^2} \right| \right). \tag{6.11}$$

Here $\bar{\Omega}_2$ is the set of all points of $\bar{\Omega}$ which are at a distance of not less than $a/8$ from the point $(0, 0)$.

Consider the difference problem

$$-L_h v^h = f \quad \text{on } D_h, \tag{6.12}$$

$$v^h = g \quad \text{on } \partial D_h \cap \Gamma_h, \tag{6.13}$$

$$v^h = w^h \quad \text{on } \partial D_h \cap \Omega_h, \tag{6.14}$$

where w^h is an arbitrary grid function defined on $\partial D_h \cap \Omega_h$. Since the difference operator L_h satisfies the maximum principle (see [113]) the homogeneous problem, corresponding to the linear algebraic system (6.12)–(6.14) has a unique solution for any right-hand side functions.

Lemma 6.1. *The estimate*

$$\max_{\bar{D}_h} |v^h - u| \leq c_3 h^2 + \max_{\partial D_h \cap \Omega_h} |w^h - u|$$

holds for (6.12)–(6.14).

PROOF. Let us represent the solution v^h in the form $v_1^h + v_2^h$ where the functions v_i^h are solutions of the following problems

$$-L_h v_1^h = 0 \quad \text{on } D_h,$$

$$v_1^h = 0 \quad \text{on } \partial D_h \cap \Gamma_h, \tag{6.15}$$

$$v_1^h = w^h - u \quad \text{on } \partial D_h \cap \Omega_h;$$

$$-L_h v_2^h = f \quad \text{on } D_h$$

$$v_2^h = g \quad \text{on } \partial D_h \cap \Gamma_h, \tag{6.16}$$

$$v_2^h = u \quad \text{on } \partial D_h \cap \Omega_h.$$

The maximum principle is valid for the first problem. From this there follows

$$\max_{\bar{D}_h} |v_1^h| \leq \max_{\partial D_h \cap \Omega_h} |w^h - u|. \tag{6.17}$$

The equalities

$$-L_h u = f + \alpha^h \quad \text{on } D_h,$$

$$u = g \qquad \text{on } \partial D_h \cap \Gamma_h.$$

hold for the exact solution u.

Therefore the difference $\varepsilon^h = u - v_2^h$ is a solution of the problem

$$-L_h \varepsilon^h = \alpha^h \quad \text{on } D_h,$$

$$\varepsilon^h = 0 \quad \text{on } \partial D_h. \tag{6.18}$$

Consider the following function on $\bar{D}_h$:

$$\rho^h(x, y) = c_1 h^2 (2b^2 - x^2 - y^2)/4 + \begin{cases} c_2 h^2, & \text{if } (x, y) \in D_h, \\ 0, & \text{if } (x, y) \in \partial D_h. \end{cases}$$

It is easy to verify that

$$\rho^h \geq 0 \qquad \text{on } \partial D_h,$$

$$-L_h \rho^h \geq \begin{cases} c_1 h^2 & \text{on } D_h^r, \\ c_2 & \text{on } D_h^{ir}. \end{cases}$$

From (6.8) and (6.9) we conclude that

$$-L_h \rho^h \geq |\alpha^h| \quad \text{on } D_h.$$

Thus by the comparison theorem (see [113]) we have

$$|\varepsilon^h| \leq \rho^h \quad \text{on } \bar{D}_h.$$

From the definition of the function ρ^h we have the estimate

$$\max_{\bar{D}_h} |\varepsilon^h| \leq b^2 c_1 h^2 / 2 + c_2 h^2.$$

Assuming

$$c_3 = c_2 + b^2 c_1 / 2,$$

we get the inequality

$$\max_{\bar{D}_h} |v^h - u| \leq \max_{\bar{D}_h} |v_2^h - u| + \max_{\bar{D}_h} |v_1^h| \leq c_3 h^2 + \max_{\partial D_h \cap \Omega_h} |w^h - u|.$$

The lemma is proved. $\square$

Consider the following problem in the sector S_1:

$$-\Delta w = f \quad \text{on } S_1,$$

$$w = g \quad \text{on } \partial S_1 \cap \Gamma, \tag{6.19}$$

$$w = u \quad \text{on } \partial S_1 \backslash \Gamma.$$

It is obvious that the function u is its solution. Now we pass to polar co-ordinates. Then (6.19) will have the form†

$$-\frac{\partial^2 w}{\partial r^2} - \frac{1}{r}\frac{\partial w}{\partial r} - \frac{1}{r^2}\frac{\partial^2 w}{\partial \varphi^2} = f, \qquad r \in (0, a), \quad \varphi \in (0, \phi), \qquad (6.20)$$

$$\left.\begin{array}{l} w(r, 0) = g(r, 0) \\ w(r, \phi) = g(r, \phi) \end{array}\right\} \qquad r \in (0, a), \qquad (6.21)$$

$$w(a, \varphi) = u(a, \varphi), \qquad \varphi \in (0, \phi). \qquad (6.22)$$

If the boundary conditions (6.21) are not homogeneous we replace the unknown function by

$$w_1(r, \varphi) = w(r, \varphi) - \left(1 - \frac{\varphi}{\phi}\right)g(r, 0) - \frac{\varphi}{\phi} g(r, \phi). \qquad (6.23)$$

Note that the function w_1 is as smooth in φ as w is. The right-hand side of equation (6.20) and the boundary condition (6.22) are transformed by this substitution to

$$-\frac{\partial^2 w_1}{\partial r^2} - \frac{1}{r}\frac{\partial w_1}{\partial r} - \frac{1}{r^2}\frac{\partial^2 w_1}{\partial \varphi^2} = f_1, \qquad r \in (0, a), \quad \varphi \in (0, \phi), \quad (6.24)$$

$$w_1(r, 0) = w_1(r, \phi) = 0, \qquad r \in (0, a), \qquad (6.25)$$

$$w_1(a, \varphi) = z_1(a, \varphi), \qquad \varphi \in (0, \phi), \qquad (6.26)$$

where

$$z_1(a, \varphi) = u(a, \varphi) - \left(1 - \frac{\varphi}{\phi}\right)g(a, 0) - \frac{\varphi}{\phi} g(a, \phi), \qquad (6.27)$$

$$f_1(r, \varphi) = f(r, \varphi) + \left(1 - \frac{\varphi}{\phi}\right)\left\{\frac{\partial^2 g}{\partial r^2}(r, 0) + \frac{1}{r}\frac{\partial g}{\partial r}(r, 0)\right\}$$

$$+ \frac{\varphi}{\phi}\left\{\frac{\partial^2 g}{\partial r^2}(r, \phi) + \frac{1}{r}\frac{\partial g}{\partial r}(r, \phi)\right\}. \qquad (6.28)$$

We apply the method of separation of variables to solve (6.24)–(6.26). Let $F_k(r)$ be the Fourier coefficients of the function $f_1(r, \varphi)$:

$$f_1(r, \varphi) = \sum_{k=1}^{\infty} F_k(r) \sin k\lambda\varphi,$$

where

$$\lambda = \pi/\phi, \qquad (6.29)$$

† When the change of variables is made we do not change notation for that function.

which are calculated from

$$F_k(r) = \frac{2}{\phi} \int_0^\phi f_1(r, \varphi) \sin k\lambda\varphi \, d\varphi. \tag{6.30}$$

We also expand the solution w_1 of (6.24)–(6.26) in a Fourier series

$$w_1(r, \varphi) = \sum_{k=1}^\infty W_k(r) \sin k\lambda\varphi. \tag{6.31}$$

The coefficients W_k are the solutions of the equation

$$-W_k'' - \frac{1}{r} W_k' + \frac{k^2\lambda^2}{r^2} W_k = F_k, \qquad r \in (0, a), \tag{6.32}$$

with boundary conditions

$$|W_k(0)| < \infty, \qquad W_k(a) = z_k, \tag{6.33}$$

where

$$z_k = \frac{2}{\phi} \int_0^\phi z(a, \varphi) \sin k\lambda\varphi \, d\varphi. \tag{6.34}$$

The solution of this boundary problem has the form

$$W_k(r) = \frac{-r^{\lambda k}}{2\lambda k a^{2\lambda k}} \int_0^a F_k(\rho)\rho^{\lambda k+1} \, d\rho + \frac{r^{\lambda k}}{2\lambda k} \int_r^a F_k(\rho)\rho^{1-\lambda k} \, d\rho$$

$$+ \frac{r^{-\lambda k}}{2\lambda k} \int_0^r F_k(\rho)\rho^{\lambda k+1} \, d\rho + \frac{r^{\lambda k}}{a^{\lambda k}} z_k. \tag{6.35}$$

We are interested only in the first M coefficients since we can use an initial segment of the Fourier

$$w_2^h(r, \varphi) = \sum_{k=1}^M W_k(r) \sin k\lambda\varphi \tag{6.36}$$

as an approximate solution of (6.24)–(6.26).

Let us clarify how the number of terms M in (6.36) will influence the degree to which w_2^h approximates the solution w_1. From the orthogonality of the functions $\sin k\lambda\varphi$ for different k it follows that

$$W_k(r) = \frac{2}{\phi} \int_0^\phi w_1(r, \varphi) \sin k\lambda\varphi \, d\varphi. \tag{6.37}$$

We note that the function w_1 can be differentiated many times with respect to φ. We will mainly be interested in the values of the functions W_k and w_1 when $r \in [a/8, a/6]$, i.e., when the function w_1 has four continuous bounded derivatives. In this case the following estimate

$$|W_k(r)| \le d_1/k^3, \qquad k = 1, 2, \ldots, \quad \forall \, r \in [a/8, a/6] \tag{6.38}$$

is guaranteed (see [37]). This inequality allows us to estimate the remainder in the Fourier series for the function w_1. We have:

$$|w_1(r, \varphi) - w_2^h(r, \varphi)| = \left| \sum_{k=M+1}^{\infty} W_k(r) \sin k\lambda\varphi \right|$$

$$\leq \sum_{k=M+1}^{\infty} |W_k(r)| \leq d_1 \sum_{k=M+1}^{\infty} k^{-3} \leq d_1/(2M^2) \quad \forall r \in [a/8, a/6]. \quad (6.39)$$

The last part of inequality (6.39) is proved using the estimate

$$\sum_{k=M+1}^{\infty} k^{-p-1} \leq 1/(pM^p), \qquad p > 0, \quad (6.40)$$

(see [36]). In order to insure that the truncated Fourier series and the difference scheme will be of the same order of accuracy it is necessary to assume that M is approximately equal to N. For simplicity we assume $M = N$. Since $h = b/N$ inequality (6.39) becomes

$$|w_1(r, \varphi) - w_2^h(r, \varphi)| \leq \frac{d_1}{2b^2} h^2. \quad (6.41)$$

The computation of the function w_2^h involves calculating integrals, which can be done via quadrature. Let us do so. We will replace the function f_1 with a piecewise-bilinear extension $\tilde{f}_1$ (see [100]) with respect to the values at the knots (ρ_i, ψ_j), where $\rho_i = ia/N$, $\psi_j = j\phi/N$, $i, j = 0, 1, \ldots, N$. We obtain within each elementary rectangle a function which is linear with respect to ρ and ψ:

$$\tilde{f}_1(\rho, \psi) = \frac{N^2}{a\phi} \{(\rho_{i+1} - \rho)[(\psi_{j+1} - \psi)f_1(\rho_i, \psi_j) + (\psi - \psi_j)f_1(\rho_i, \psi_{j+1})]$$

$$+ (\rho - \rho_i)[(\psi_{j+1} - \psi)f_1(\rho_{i+1}, \psi_j) + (\psi - \psi_j)(\rho_{i+1}, \psi_{j+1})]\}, \quad (6.42)$$

if $\psi \in [\psi_j, \psi_{j+1}]$, $\rho \in [\rho_i, \rho_{i+1}]$.

After substituting $\tilde{f}_1$ for f_1 in (6.30) we have

$$\tilde{F}_k(\rho) = \frac{2}{\phi} \int_0^{\phi} \tilde{f}_1(\rho, \psi) \sin k\lambda\psi \, d\psi. \quad (6.43)$$

This approximation is exact at the knots ρ_i since

$$\tilde{F}_k(\rho) = [(\rho_{i+1} - \rho)\tilde{F}_k(\rho_i) + (\rho - \rho_i)\tilde{F}_k(\rho_{i+1})]N/a$$

if $\rho \in [\rho_i, \rho_{i+1}]$. Therefore, we should calculate the integrals (6.43) only for $\rho = \rho_j$, where $j = 0, 1, \ldots, N$.

Similarly in (6.34) replace the function z by the piecewise-linear extension

$$\tilde{z}(\psi) = \frac{N}{\phi} [(\psi_{j+1} - \psi)z(\psi_j) + (\psi - \psi_j)z(\psi_{j+1})]. \quad (6.44)$$

Then the formula

$$\tilde{z}_k = \frac{2}{\phi} \int_0^\phi \tilde{z}(\psi) \sin k\lambda\psi \, d\psi \qquad (6.45)$$

is (as (6.42)) a quadrature in the form of the trapezoidal rule with weights $\sin k\lambda\psi$.

An approximate calculation of the Fourier coefficients can thus be made:

$$\tilde{W}_k(r) = \frac{-r^{\lambda k}}{2\lambda k a^{2\lambda k}} \int_0^a \tilde{F}_k(\rho)\rho^{\lambda k+1} \, d\rho + \frac{r^{\lambda k}}{2\lambda k} \int_r^a \tilde{F}_k(\rho)\rho^{1-\lambda k} \, d\rho$$

$$+ \frac{r^{-\lambda k}}{2\lambda k} \int_0^r \tilde{F}_k(\rho)\rho^{\lambda k+1} \, d\rho + \frac{r^{\lambda k}}{a^{\lambda k}} \tilde{z}_k. \qquad (6.46)$$

Using these coefficients we get the function

$$w_3^h(r, \varphi) = \sum_{k=1}^{N} \tilde{W}_k(r) \sin k\lambda\varphi. \qquad (6.47)$$

instead of w_2^h. Let us estimate the difference between w_2^h and w_3^h.

From [14, 100] we know that the linear and bilinear interpolations given have second-order accuracy for functions with bounded second derivatives. Thus

$$|f_1(\rho, \psi) - \tilde{f}_1(\rho, \psi)| \le d_2 h^2 \quad \forall \, (\rho, \psi) \in [0, a] \times [0, \phi],$$

$$|z(\psi) - \tilde{z}(\psi)| \le d_3 h^2 \quad \forall \, \psi \in [0, \phi].$$

From (6.34) and (6.45) it follows that

$$|z_k - \tilde{z}_k| \le d_3 h^2 \frac{2}{\phi} \int_0^\phi |\sin k\lambda\psi| \, d\psi \le 2d_3 h^2, \qquad k = 1, 2, \ldots$$

Similarly,

$$|F_k(\rho) - \tilde{F}_k(\rho)| \le 2d_2 h^2, \qquad \rho \in [0, a], \quad k = 1, 2, \ldots$$

Let us subtract (6.46) from (6.35) and use the estimates we have obtained. We have

$$|W_k(r) - \tilde{W}_k(r)| \le \frac{d_2 h^2}{2\lambda k} \left(\frac{r^{\lambda k}}{a^{2\lambda k}} \int_0^a \rho^{\lambda k+1} \, d\rho \right.$$

$$\left. + r^{\lambda k} \int_r^a \rho^{1-\lambda k} \, d\rho + r^{-\lambda k} \int_0^r \rho^{\lambda k+1} \, d\rho \right) + \frac{r^{\lambda k}}{a^{\lambda k}} 2d_3 h^2.$$

Let us further estimate this difference for $r \in [a/8, a/6]$. The first term in the brackets is estimated simply

$$\frac{r^{\lambda k}}{a^{2\lambda k}} \int_0^a \rho^{\lambda k+1} \, d\rho = r^{\lambda k} a^{2-\lambda k}/(\lambda k + 2) \le a^2/(k\lambda 6^{\lambda k}) \le a^2/(k\lambda).$$

When estimating the second term two cases arise. If $k = 2/\lambda$, then

$$r^{\lambda k} \int_r^a \rho^{1-\lambda k} \, d\rho = r^2(\ln a - \ln r) \le (a^2 \ln 8)/36.$$

Otherwise $\lambda k \ne 2$, and therefore

$$r^{\lambda k} \int_r^a \rho^{1-\lambda k} \, d\rho = r^{\lambda k} |a^{2-\lambda k} - r^{2-\lambda k}|/|2 - \lambda k| \le a^2/|2 - \lambda k|.$$

The third term is estimated quite simply:

$$r^{-\lambda k} \int_0^r \rho^{\lambda k+1} \, d\rho = r^2/(\lambda k + 2) \le a^2/(\lambda k).$$

Thus we see that for all $k \le 4/\lambda$ the estimate

$$|W_k(r) - \tilde{W}_k(r)| \le d_4 h^2$$

holds. In the opposite case $\lambda k \ge 4$, $|2 - \lambda k| \ge \lambda k/2$ and therefore

$$|W_k(r) - \tilde{W}_k(r)| \le d_5 h^2/k^2 + 2d_3 h^2 6^{-\lambda k}, \qquad d_5 = 4a^2 d_2/\lambda^2.$$

Taking these estimates into account it follows from (6.36) and (6.47) that

$$|w_2^h(r, \varphi) - w_3^h(r, \varphi)| \le \sum_{k=1}^N |\tilde{W}_k(r) - W_k(r)|$$

$$\le \sum_{k=1}^{[4/\lambda]} d_4 h^2 + \sum_{k=[4/\lambda]+1}^N (d_5 h^2/k^2 + 2d_3 h^2 6^{-\lambda k})$$

$$\le 4d_4 h^2/\lambda + d_5 h^2 \sum_{k=1}^\infty k^{-2} + 2d_3 h^2 \sum_{k=1}^\infty 6^{-\lambda k}.$$

In this expression we estimate the first sum using (6.40) and calculate the second sum explicitly (it is a geometric progression). Finally, we estimate the difference

$$|w_2^h(r, \varphi) - w_3^h(r, \varphi)| \le (4d_4/\lambda + 2d_5 + 2d_3 6^\lambda/(6^\lambda - 1))h^2$$
$$= d_6 h^2 \quad \forall r \in [a/8, a/6], \quad \forall \varphi \in [0, \phi]. \quad (6.48)$$

Adding the polynomial we subtracted earlier to function w_3^h we have

$$w_4^h(r, \varphi) = w_3^h(r, \varphi) + \left(1 - \frac{\varphi}{\phi}\right)g(r, 0) + \frac{\varphi}{\phi} g(r, \phi), \qquad (6.49)$$

which is an approximate solution of the original problem (6.20)–(6.22). Combining estimates (6.41) and (6.48) we obtain the following inequality which characterizes the accuracy of w_4^h in $[a/8, a/6] \times [0, \phi]$:

$$|w_4^h(r, \varphi) - u(r, \varphi)| \le (d_6 + d_1/(2b^2))h^2 = d_7 h^2. \qquad (6.50)$$

We will apply (6.42)–(6.47) to the problem

$$\Delta w_5 = 0 \quad \text{on } S_1,$$
$$w_5 = 0 \quad \text{on } \partial S_1 \cap \Gamma, \tag{6.51}$$
$$w_5 = z_2 \quad \text{on } \partial S_1 \backslash \Gamma.$$

Let us pass to polar coordinates. Note that it is sufficient to specify the values of function z_2 at the knots (a, φ_i), $i = 0, 1, \ldots, N$. Formulas (6.42)–(6.47) for (6.51) result in the following sequence of calculations:

$$\check{Z}_k = \frac{2}{\phi} \int_0^\phi \tilde{z}_2(a, \psi) \sin k\lambda\psi \, d\psi,$$
$$\tilde{W}_k(r) = (r/a)^{\lambda k} \check{Z}_k, \qquad k = 1, \ldots, N, \tag{6.52}$$

where $\tilde{z}_2$ is a piecewise-linear interpolant of the function z_2, and

$$w_6^h(r, \varphi) = \sum_{k=1}^N \tilde{W}_k(r) \sin k\lambda\varphi. \tag{6.53}$$

We are interested in the rate of decrease of the modulus of the approximate solution w_6^h as the distance of that part of the boundary with non-zero boundary conditions decreases. We will introduce the notation

$$\|\tilde{z}_2\|_C \equiv \max_{\varphi \in [0, \phi]} |\tilde{z}_2(a, \varphi)| = \max_{0 \le j \le N} |z_2(a, \varphi_j)|.$$

From (6.52) it follows that

$$|\tilde{W}_k(r)| \le (r/a)^{\lambda k} \frac{2}{\phi} \int_0^\phi |\sin k\lambda\varphi| \, d\varphi \|\tilde{z}_2\|_C \le (r/a)^{\lambda k} \frac{4}{\pi} \|\tilde{z}_2\|_C.$$

We investigate those values of $\tilde{W}_k(r)$ with $r \in [a/8, a/6]$ assuming that $\lambda \ge \frac{1}{2}$. We have

$$|\tilde{W}_k(r)| \le \frac{4}{\pi} 6^{-k/2} \|\tilde{z}_2\|_C \quad \forall \, r \in [a/8, a/6].$$

Let us use this estimate in equality (6.53). We get

$$|w_6^h(r, \varphi)| \le \sum_{k=1}^N |\tilde{W}_k(r)| |\sin k\lambda\varphi| \, d\varphi$$
$$\le \sum_{k=1}^\infty \frac{4}{\pi} 6^{-k/2} \|\tilde{z}_2\|_C = \frac{4}{\pi(\sqrt{6} - 1)} \|\tilde{z}_2\|_C.$$

Let us define

$$q = \frac{4}{\pi(\sqrt{6} - 1)}. \tag{6.54}$$

Then

$$\max_{\substack{r \in [a/8, a/6] \\ \varphi \in [0, \phi]}} |w_6^h(r, \varphi)| \leq q \|\tilde{z}_2\|_C. \tag{6.55}$$

We will use estimates (6.50), (6.55) to obtain an approximate solution to the problem

$$\begin{aligned} \Delta w_7 &= f \quad \text{on } S_1, \\ w_7 &= g \quad \text{on } \partial S_1 \cap \Gamma, \\ w_7 &= z \quad \text{on } \partial S_1 \backslash \Gamma, \end{aligned} \tag{6.56}$$

where z is a certain function given on $\partial S_1 \backslash \Gamma$.

Lemma 6.2. *Assume that (6.56) has been solved by the method (6.42)–(6.47). Then the approximate solution w_8^h obtained satisfies*

$$|w_8^h(r, \varphi) - u(r, \varphi)| \leq d_7 h^2 + q \max_{0 \leq j \leq N} |z(a, \varphi_j) - u(a, \varphi_j)|, \tag{6.57}$$

if $r \in [a/8, a/6]$, where the constant d_7 is taken from (6.50) and q is as in (6.54).

Let us solve the problem (6.1), (6.2) as a whole by applying the alternating Schwarz process (see [45]). We will consider an iterative process each step of which consists of three stages.

Stage I. Assume that the values $z_{i-1}^h(a, \varphi_j), j = 0, \ldots, N$ are known from the $(i - 1)$th step. Find an approximate solution of

$$\begin{aligned} \Delta v_i &= f \quad \text{on } S_1, \\ v_i &= g \quad \text{on } \partial S_1 \cap \Gamma, \\ v_i &= z_{i-1}^h \quad \text{on } \partial S_1 \backslash \Gamma, \end{aligned} \tag{6.58}$$

by method (6.42)–(6.47). Calculate an approximate solution v_i^h only at the knots $\partial D_h \cap \Omega_h$. From the results of Lemma 6.2 we have

$$\max_{\partial D_h \cap \Omega_h} |v_i^h - u| \leq d_7 h^2 + q \max_{0 \leq j \leq N} |z_{i-1}^h(a, \varphi_j) - u(a, \varphi_j)|. \tag{6.59}$$

At this step it is necessary to have of N^3 order arithmetic operations if the right-hand side is nonzero, and of N^2 order arithmetic operations if it is equal to zero. Therefore, it seems reasonable to deal with the right-hand side before the beginning the iterative process. In fact, the number of operations can be reduced to order $N^2 \ln N$ by using various modifications of the fast Fourier transform (see [87]).

Stage II. Since the values of v_i^h on $\partial D_h \cap \Omega_h$ are known one may pose the difference problem

$$-L_h u_i^h = f \quad \text{on } D_h,$$
$$u_i^h = g \quad \text{on } \partial D_h \cap \Gamma_h, \qquad\qquad (6.60)$$
$$u_i^h = v_i^h \quad \text{on } \partial D_h \cap \Omega_h.$$

The estimate

$$\max_{\bar{D}_h} |u_i^h - u| \le c_3 h^2 + \max_{\partial D_h \cap \Omega_h} |v_i^h - u| \qquad\qquad (6.61)$$

is valid for its solution by virtue of Lemma 6.1.

Stage III. Since we need approximate values of u at the knots (a, φ_j) of the polar grid to continue the process, we will use a simple piecewise-linear interpolation of the values of the function u_i^h using the three nearest knots of the rectangular net $\bar{D}_h$. We have

$$z_i^h(a, \varphi_j) = \sum_{x \in \bar{D}_h} G_j(x) u_i^h(x); \qquad j = 0, 1, \ldots, N,$$

where the weights $G_j(x)$, for any j, can differ from zero only at three knots of $\bar{D}_h$, where they are positive. Since the line $r = a$ is situated at a nonzero distance from $(0, 0)$ the function u is sufficiently smooth in a neighborhood of this point and we have the estimate

$$\left| u(a, \varphi_j) - \sum_{x \in \bar{D}_h} G_j(x) u(x) \right| \le d_8 h^2.$$

Using the property of the interpolation coefficients:

$$\left| \sum_{x \in \bar{D}_h} G_j(x) \right| = 1,$$

we get

$$|z_i^h(a, \varphi_j) - u(a, \varphi_j)| \le \sum_{x \in \bar{D}_h} G_j(x) |u_i^h(x) - u(x)|$$
$$+ \left| u(a, \varphi_j) - \sum_{x \in \bar{D}_h} G_j(x) u(x) \right|.$$

whence

$$\max_{1 \le j \le N} |z_i^h(a, \varphi_j) - u(a, \varphi_j)| \le d_8 h^2 + \max_{x \in \bar{D}_h} |u_i^h - u|. \qquad (6.62)$$

Thus the ith step of the iterative process has now been defined completely. It is easy to see that (6.59), (6.61), and (6.62) lead to the estimate

$$\max_{0 \le j \le N} |z_i^h(a, \varphi_j) - u(a, \varphi_j)| \le (c_3 + d_7 + d_8) h^2$$

$$+ q \max_{0 \le j \le N} |z_{i-1}^h(a, \varphi_j) - u(a, \varphi_j)|. \quad (6.63)$$

To begin the process we take $z_0^h \equiv 0$. Repeating Stages I–III we obtain an approximate solution with second-order accuracy.

Theorem 6.3. *When conditions (6.3)–(6.5) for (6.1), (6.2) are satisfied the iterative process* I–III *converges, the estimate*

$$\max_{\bar{D}_h} |u_i^h - u| \leq d_9 h^2 + q^i \max_{\bar{\Omega}} |u| \qquad (6.64)$$

being valid at the initial approximation $z_0^h \equiv 0$, *where* q *is defined by relation* (6.54).

PROOF. Since $z_0^h \equiv 0$, then from estimate (6.63) it follows that

$$\max_{0 \leq j \leq N} |z_1^h(a, \varphi_j) - u(a, \varphi_j)| \leq (c_3 + d_7 + d_8)h^2 + q \max_{\bar{\Omega}} |u|. \qquad (6.65)$$

Using estimate (6.63) $(i - 2)$ times we will have

$$\max_{0 \leq j \leq N} |z_{i-1}^h(a, \varphi_j) - u(a, \varphi_j)| \leq (d_7 + d_8 + c_3)h^2 \sum_{i=0}^{i-2} q^i + q^{i-1} \max_{\bar{\Omega}} |u|$$

$$\leq h^2(c_3 + d_7 + d_8)/(1 - q) + q^{i-1} \max_{\bar{\Omega}} |u|. \qquad (6.66)$$

Because for (6.58) we have (6.59), and for (6.60) there is (6.61) it follows from (6.66) that

$$\max_{\bar{D}_h} |u_i^h - u| \leq (c_3 + d_7 + q(c_3 + d_7 + d_8)/(1 - q))h^2 + q^i \max_{\bar{\Omega}} |u|.$$

Assuming

$$d_9 = c_3 + d_7 + q(c_3 + d_7 + d_8)/(1 - q),$$

we come to (6.64). The theorem is proved. $\qquad\qquad\square$

Thus to insure that the orders of the terms in (6.64) coincide it suffices to perform l iterations:

$$l = 2[\ln h/\ln q]. \qquad (6.67)$$

Finally, we have a numerical solution at the knots of the grid D_h with h^2 order accuracy.

We now present a numerical example. Let the domain Ω be the interior of a square with side $b = 2.4$ and center at the origin, and cut from $(0, 0)$ to $(b/2, 0)$ (Figure 4.10). On this domain Ω we seek a solution of Laplace's equation

$$-\Delta u = 0, \qquad (6.68)$$

satisfying the boundary condition

$$u = 0 \qquad (6.69)$$

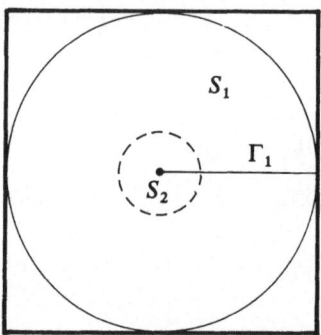

Figure 4.10. The cut square.

on the cut Γ_1 and the condition

$$u(x, y) = g(x, y) \tag{6.70}$$

on the sides of square Γ_2. Here

$$g(x, y) = \sqrt{\sqrt{x^2 + y^2} - x}.$$

It is easy to prove that the solution of (6.68)–(6.70) is

$$u(x, y) = g(x, y) \quad \text{on } \overline{\Omega}.$$

Problem (6.68)–(6.70) was solved using two different approaches, for several parameters h. One of the approaches used the ordinary fivepoint difference scheme of the form (6.7). The other used the method developed in this section. The singularity at the point $(0, 0)$ was taken into account. We note that the domain Ω has six corners. But the angles are right angles and they satisfy conditions (3.6). From these conditions and [137, 43] it follows that the fourth derivatives of the solution u of (6.68)–(6.70) are bounded near these corners. Therefore, it is unnecessary to apply the method of separation of singularities for these six corners. We have constructed Table 4.1 giving the maximum error of the solutions on the grid Ω_h using these two approaches.

We should note that the maximum error of the solution given by the iterative process I–III decreases proportional to h^2, as was expected, while the error of a standard fivepoint difference scheme decreases only slightly.

Table 4.1

h	0.2	0.15	0.1	0.075	0.05	0.0375
The iterative process I–III	0.01444	0.00703	0.00367	0.00206	0.00091	0.00052
Fivepoint difference scheme	0.10787	0.09636	0.08158	0.07168	0.05905	0.05147

CHAPTER 5
Nonstationary Problems

Nonstationary equations tend to be more complicated than stationary ones. They have many applications in various fields of science and technology. When constructing numerical algorithms one must deal with spatial variables as well as with the time variable. The question of finding a corrector for an approximate solution on a sequence of nets is considerably more complicated. Nevertheless in a great number of cases such correctors can be found. These questions will now occupy us. In this book we will not treat high order schemes which use another method, nor will we discuss other algorithms even though they also give successful results. The reader is referred to [4, 131, 51, 87, 112, 114, 42, 146], for further discussion.

5.1. The Simplest Type of Parabolic Equation

Let us first consider the simplest parabolic equation, the one-dimensional heat equation:

$$\frac{\partial u}{\partial t} = \frac{\partial}{\partial x}\left(p\,\frac{\partial u}{\partial x}\right) + f, \qquad p(x, t) \geq c_0 > 0, \tag{1.1}$$

where $x \in (0, 1)$, $t \in (0, T)$. We impose initial conditions and boundary conditions on the function u

$$u(x, 0) = 0, \qquad x \in [0, 1], \tag{1.2}$$

$$\begin{aligned} u(0, t) &= u_0(t), \\ u(1, t) &= u_1(t), \end{aligned} \qquad t \in [0, T]. \tag{1.3}$$

Let Q be the rectangle $(0, 1) \times (0, T)$, $\bar{Q}$ its closure, and l an arbitrary positive nonintegral number. In order to describe the smoothness of the initial data, the boundary conditions and the solution of (1.1)–(1.3) let us introduce the following class of functions (see [71]). Denote by $H^l(\bar{Q})$ the Banach space of functions $u(x, t)$, continuous on $\bar{Q}$ along with all their derivatives of the form:

$$\frac{\partial^{r+s} u}{\partial t^r \, \partial x^s}, \qquad 2r + s < l$$

and with finite norm

$$\|u\|_{H^l(\bar{Q})} = \langle u \rangle_Q^{(l)} + \sum_{j=0}^{[l]} \langle u \rangle_Q^{(j)}, \tag{1.4}$$

where

$$\langle u \rangle_Q^{(j)} = \sum_{2r+s=j} \max_Q \left| \frac{\partial^{r+s} u}{\partial t^r \, \partial x^s} \right|, \qquad j = 0, 1, \ldots, [l].$$

$$\langle u \rangle_Q^{(l)} = \langle u \rangle_x^{(l)} + \langle u \rangle_t^{(l/2)},$$

$$\langle u \rangle_x^{(l)} = \sum_{2r+s=[l]} \left\langle \frac{\partial^{r+s} u}{\partial t^r \, \partial x^s} \right\rangle_x^{l-[l]},$$

$$\langle u \rangle_t^{(l/2)} = \sum_{0 < l-2r-s < 2} \left\langle \frac{\partial^{r+s} u}{\partial t^r \, \partial x^s} \right\rangle_t^{(l-2r-s)/2},$$

$$\langle u \rangle_x^\alpha = \sup_{(x, t), (x', t) \in \bar{Q}} |u(x, t) - u(x', t)| / |x - x'|^\alpha, \qquad \alpha \in (0, 1),$$

$$\langle u \rangle_t^\beta = \sup_{(x, t), (x, t') \in Q} |u(x, t) - u(x, t')| / |t - t'|^\beta, \qquad \beta \in (0, 1).$$

For the solution to be continuous up to the boundary, it is necessary that

$$u_0(0) = u_1(0) = 0. \tag{1.5}$$

The conditions (1.5) will be referred to as the coherence conditions of zero order. If one needs a solution whose derivatives $\partial u / \partial t$ and $\partial^2 u / \partial x^2$ are continuous up to the boundary then it is necessary that

$$\frac{\partial u_0}{\partial t}(0) = f(0, 0), \qquad \frac{\partial u_1}{\partial t}(0) = f(1, 0) \tag{1.6}$$

hold. These relations will be referred to as the coherence conditions of the first order. Differentiating the equation, assuming that all derivatives thus obtained are continuous up to the boundary, we have

$$\frac{\partial^2 u}{\partial t^2} = \frac{\partial}{\partial t}\left(\frac{\partial}{\partial x} p \frac{\partial u}{\partial x} + f \right) = \frac{\partial}{\partial x}\left(\frac{\partial p}{\partial t} \frac{\partial u}{\partial x} \right) + \frac{\partial}{\partial x} p \frac{\partial u}{\partial x} \frac{\partial u}{\partial t} + \frac{\partial f}{\partial t}$$

$$= \frac{\partial}{\partial x} p \frac{\partial}{\partial x}\left(\frac{\partial}{\partial x} p \frac{\partial u}{\partial x} + f \right) + \frac{\partial f}{\partial t} = \frac{\partial}{\partial x} p \frac{\partial f}{\partial x} + \frac{\partial f}{\partial t} \qquad \text{at } t = 0. \tag{1.7}$$

The derivatives of the function u with respect to the variable x are zero because of the uniform initial condition. Since the values of $\partial^2 u/\partial t^2$ at the corners $(0, 0)$ and $(1, 0)$ can be expressed in terms of known values then from (1.7) there follows

$$\frac{d^2 u_0}{dt^2}(0) = \frac{\partial}{\partial x} p \frac{\partial f}{\partial x}(0, 0) + \frac{\partial f}{\partial t}(0, 0),$$

$$\frac{d^2 u_1}{dt^2}(0) = \frac{\partial}{\partial x} p \frac{\partial f}{\partial x}(1, 0) + \frac{\partial f}{\partial t}(1, 0).$$

(1.8)

We will refer to (1.8) as the coherence conditions of order 2. Using a procedure similar to the one above we may obtain the coherence conditions of any higher order. The conditions are turn out not only to be necessary for the continuity of the corresponding derivatives up to the boundary (see [71]) but also sufficient when the data of the problem are smooth. Let us formulate this result in the form of a theorem.

Theorem 1.1 [71]. *Let $k > 0$, k not an integer, and let*

$$f \in H^k(\bar{Q}), \qquad p \in H^{k+1}(\bar{Q}), \qquad u_0, u_1 \in C^{k/2+1}[0, T].$$

Then (1.1)–(1.3) has a unique solution when the coherence conditions of orders $0, 1, \ldots, [k/2] + 1$ hold.

We will solve (1.1)–(1.3) when $u_0 = u_1 = 0$ on $[0, T]$ and the assumptions of Theorem 1.1 with $l \in (3, 4)$ hold. The condition that the boundary values be uniform (as well as the initial values) is a rather weak condition. If the function u had nonzero values $u(x, 0) = v(x) \; \forall \; x \in [0, 1]$, at the beginning of time then replace the unknown function u by $w(x, t) = u(x, t) - v(x)$. This leads to a uniform initial condition. If (1.2) is satisfied then (1.3) can be made uniform by assuming

$$w(x, t) = u(x, t) - u_0(t)(1 - x) - xu_1(t).$$

For the function w conditions (1.2), (1.3) are now uniform.

Fix an integer $M \geq 2$ and assume $\tau = 1/M$. Introduce the following notation

$$\bar{\omega}_\tau = \{t_j = j\tau; j = 0, 1, \ldots, M\},$$

$$\omega_\tau = \{t_j = j\tau; j = 1, 2, \ldots, M\}.$$

Let us choose a uniform net in the spatial variable. Let $h = 1/N$; then

$$\bar{\omega}_h = \{x_i = ih; i = 0, 1, \ldots, N\},$$

$$\omega_h = \{x_i = ih; i = 1, 2, \ldots, N - 1\}.$$

We form two-dimensional rectangular nets as the Cartesian products of one-dimensional ones:

$$\bar{Q}_h^\tau = \bar{\omega}_h \times \bar{\omega}_\tau, \qquad Q_h^\tau = \omega_h \times \omega_\tau. \tag{1.9}$$

We seek an approximate solution at the points of $\bar{Q}_h^\tau$ and construct our difference equations at the points of the net Q_h^τ using the implicit scheme (see [87, 114, 146]):

$$u_i^\tau = (p u_{\bar{x}}^\tau)_{\hat{x}} + f \quad \text{on } Q_h^\tau. \tag{1.10}$$

Since the number of equations is less than the number of unknowns we add initial conditions and boundary conditions (for simplicity we will write u^τ without the second index h):

$$u^\tau(x, 0) = 0, \qquad x \in \omega_h, \tag{1.11}$$

$$u^\tau(0, t) = u^\tau(1, t) = 0, \qquad t \in \omega_\tau. \tag{1.12}$$

In [114] an estimate is made, which characterizes the stability of the solution

$$\max_{x \in \bar{\omega}_h} |u^\tau(x, t_{j+1})| \leq \max_{x \in \bar{\omega}_h} |u^\tau(x, 0)| + \sum_{j'=0}^{j} \tau \max_{x \in \omega_h} |f(x, t_{j'})|.$$

We use the following simple corollary

$$\max_{\bar{Q}_h^\tau} |u^\tau| \leq T \max_{Q_h^\tau} |f|. \tag{1.13}$$

It should be kept in mind that using the factorization method one can compute the approximate solution u^τ, passing from one value of the time parameter to the next. The system (1.10)–(1.12) is linear in u^τ, and therefore the uniqueness of its solution is related to the number of solutions of the homogeneous problem. But the homogeneous problem (with the right-hand side f being zero), according to (1.13), has only the zero solution. Thus the solution u^τ is unique.

Theorem 1.2. *Suppose that the conditions of Theorem 1.1, with $l \in (3, 4)$, hold for (1.1)–(1.3). Then the solution of the difference problem (1.10)–(1.12) can be written as*

$$u^\tau = u + h^2 v + \tau w + (h^l + \tau^{l/2}) \eta^\tau \quad \text{on } \bar{Q}_h^\tau, \tag{1.14}$$

where the functions v and w are continuous on $\bar{Q}$ and are independent of τ and h, and the net function η^τ is uniformly bounded:

$$|\eta^\tau| \leq c_1 \quad \text{on } \bar{Q}_h^\tau. \tag{1.15}$$

PROOF. Choose the function v as the solution of

$$\frac{\partial v}{\partial t} = \frac{\partial}{\partial x} p \frac{\partial v}{\partial x} + g_1 \quad \text{on } Q,$$

$$v(x, 0) = 0, \qquad x \in [0, 1], \qquad v(1, t) = v(0, t) = 0, \qquad t \in [0, T], \tag{1.16}$$

where

$$g_1 = \frac{1}{24}\left(\frac{\partial^3}{\partial x^3} p \frac{\partial u}{\partial x} + \frac{\partial}{\partial x} p \frac{\partial^3 u}{\partial x^3}\right).$$

Note that the data of problem (1.16) satisfy the conditions of Theorem 1.1 where the number k should be replaced by $l - 2$. In particular, the function g_1 is continuous and becomes zero at the corners $(0, 0)$ and $(1, 0)$. Therefore, the coherence conditions of order 1 hold for (1.16). The coherence conditions of order 0 hold automatically from the uniformity of the boundary data and initial data. Thus $v \in H^l(\bar{Q})$.

Choose the function w as the solution of

$$\frac{\partial w}{\partial t} = \frac{\partial}{\partial x} p \frac{\partial w}{\partial x} + g_2 \quad \text{on } Q,$$

$$w(x, 0) = 0, \qquad x \in [0, 1], \qquad w(0, t) = w(1, t) = 0, \qquad t \in [0, T], \quad (1.17)$$

where $g_2 = -\frac{1}{2}(\partial^2 u/\partial t^2)$. Since $g_2 \in H^{l-2}(\bar{Q})$ and the coherence conditions of orders 0 and 1 hold then by Theorem 1.1 we have that $w \in H^l(\bar{Q})$.

The functions u^τ, u, v, w are defined unambiguously at the net points of Q_h^τ. Therefore, one can consider (1.14) as defining the net function η^τ. To complete the proof of the theorem we must obtain (1.15). Substitute (1.14) in (1.10) for u^τ. We have

$$u_{\bar{t}} + h^2 v_{\bar{t}} + \tau w_{\bar{t}} + (h^l + \tau^{l/2})\eta_{\bar{t}}^\tau = (pu_{\hat{x}})_{\hat{x}} + h^2(pv_{\hat{x}})_{\hat{x}}$$
$$+ \tau(pw_{\hat{x}})_{\hat{x}} + (h^l + \tau^{l/2})(p\eta_{\hat{x}}^\tau)_{\hat{x}} + f \quad \text{on } Q_h^\tau. \quad (1.18)$$

Using Lemmas 1.1 and 1.2 of §7.1 we have

$$u_{\bar{t}} = \frac{\partial u}{\partial t} + \frac{\tau}{2}\frac{\partial^2 u}{\partial t^2} + \tau^{l/2}\xi_1,$$

$$v_{\bar{t}} = \frac{\partial v}{\partial t} + \tau^{l/2-1}\xi_2,$$

$$w_{\bar{t}} = \frac{\partial w}{\partial t} + \tau^{l/2-1}\xi_3,$$

$$(pu_{\hat{x}})_{\hat{x}} = \frac{\partial}{\partial x} p \frac{\partial u}{\partial x} + h^2\left(\frac{1}{24}\frac{\partial^3}{\partial x^3} p \frac{\partial u}{\partial x} + \frac{1}{24}\frac{\partial}{\partial x} p \frac{\partial^3 u}{\partial x^3}\right) + h^l\xi_4,$$

$$(pv_{\hat{x}})_{\hat{x}} = \frac{\partial}{\partial x} p \frac{\partial v}{\partial x} + h^{l-2}\xi_5,$$

$$(pw_{\hat{x}})_{\hat{x}} = \frac{\partial}{\partial x} p \frac{\partial w}{\partial x} + h^{l-2}\xi_6,$$

where

$$|\xi_i| \le c_i, \qquad i = 1, \ldots, 6.$$

and the constants c_i depend neither on the points of Q_h^τ nor on h and τ. Substituting the terms of equation (1.18) for expansions obtained we get

$$
\frac{\partial u}{\partial t} + h^2 \frac{\partial v}{\partial t} + \tau\left(\frac{1}{2}\frac{\partial^2 u}{\partial t^2} + \frac{\partial w}{\partial t}\right) + (h^2\tau^{l/2-1} + \tau^{l/2})\xi_7 + (h^l + \tau^{l/2})\eta_{\bar t}^\tau
$$

$$
= \frac{\partial}{\partial x} p \frac{\partial u}{\partial x} + h^2\left(\frac{1}{24}\frac{\partial^3}{\partial x^3} p \frac{\partial u}{\partial x} + \frac{1}{24}\frac{\partial}{\partial x} p \frac{\partial^3 u}{\partial x^3} + \frac{\partial}{\partial x} p \frac{\partial v}{\partial x}\right)
$$

$$
+ \tau \frac{\partial}{\partial x} p \frac{\partial w}{\partial x} + (h^l + h^{l-2}\tau)\xi_8
$$

$$
+ (h^l + \tau^{l/2})(p\eta_{\bar x}^\tau)_{\hat x} + f \quad \text{on } Q_h^\tau. \tag{1.19}
$$

Here

$$
|\xi_7| \le c_1 + c_2 + c_3, \qquad |\xi_8| \le c_4 + c_5 + c_6.
$$

The terms in h^0, τ^0 can be dropped since the function u satisfies equation (1.1). The coefficients of h^2 on both sides of (1.19) cancel because from (1.16) we have

$$
\frac{\partial v}{\partial t} = \frac{\partial}{\partial x} p \frac{\partial v}{\partial x} + \frac{1}{24}\frac{\partial^3}{\partial x^3} p \frac{\partial u}{\partial x} + \frac{1}{24}\frac{\partial}{\partial x} p \frac{\partial^3 u}{\partial x^3}.
$$

The function w satisfies

$$
\frac{\partial w}{\partial t} + \frac{1}{2}\frac{\partial^2 u}{\partial t^2} = \frac{\partial}{\partial x} p \frac{\partial w}{\partial x},
$$

which allows us to cancel the terms in τ. Write the remaining terms in formula (1.19) as:

$$
(h^l + \tau^{l/2})\eta_{\bar t}^\tau = (h^l + \tau^{l/2})(p\eta_{\bar x}^\tau)_{\hat x} + (h^l + h^{l-2}\tau)\xi_8
$$
$$
+ (h^2\tau^{l/2-1} + \tau^{l/2})\xi_7 \quad \text{on } Q_h^\tau. \tag{1.20}
$$

Use Young's inequality (see [73]): if $a \ge 0, b \ge 0$, then

$$
ab \le \frac{1}{p} a^p + \frac{1}{p'} b^{p'},
$$

where

$$
1/p + 1/p' = 1.
$$

For the products $h^{l-2}\tau$ and $h^2\tau^{l/2-1}$ this gives

$$
h^{l-2}\tau \le \left(1 - \frac{2}{l}\right)h^l + \frac{2}{l}\tau^{l/2} \le h^l + \tau^{l/2},
$$

$$
h^2\tau^{l/2-1} \le \frac{2}{l}h^l + \left(1 - \frac{2}{l}\right)\tau^{l/2} \le h^l + \tau^{l/2},
$$

respectively. Dividing both sides of (1.20) by $h^l + \tau^{l/2}$ and using two last inequalities we have

$$\eta_{\bar{t}} = (p\eta_{\bar{x}}^{\tau})_{\hat{x}} + \xi_9 \quad \text{on } Q_h^{\tau}, \tag{1.21}$$

where

$$|\xi_9| \leq 2\sum_{i=1}^{6} c_i = c_7. \tag{1.22}$$

In order to use (1.13) it is necessary to specify boundary conditions and initial conditions for the net function η^{τ}. Since for $t = 0$ we have

$$u(x, 0) = v(x, 0) = w(x, 0) = 0 \quad \forall\, x \in [0, 1], \qquad u^{\tau}(x, 0) = 0 \quad \forall\, x \in \bar{\omega}_h$$

then from (1.14) it follows that

$$\eta^{\tau}(x, 0) = 0 \quad \forall\, x \in \bar{\omega}_h.$$

In the similar way using corresponding boundary conditions we make sure that

$$\eta^{\tau}(0, t) = \eta^{\tau}(1, t) = 0 \quad \forall\, t \in \omega_{\tau}.$$

Thus estimate (1.13) can be applied to (1.21) and yields

$$\max_{\bar{Q}_h^{\tau}} |\eta^{\tau}| \leq T \max_{Q_h^{\tau}} |\xi_9| \leq c_7 T.$$

This means that estimate (1.15) holds with constant $c_1 = c_7 T$ and the proof of the theorem is complete. $\qquad\square$

Note that the degree of smoothness which the solution possesses is different in x and in t. This leads to the realization that different orders of accuracy can be obtained by adjusting the mesh-sizes τ and h. If the amount of smoothness of the solution in x dictates that the maximal accuracy of the approximate solution is of order h^k, then an analysis of the time derivatives shows that the maximal accuracy of the approximate solution, not considering the spatial derivatives, is of order $\tau^{k/2}$. Thus the fact that (1.14) has τ equal to h^2 is quite natural.

Now we describe a method for increasing the accuracy using (1.14). Assume the conditions of Theorem 1.2 hold. Choose integers $M \geq 2$ and $N \geq 2$. Construct the difference net with time step $\tau = 1/M$ and spatial mesh-size $h = 1/N$. Find the solution u^{τ} of (1.10)–(1.12) on the net $\bar{Q}_h^{\tau}$.

Then construct the difference net $\bar{Q}_{h/2}^{\tau/4}$ with time step $\tau/4$ and spatial mesh-size $h/2$. Solve (1.10)–(1.12) again, and denote its solution $u^{\tau/4}$. Thus on the net points $\bar{Q}_h^{\tau}$ we now have two approximate solutions u^{τ} and $u^{\tau/4}$. Form the linear combination

$$U = \tfrac{4}{3}u^{\tau/4} - \tfrac{1}{3}u^{\tau} \quad \text{on } \bar{Q}_h^{\tau}. \tag{1.23}$$

We will show that the corrected difference solution U approximates the exact solution with $(h^l + \tau^{l/2})$ order accuracy in the uniform metric.

At each net point $(x, t) \in \bar{Q}_h^\tau$ we have

$$u^\tau = u + h^2 v + \tau w + (h^l + \tau^{l/2})\eta^\tau,$$

$$u^{\tau/4} = u + \frac{h^2}{4} v + \frac{\tau}{4} w + \frac{h^l + \tau^{l/2}}{2^l} \eta^{\tau/4}.$$

Since the functions u, v, w are independent of τ and h cancellation of the addends containing v and w in our linear combination will take place:

$$U = u + (h^l + \tau^{l/2})\tfrac{1}{3}(2^{2-l}\eta^{\tau/4} - \eta^\tau).$$

Using (1.15) we have

$$|U - u| \le \tfrac{1}{3}(2^{2-l} + 1)c_1(h^l + \tau^{l/2}) \le c_1(h^l + \tau^{l/2})/2 \qquad (1.24)$$

for all $(x, t) \in \bar{Q}_h^\tau$.

Thus the difference solution (1.23) obtained by taking a linear combination of the approximate solutions with accuracy of $\tau + h^2$ order approximates the exact solution u at the points of $\bar{Q}_h^\tau$ with accuracy of $h^l + \tau^{l/2}$ order, $l \in (3, 4)$. We should note that one can guarantee an accuracy in (1.23) of $h^4 + \tau^2$ order by slightly increasing the smoothness of the functions f, p, for example, $f \in H^{4+\alpha}(\bar{Q})$, $p \in H^{5+\alpha}(\bar{Q})$, where $\alpha \in (0, 1)$. We need not impose additional coherence conditions beyond those of Theorem 1.1. This is explained by the fact that the derivatives taking part which occur in the reminder of h^4 and τ^2 order are bounded at the corners $(0, 0)$ and $(1, 0)$. Justification of this fact in the general case requires a large number of manipulations not related to our particular difference scheme.

If we require that the data be smoother and that more coherence conditions are satisfied then the solution (1.1)–(1.3) will have a higher order of smoothness. But without changing the differential equation it is impossible to obtain expansions like (1.14) in powers of h^2 and τ with the order of the remainder greater than $h^4 + \tau^2$. This is because the auxiliary problems which give us the coefficients of the expansion obey only two coherence conditions. In order to have more coherence conditions satisfied it is necessary to make substitution

$$u(x, t) = z(x, t) + tx(1 - x)R(x, t).$$

Here z is a new unknown function, and R is a polynomial in x, t providing the coherence of the auxiliary problems. Note that such a substitution effects neither the smoothness of the solution nor the uniformity of the initial conditions and boundary conditions.

Using the difference schemes from Chapter 2 to approximate the spatial derivatives one easily succeeds in generalizing our results to the case of discontinuous coefficients (with fixed points of discontinuity) to the third boundary problem and to the case where the coefficients depend on the solution. Note that the useful estimate (1.13) in [114] can also be extended to these cases.

5.2. Increasing the Accuracy of the Splitting-Up Method

We now present a method for increasing accuracy using the abstract evolutionary equation (see [66]).

$$\frac{du}{dt} + Au = f, \qquad t \in (0, T), \tag{2.1}$$

with the initial condition

$$u(0) = u_0 \tag{2.2}$$

as our model. Here the functions $u(t)$ and $f(t)$ are elements of a Hilbert space H with scalar product (u, v) and norm

$$\|u\| = (u, u)^{1/2}.$$

The linear operator A is independent of time (dependence on t or even on u being irrelevant) and is nonnegative definite, i.e.,

$$(Au, u) \geq 0 \quad \forall u \in H.$$

Let us assume also that the solution u and the right-hand side f have a sufficient number of bounded derivatives with respect to t so that subsequent steps are justified.

We will consider two approaches to the questions of convergence and order of accuracy for one of the splitting-up methods.

Let the operator A be represented as the sum of two nonnegative definite operators: $A = A_1 + A_2$. Divide the interval $[0, T]$ into M equal parts and introduce the following notation:

$$\bar{\omega}_\tau = \{t_j = j\tau; j = 0, 1, \ldots, M\}, \qquad \omega_\tau = \{t \in \bar{\omega}_\tau, t \neq 0\}.$$

Consider the scheme

$$(u^*(t) - u^\tau(t - \tau))/\tau + A_1 u^*(t) = f(t),$$

$$(u^\tau(t) - u^*(t))/\tau + A_2 u^\tau(t) = 0, \qquad t \in \omega_\tau, \tag{2.3}$$

with the initial condition

$$u^\tau(0) = u_0. \tag{2.4}$$

This scheme is absolutely stable and there is an *a priori* estimate

$$\max_{\bar{\omega}_\tau} \|u^\tau\| \leq \|u_0\| + T \max_{\omega_\tau} \|f\| \tag{2.5}$$

which guarantees the uniqueness and boundedness of the approximate solution (see [86]).

The first approach to justifying this method as a means of increasing the accuracy of the solution excludes the intermediate value and inquires as to the approximating properties and convergence of the reduced problem:

$$(I + \tau A_1)(I + \tau A_2)u^\tau(t) = u^\tau(t - \tau) + \tau f(t), \qquad t \in \omega_\tau,$$

$$u^\tau(0) = u_0. \tag{2.6}$$

Here I is the identity operator in the space H. First assume that we can write

$$u^\tau = u + \tau v_1 + \tau^2 v_2 + \tau^3 \eta^\tau \quad \text{on } \bar{\omega}_\tau \tag{2.7}$$

with the functions v_1, v_2 independent of τ and with bounded net function η^τ. Substitute u^τ into (2.6) using this expansion. We have

$$(I + \tau A_1)(I + \tau A_2)(u + \tau v_1 + \tau^2 v_2 + \tau^3 \eta^\tau)/_t$$
$$= (u + \tau v_1 + \tau^2 v_2 + \tau^3 \eta^\tau)/_{t-\tau} + \tau f(t).$$

Use the Taylor formula for each of the functions u, v_1, v_2 supposing them to be sufficiently smooth:

$$u(t - \tau) = u(t) - \tau \frac{du}{dt}(t) + \frac{\tau^2}{2}\frac{d^2u}{dt^2}(t) - \frac{\tau^3}{6}\frac{d^3u}{dt^3}(t) + \tau^4 \xi_1,$$

$$v_1(t - \tau) = v_1(t) - \tau \frac{dv_1}{dt}(t) + \frac{\tau^2}{2}\frac{d^2v_1}{dt^2}(t) + \tau^3 \xi_2,$$

$$v_2(t - \tau) = v_2(t) - \tau \frac{dv_2}{dt}(t) + \tau^2 \xi_3, \tag{2.8}$$

where

$$\|\xi_i\| \le c_1 \quad \forall\, t \in \omega_\tau, \quad i = 1, 2, 3.$$

Then we have

$$(I + \tau A_1)(I + \tau A_2)(u + \tau v_1 + \tau^2 v_2 + \tau^3 \eta^\tau)|_t$$

$$= u(t) + \tau\left(-\frac{du}{dt} + v_1\right)\bigg|_t + \tau^2\left(\frac{1}{2}\frac{d^2u}{dt^2} - \frac{dv_1}{dt} + v_2\right)\bigg|_t$$

$$+ \tau^3\left(-\frac{1}{6}\frac{d^3u}{dt^3} + \frac{1}{2}\frac{d^2v_1}{dt^2} - \frac{dv_2}{dt}\right)\bigg|_t + \tau^3\eta^\tau(t - \tau)$$

$$+ \tau f(t) + \tau^3(\xi_1 + \xi_2 + \xi_3)|_t.$$

Multiplying out the expressions in brackets and collecting terms we have

$$\tau^2(Av_1 + A_1A_2u)|_t + \tau^3(Av_2 + A_1A_2v_1)|_t$$

$$+ \tau^4 A_1A_2 v_2(t) + \tau^3(I + \tau A_1)(I + \tau A_2)\eta^\tau(t)$$

$$= \tau^2\left(\frac{1}{2}\frac{d^2u}{dt^2} - \frac{dv_1}{dt}\right)\bigg|_t + \tau^3\left(-\frac{1}{6}\frac{d^3u}{dt^3} + \frac{1}{2}\frac{d^2v_1}{dt^2} - \frac{dv_2}{dt}\right)\bigg|_t$$

$$+ \tau^3\eta^\tau(t - \tau) + \tau^4(\xi_1 + \xi_2 + \xi_3)|_t. \tag{2.9}$$

Equate the coefficients in τ^2:

$$Av_1 + A_1 A_2 u = \frac{1}{2} \frac{d^2 u}{dt^2} - \frac{dv_1}{dt}, \qquad t \in \omega_\tau.$$

Suppose that the product $A_1 A_2 u$ is bounded for any $t \in [0, T]$. Then the function v_1 is a solution of

$$\frac{dv_1}{dt} + Av_1 = \frac{1}{2} \frac{d^2 u}{dt^2} - A_1 A_2 u, \qquad t \in [0, T],$$

$$v_1(0) = 0. \tag{2.10}$$

The initial condition of the above has been obtained as follows. For (2.7) to hold on $\bar{\omega}_\tau$ the relation

$$u^\tau(0) = u(0) + \tau v_1(0) + \tau^2 v_2(0) + \tau^3 \eta^\tau(0) \tag{2.11}$$

should be valid at the beginning of time. Since $u^\tau(0) = u(0) = u_0$ then after canceling like terms on both sides of (2.11) we have

$$\tau v_1(0) + \tau^2 v_2(0) + \tau^3 \eta^\tau(0) = 0. \tag{2.12}$$

Equating the coefficient of τ to zero we obtain the initial condition for (2.10).

Choosing the function v_1 as the solution of (2.10) we rewrite (2.9) in the following form:

$$(Av_2 + A_1 A_2 v_1 + \tau A_1 A_2 v_2)|_t + (I + \tau A_1)(I + \tau A_2)\eta^\tau(t)$$

$$= -\frac{1}{6} \frac{d^3 u}{dt^3}(t) + \frac{1}{2} \frac{d^2 v_1}{dt^2}(t) - \frac{dv_2}{dt}(t)$$

$$+ \eta^\tau(t - \tau) + \tau(\xi_1 + \xi_2 + \xi_3)|_t. \tag{2.13}$$

In order that the function η^τ be bounded as in (2.5) it is sufficient to choose v_2 in such a way that only terms of order τ and terms containing η^τ should remain in (2.13). Such a choice results in

$$Av_2 + A_1 A_2 v_1 = -\frac{1}{6} \frac{d^3 u}{dt^3} + \frac{1}{2} \frac{d^2 v_1}{dt^2} - \frac{dv_2}{dt}, \qquad t \in \omega_\tau.$$

Let us require that

$$\frac{dv_2}{dt} + Av_2 = -\frac{1}{6} \frac{d^3 u}{dt^3} + \frac{1}{2} \frac{d^2 v_1}{dt^2} - A_1 A_2 v_1, \qquad t \in [0, T],$$

$$v_2(0) = 0 \tag{2.14}$$

hold for v_2.

Note that the uniform initial condition in this initial-value problem guarantees the boundedness of $\eta^\tau(0)$. Thus, supposing $A_1 A_2 v_1$ to be bounded we can now uniquely define v_2.

This choice of v_1, v_2 brings (2.13) to the form

$$(I + \tau A_1)(I + \tau A_2)\eta^\tau(t) = \eta^\tau(t - \tau) + \tau(\xi_1 + \xi_2 + \xi_3)|_t$$
$$- \tau A_1 A_2 v_2(t), \qquad t \in \omega_\tau. \qquad (2.15)$$

Thus the functions u^τ, u, v_1, v_2 are uniquely defined in (2.7). Therefore η^τ is uniquely defined from (2.7). Let us prove that it is bounded. From (2.12) we have

$$\eta^\tau(0) = 0. \qquad (2.16)$$

Due to the boundedness of the product $A_1 A_2 v_2$ and because of (2.5) we have

$$\|\eta^\tau\| \le c_2 \quad \forall\, t \in \bar{\omega}_\tau. \qquad (2.17)$$

Thus the functions v_1 and v_2 have been found and are independent of τ. The net function η^τ is bounded.

We will now show how this expansion can be applied. Take some natural number $M > 2$ and solve (2.3), (2.4) with mesh-sizes $\tau = T/M$, $\tau/2$, $\tau/3$. On the net $\bar{\omega}_\tau$ of mesh-size τ we have the three expansions:

$$u^\tau = u + \tau v_1 + \tau^2 v_2 + \tau^3 \eta^\tau,$$

$$u^{\tau/2} = u + \frac{\tau}{2} v_1 + \frac{\tau^2}{4} v_2 + \frac{\tau^3}{8} \eta^{\tau/2},$$

$$u^{\tau/3} = u + \frac{\tau}{3} v_1 + \frac{\tau^2}{9} v_2 + \frac{\tau^3}{27} \eta^{\tau/3}.$$

Summing these up with weights $\gamma_1 = \frac{1}{2}, \gamma_2 = -4, \gamma_3 = \frac{9}{2}$, respectively, we have

$$U = \tfrac{1}{2}u^\tau - 4u^{\tau/2} + \tfrac{9}{2}u^{\tau/3} = u + \tau^3(\tfrac{1}{2}\eta^\tau - \tfrac{1}{2}\eta^{\tau/2} + \tfrac{1}{6}\eta^{\tau/3}), \qquad t \in \bar{\omega}_\tau.$$

Since γ_i solve the system

$$\gamma_1 + \gamma_2 + \gamma_3 = 1,$$

$$\gamma_1 + \tfrac{1}{2}\gamma_2 + \tfrac{1}{3}\gamma_3 = 0,$$

$$\gamma_1 + \tfrac{1}{4}\gamma_2 + \tfrac{1}{9}\gamma_3 = 0$$

the terms in τ and τ^2 in the expression for U are zero. Using (2.17) for the remainder terms we have

$$\|U(t) - u(t)\| \le \tfrac{7}{6}c_2 \tau^3 \quad \forall\, t \in \bar{\omega}_\tau. \qquad (2.18)$$

This inequality reflects the effect of our correction to the approximate solution.

This method can be used when A is a matrix or an integral operator. In certain cases this method can be generalized to differential operators. One application of the this method to the equation of motion will be presented in §5.4.

This method fails in a more general setting (as has been emphasized many times) because the product $A_1 A_2 w$ is not, in general, bounded. For differential operators giving rise to boundary-value problems, $A_1 A_2 w$ is most often usually not even well defined. Discretizing the problem in the space variables fails to improve the matter. Though we now have a matrix A, and the formal manipulations can be carried out all constants characterizing the goodness of the approximation and the rate of convergence depend on the spatial mesh-sizes. In a number of cases this often degrades or eliminates convergence (see [146]).

In order to avoid having to deal with the product $A_1 A_2 w$ we will not drop the intermediate value $u^*(t)$ from (2.3) but rather consider it as some sort of approximate solution. This point of view is the basis of the method of additive approximation (see [114]) and leads to the an algorithm for obtaining increased accuracy. Consider the following expansions for u^τ and u^*:

$$u^\tau = u + \tau v_1 + \tau^2 v_2 + \tau^3 \eta^\tau \quad \text{on } \bar{\omega}_\tau, \qquad (2.19)$$

$$u^* = u + \tau w_1 + \tau^2 w_2 + \tau^3 \zeta^\tau \quad \text{on } \omega_\tau. \qquad (2.20)$$

Here the functions v_1, v_2, w_1, w_2 are independent of τ, and η^τ and ζ^τ are bounded on ω_τ.

In order to justify the above expansions we use them for u^τ and u^* in (2.3). (In what follows we will only show arguments which differ from t):

$$(u - u(t - \tau))/\tau + w_1 + \tau w_2 + \tau^2 \zeta^\tau - v_1(t - \tau) - \tau v_2(t - \tau) - \tau^2 \eta^\tau(t - \tau)$$

$$+ A_1 u + \tau A_1 w_1 + \tau^2 A_1 w_2 + \tau^3 A_1 \zeta^\tau = f, \qquad (2.21)$$

$$v_1 + \tau v_2 + \tau^2 \eta^\tau - w_1 - \tau w_2 - \tau^2 \zeta^\tau$$

$$+ A_2 u + \tau A_2 v_1 + \tau^2 A_2 v_2 + \tau^3 A_2 \eta^\tau = 0. \qquad (2.22)$$

Use the Taylor-series expansions of the functions u, v_1, v_2 from (2.8) in (2.21) to get

$$\frac{du}{dt} - \frac{\tau}{2}\frac{d^2 u}{dt^2} + \frac{\tau^2}{6}\frac{d^3 u}{dt^3} + w_1 - v_1 + \tau \frac{dv_1}{dt} - \tau^2 \frac{d^2 v_1}{dt^2}$$

$$+ \tau w_2 - \tau v_2 + \tau^2 \frac{dv_2}{dt} + \tau \zeta^\tau - \tau^2 \eta^\tau(t - \tau) + A_1 u + \tau A_1 w_1$$

$$+ \tau^2 A_1 w_2 + \tau^3 (\zeta_1 + \xi_2 + \xi_3) + \tau^3 A_1 \xi^\tau = f, \qquad t \in \omega_\tau. \qquad (2.23)$$

Equating the coefficients of τ^0 in (2.22) and (2.23) we get the system of equations:

$$v_1 - w_1 + A_2 u = 0,$$

$$\frac{du}{dt} + w_1 - v_1 + A_1 u = f, \qquad t \in \omega_\tau. \qquad (2.24)$$

This system is singular in the unknowns v_1, w_1; eliminating $v_1 - w_1$ from the second equation with the help of the first one we have:

$$\frac{du}{dt} + Au = f.$$

Thus (2.24) imposes only a single condition on the functions v_1 and w_1:

$$w_1 - v_1 = A_2 u, \qquad t \in (0, T). \tag{2.25}$$

Now equate the coefficients of τ. We have

$$v_2 - w_2 + A_2 v_1 = 0,$$

$$-\frac{1}{2}\frac{d^2u}{dt^2} + \frac{dv_1}{dt} + w_2 - v_2 + A_1 w_1 = 0, \qquad t \in \omega_\tau. \tag{2.26}$$

Though the system again appears to be singular with respect to v_2 and w_2 it will have a solution if the consistency condition

$$-\frac{1}{2}\frac{d^2u}{dt^2} + \frac{dv_1}{dt} + A_2 v_1 + A_1 w_1 = 0, \qquad t \in \omega_\tau,$$

is realized. We require that these equalities hold true not simply at the net points but over the entire interval $(0, T]$:

$$\frac{dv_1}{dt} + A_2 v_1 + A_1 w_1 = \frac{1}{2}\frac{d^2u}{dt^2}, \qquad t \in (0, T]. \tag{2.27}$$

Note that the system (2.25), (2.27) is a differential-algebraic problem. For (2.19) to be valid at all net points of $\bar{\omega}_\tau$ the following representation

$$u^\tau(0) = u(0) + \tau v_1(0) + \tau^2 v_2(0) + \tau^3 \eta^\tau(0) \tag{2.28}$$

should hold at the initial moment of time. Since $u^\tau(0) = u(0) = u_0$, then

$$\tau v_1(0) + \tau^2 v_2(0) + \tau^3 \eta^\tau(0) = 0. \tag{2.29}$$

Equating the coefficient of τ to zero we have the initial condition: $v_1(0) = 0$.
Suppose that

$$\frac{dv_1}{dt} + A_2 v_1 + A_1 w_1 = \frac{1}{2}\frac{d^2u}{dt^2}, \qquad t \in (0, T],$$

$$w_1 - v_1 = A_2 u, \qquad t \in (0, T],$$

$$v_1(0) = 0 \tag{2.30}$$

has at least one solution. (We leave the questions of existence and smoothness open; we will return to these in some concrete cases.)

The question of uniqueness is rather simple. Indeed, the uniqueness of the solution of (2.30) is related to the nonexistence of nontrivial solutions of

$$\frac{dv_1}{dt} + A_2 v_1 + A_1 w_1 = 0 \quad \text{on } (0, T],$$

$$w_1 - v_1 = 0 \quad \text{on } (0, T],$$

$$v_1(0) = 0.$$

But there cannot be any nontrivial solutions of this problem, since eliminating w_1 leads to

$$\frac{dv_1}{dt} + A v_1 = 0 \quad \text{on } (0, T],$$

$$v_1(0) = 0$$

which has only the trivial solution (see [66]).

Let us choose the functions v_1 and w_1 from (2.30) thus eliminating terms of order τ^2 from (2.22) and (2.23). We divide what remains by τ^3 and implement the usual splitting-up scheme with implicit local steps:

$$\frac{\eta^\tau - \zeta^\tau}{\tau} + A_2 \eta^\tau = -\frac{1}{\tau} A_2 v_2, \tag{2.31}$$

$$\frac{\zeta^\tau - \eta^\tau(t - \tau)}{\tau} + A_1 \zeta^\tau$$

$$= \frac{1}{\tau}\left(-\frac{1}{6}\frac{d^3 u}{dt^3} + \frac{1}{2}\frac{d^2 v_1}{dt^2} - \frac{dv_2}{dt} - A_1 w_2\right)$$

$$+ \xi_1 + \xi_2 + \xi_3, \qquad t \in \omega_\tau. \tag{2.32}$$

The conditions

$$A_1 v_2 = 0,$$

$$-\frac{1}{6}\frac{d^3 u}{dt^3} + \frac{1}{2}\frac{d^2 v_1}{dt^2} - \frac{dv_2}{dt} - A_1 w_2 = 0$$

are sufficient for the functions η^τ and ζ^τ to be bounded. But then the problem of finding the functions v_2, w_2 becomes overdetermined (in addition to the first equation of (2.26) there are three other equations for the two unknowns) and inconsistent. Let us use the method of additive approximation. For η^τ and ζ^τ to be bounded it is sufficient that the right-hand side be a bounded function. We have

$$-\frac{1}{6}\frac{d^3 u}{dt^3} + \frac{1}{2}\frac{d^2 v_1}{dt^2} - \frac{dv_2}{dt} - A_1 w_2 - A_2 v_2 = 0.$$

Extend this relation to the entire interval $(0, T]$:

$$\frac{dv_2}{dt} + A_1 w_2 + A_2 v_2 = -\frac{1}{6}\frac{d^3 u}{dt^3} + \frac{1}{2}\frac{d^2 v_1}{dt^2}, \qquad t \in (0, T]. \qquad (2.33)$$

In order to find the initial condition for the function v_2 we return to (2.29). Since $v_1(0) = 0$, then

$$v_2(0) + \tau \eta^\tau(0) = 0.$$

This means that $v_2(0) = 0$. Adding this condition to equation (2.33) and the first equation in (2.26) we have the system

$$\frac{dv_2}{dt} + A_1 w_2 + A_2 v_2 = -\frac{1}{6}\frac{d^3 u}{dt^3} + \frac{1}{2}\frac{d^2 v_1}{dt^2}, \qquad t \in (0, T],$$

$$w_2 - v_2 = A_2 v_1, \qquad\qquad\qquad t \in (0, T],$$

$$v_2(0) = 0. \qquad\qquad\qquad\qquad\qquad (2.34)$$

Suppose that the solution of this problem exists and that the function v_2 is sufficiently smooth. The uniqueness of the solution is proved as for system (2.30). Therefore our construction of the functions v_1, v_2, w_1, w_2 is complete. Now let us investigate the system (2.31), (2.32). Add the uniform initial condition which is a consequence of (2.29) and the conditions $v_1(0) = v_2(0) = 0$. We have

$$\frac{\eta^\tau - \zeta^\tau}{\tau} + A_2 \eta^\tau = g_1 \equiv \frac{1}{\tau}\psi_1, \qquad t \in \omega_\tau,$$

$$\frac{\zeta^\tau - \eta^\tau(t - \tau)}{\tau} + A_1 \zeta^\tau = g_2 \equiv \frac{1}{\tau}\psi_2 + \psi_3, \qquad t \in \omega_\tau,$$

$$\eta^\tau(0) = 0. \qquad\qquad\qquad\qquad (2.35)$$

Here all values of ψ_i are bounded:

$$\|\psi_i\| \le c_3, \qquad t \in \omega_\tau, \qquad i = 1, 2, 3. \qquad (2.36)$$

Let us show that the solution of (2.35) is bounded in spite of the fact that the right-hand side is of order τ^{-1}. Consider

$$\frac{\rho^\tau - \sigma^\tau}{\tau} = \frac{1}{\tau}\psi_1, \qquad t \in \omega_\tau,$$

$$\frac{\sigma^\tau - \rho^\tau(t - \tau)}{\tau} = \frac{1}{\tau}\psi_2, \qquad t \in \omega_\tau,$$

$$\rho^\tau(0) = 0. \qquad\qquad\qquad\qquad (2.37)$$

Its solution can be constructed explicitly:

$$\rho^\tau = 0, \qquad t \in \bar{\omega}_\tau,$$
$$\sigma^\tau = \psi_2, \qquad t \in \omega_\tau. \qquad (2.38)$$

Here we use the fact that from (2.33) we have

$$\psi_1 + \psi_2 = 0, \qquad t \in \omega_\tau.$$

Rewrite (2.37) in the form

$$\frac{\rho^\tau - \sigma^\tau}{\tau} + A_2 \rho^\tau = \frac{1}{\tau} \psi_1, \qquad t \in \omega_\tau,$$

$$\frac{\sigma^\tau - \rho^\tau(t - \tau)}{\tau} + A_1 \sigma^\tau = \frac{1}{\tau} \psi_2 + A_1 \psi_2, \qquad t \in \omega_\tau,$$

$$\rho^\tau(0) = 0. \qquad (2.39)$$

Assuming $\eta^\tau - \rho^\tau = \varepsilon^\tau$, $\zeta^\tau - \sigma^\tau = \delta^\tau$ and subtracting equations (2.39) from the corresponding equations (2.35) we come to

$$\frac{\varepsilon^\tau - \delta^\tau}{\tau} + A_2 \varepsilon^\tau = 0, \qquad t \in \omega_\tau,$$

$$\frac{\delta^\tau - \varepsilon^\tau(t - \tau)}{\tau} + A_1 \delta^\tau = \psi_3 - A_1 \psi_2, \qquad t \in \omega_\tau,$$

$$\varepsilon^\tau(0) = 0. \qquad (2.40)$$

Since the function ψ_2 is known, it is not difficult to show that $A_1 \psi_2$ is bounded. Therefore, the right-hand side of (2.40) is bounded and an application of (2.5) leads to

$$\max_{\bar{\omega}_\tau} \|\varepsilon^\tau\| \le c_4, \qquad \max_{\omega_\tau} \|\delta^\tau\| \le c_4.$$

Taking into account that

$$\eta^\tau = \rho^\tau + \varepsilon^\tau, \qquad \zeta^\tau = \sigma^\tau + \delta^\tau$$

we have

$$\max_{\bar{\omega}_\tau} \|\eta^\tau\| \le \max_{\bar{\omega}_\tau} \|\rho^\tau\| + \max_{\bar{\omega}_\tau} \|\varepsilon^\tau\| \le c_4,$$

$$\max_{\omega_\tau} \|\zeta^\tau\| \le \max_{\omega_\tau} \|\sigma^\tau\| + \max_{\omega_\tau} \|\delta^\tau\| \le c_3 + c_4.$$

Thus the validity of (2.19) is proved. Using expansion (2.20) for the same purpose as (2.19) is theoretically possible. However, the functions w_1, w_2 are usually less smooth than v_1 and v_2, respectively, and have lesser accuracy. This means that the corrected solution obtained from (2.19) is more accurate than the solution which uses (2.20). This difference is especially noticeable near the boundary for partial differential equations.

5.3. The Two-Dimensional Heat Equation

The method described in the previous section can be illustrated with the heat equation.

Let Ω be a two-dimensional bounded region with Γ its boundary. Denote by Q an open cylinder $\Omega \times (0, T)$ with surface $S = \Gamma \times [0, T]$. Consider the equation

$$\frac{\partial u}{\partial t} = \Delta u + f \quad \text{in } Q. \tag{3.1}$$

We put impose initial conditions and boundary conditions on u:

$$u(x, 0) = 0, \qquad x \in \Omega, \tag{3.2}$$

$$u(x, t) = 0, \qquad (x, t) \in S. \tag{3.3}$$

For the one-dimensional case we have already introduced the function classes which characterize the solution. For the two-dimensional case these classes are similar to those in [71]. For any positive nonintegral l we denote by $H^l(\bar{Q})$ the Banach space of functions $v(x, t)$ which are continuous on Q together with all derivatives of the form

$$\frac{\partial^{r+s_1+s_2} v}{\partial t^2 \, \partial x^{s_1} \, \partial y^{s_2}} \quad \text{at } 2r + s_1 + s_2 < l$$

and with finite norm

$$\|u\|_{H^l(\bar{Q})} = \langle u \rangle_Q^{(l)} + \sum_{j=0}^{[l]} \langle u \rangle_Q^{(j)}, \tag{3.4}$$

where

$$\langle u \rangle_Q^{(j)} = \sum_{2r+s_1+s_2=j} \max_{\bar{Q}} \left| \frac{\partial^{r+s_1+s_2} u}{\partial t^r \, \partial x^{s_1} \, \partial y^{s_2}} \right|, \qquad j = 0, 1, \ldots, [l],$$

$$\langle u \rangle_Q^{(l)} = \langle u \rangle_x^l + \langle u \rangle_t^{(l/2)},$$

$$\langle u \rangle_x^{(l)} = \sum_{2r+s_1+s_2=[l]} \left\langle \frac{\partial^{r+s_1+s_2} u}{\partial t^r \, \partial x^{s_1} \, \partial y^{s_2}} \right\rangle_x^{l-[l]},$$

$$\langle u \rangle_t^{(l/2)} = \sum_{0 < l-2r-s_1-s_2 < 2} \left\langle \frac{\partial^{r+s_1+s_2} u}{\partial t^r \, \partial x^{s_1} \, \partial y^{s_2}} \right\rangle_t^{(l-2r-s_1-s_2)/2}$$

The values $\langle u \rangle_x^\alpha$ and $\langle u \rangle_t^\beta$ for $\alpha, \beta \in (0, 1)$ are as in §5.1.

The coherence condition of order 0 is automatically satisfied because of the homogeneity of the initial values and boundary values: if $(x, y) \in \Gamma$, then

$$\lim_{t \to 0} u(x, y, t) = \lim_{(x', y') \to (x, y)} u(x', y', 0). \tag{3.5}$$

The necessary coherence condition of order 1 for the solution with derivatives $\partial u/\partial t$, $\partial^2 u/\partial x^2$, $\partial^2 u/\partial y^2$ continuous on $\bar{Q}$ takes the form

$$f(x, y, 0) = 0 \quad \forall\, (x, y) \in \Gamma. \tag{3.6}$$

The second-order derivative $\partial^2 u/\partial t^2$ is equal to zero on S due to homogeneity of the boundary condition (3.3). Let us calculate this derivative using (3.1):

$$\frac{\partial^2 u}{\partial t^2} = \frac{\partial}{\partial t}(\Delta u + f) = \Delta\left(\frac{\partial u}{\partial t}\right) + \frac{\partial f}{\partial t} = \Delta(\Delta u + f) + \frac{\partial f}{\partial t}$$

$$= \Delta\Delta u + \Delta f + \frac{\partial f}{\partial t}.$$

The term $\Delta\Delta u$ is zero at $t = 0$ because of the initial condition (3.2). Therefore for $\partial^2 u/\partial t^2$ and $\Delta\Delta u$ to exist and be continuous it is necessary that the coherence condition of order 2

$$\Delta f(x, 0) + \frac{\partial f}{\partial t}(x, 0) = 0 \quad \forall\, x \in \Gamma \tag{3.7}$$

hold. The higher-order conditions are constructed similarly (see [71]). As in the one-dimensional case these conditions are not only necessary but also sufficient for the corresponding derivatives to be continuous on $\bar{Q}$ when f is smooth.

Theorem 3.1 (see [71]). *Let $l > 0$, be nonintegral $f \in H^l(\bar{Q})$, $\Gamma \in C^{l+2}$. Then (3.1)–(3.3) has a unique solution $u \in H^{l+2}(\bar{Q})$ when the coherence conditions of order $0, 1, \ldots, [l/2] + 1$ are satisfied.*

Let us construct the difference scheme. Using Richardson extrapolation in the case of a locally-one-dimensional scheme [114] results in an approximate solution of $\tau^2 + h^3$ order accuracy, since one fails to reduce the contribution of the irregular error terms appearing near the boundary when one uses a threepoint scheme. Improving this scheme is of great interest; we will formulate it more precisely at the end of the section.

In §4.2 we discussed the problem of reducing the irregular boundary error and gave two approaches to solving this problem. The first one used a special multipoint approximation of the boundary conditions near the boundary. The second one uses special multipoint approximations of the second partial derivatives on knots placed nonuniformly near the boundary. In both cases only the standard fivepoint scheme was used inside the region. The second approach appears to be more profitable for the parabolic equation (in order to reach at least h^3 or h^4 accuracy).

Let us first construct the spatial difference net. Suppose a region Ω is contained in the square $\{-b < x < b, -b < y < b\}$. Cover this square with a rectangular uniform net with mesh-size $h = b/N$, formed by the lines $x_i = ih$ and $y_j = jh$, where $i, j = -N, \ldots, N$. Denote by Ω_h the set of net points inside Ω, which we will refer to as the set of inner net points. As in §§4.2.1 and 4.2.3 we will introduce the sets of boundary, regular, and irregular net points in the x- and y-directions and denote them by $\Gamma_{h,x}$, $\Omega^r_{h,x}$, $\Omega^{ir}_{h,x}$, $\Gamma_{h,y}$, $\Omega^r_{h,y}$, $\Omega^{ir}_{h,y}$, respectively. To form the difference approximation for the operators $L_1 = \partial^2/\partial x^2$ and $L_2 = \partial^2/\partial y^2$ we use the threepoint formula at the regular net points and the fourpoint formula at the irregular net points. Let (x, y) be a net point of Ω_h. When we replace the operators L_i by the difference operators L^h_i four cases are possible.

(a) The net point (x, y) is regular in the x-direction. Then

$$L^h_1 u(x, y) = u_{\bar{x}\hat{x}}(x, y) = \{u(x - h, y) - 2u(x, y) + u(x + h, y)\}/h^2. \quad (3.8)$$

(b) The net point (x, y) is irregular in the x-direction. Then by moving in the Ox direction one can find the point $(\xi, y) \in \Gamma_{h,x}$, nearest to the net point (x, y) at a distance less than h. Let this distance be δh, where $\delta \in (0, 1)$. Assume

$$L^h_1 u(x, y) = \frac{6}{\delta(\delta + 1)(\delta + 2)h^2} u(\xi, y) - \frac{3 - \delta}{\delta h^2} u(x, y)$$

$$+ \frac{4 - 2\delta}{(1 + \delta)h^2} u(x \pm h, y) - \frac{1 - \delta}{(2 + \delta)h^2} u(x \pm 2h, y). \quad (3.9)$$

The minus sign is taken when $\xi > x$ (see Figure 4.2). The sign plus should be taken otherwise. It is possible that the point $(x \pm 2h, y)$ lies outside the region and the approximation becomes meaningless. In §4.2.3 we showed that in order to extricate ourselves from this situation we should treat several of the regular points as irregular points. We must take into account the fact that in some cases we have $1 < \delta < \frac{3}{2}$. Note that δ is defined at every point of $\Omega^{ir}_{h,x}$. Thus we have constructed a positive function $\delta(x)$ on $\Omega^{ir}_{h,x}$.

(c) The net point (x, y) is regular in the y-direction. Then

$$L^h_2 u(x, y) = u_{\bar{y}\hat{y}}(x, y) = \{u(x, y - h) - 2u(x, y) + u(x, y + h)\}/h^2. \quad (3.10)$$

(d) The net point (x, y) is irregular in the y-direction. Then one can find a point $(x, \eta) \in \Gamma_{h,y}$, which is nearest to the net point (x, y) in the Oy direction, at a distance ρh, where $\rho \in (0, 1)$. Assume

$$L^h_2 u(x, y) = \frac{6}{\rho(\rho + 1)(\rho + 2)h^2} u(x, \eta) - \frac{3 - \rho}{\rho h^2} u(x, y)$$

$$+ \frac{4 - 2\rho}{(1 + \rho)h^2} u(x, y \pm h) - \frac{1 - \rho}{(2 + \rho)h^2} u(x, y \pm 2h). \quad (3.11)$$

The minus sign is taken when $\eta > y$. Otherwise one should take the plus sign. It is possible that the interval with endpoints (x, η) and $(x, y \pm 2h)$ does not belong to $\bar{\Omega}$. An application of the method suggested in §4.2.3

gives a value of ρ greater than 1 and less than $\frac{3}{2}$, for some net points $x \in \Omega_{h,y}^{ir}$. Thus we have constructed a function $\rho(x)$ defined at every point $x \in \Omega_{h,y}^{ir}$.

Let us now describe the splitting-up scheme. Introduce on the interval $[0, T]$ the net

$$\bar{\omega}_\tau = \{t_k = k\tau; k = 0, 1, \ldots, M\}$$

with mesh-size $\tau = T/M$. Assume

$$\omega_\tau = \{t_k \in \bar{\omega}_\tau, k \neq 0\}.$$

We introduce a difference net on the cylinder $Q = \Omega \times (0, T)$ as the Cartesian product: $Q_h^\tau = \Omega_h \times \omega_\tau$. Define

$$\bar{Q}_h^\tau = \bar{\Omega}_h \times \bar{\omega}_\tau, \qquad S_h^\tau = \Gamma_h \times \bar{\omega}_\tau.$$

We will study the difference scheme consisting of cyclicly repeated spatially-one-dimensional equations. We omit the arguments which are not differentiated we have:

$$\frac{u^* - u^\tau(x, y, t - \tau)}{\tau} - L_1^h u^* = f, \qquad (x, y, t) \in Q_h^\tau,$$

$$u^* = 0, \qquad (x, y, t) \in \Gamma_{h,x} \times \omega_\tau, \qquad (3.12)$$

$$\frac{u^\tau - u^*}{\tau} - L_2^h u^\tau = 0, \qquad (x, y, t) \in Q_h^\tau,$$

$$u^\tau = 0, \qquad (x, y, t) \in \Gamma_{h,y} \times \omega_\tau. \qquad (3.13)$$

The following condition is imposed at $t = 0$:

$$u^\tau(x, y, 0) = 0, \qquad (x, y) \in \Omega_h. \qquad (3.14)$$

We will exhibit a method for finding the solution of (3.12)–(3.14). Suppose $u^\tau(x, y, t - \tau)$ has been computed for $(x, y) \in \Omega_h$. Then the solution is constructed by first finding $u^*(x, y, t)$ $\forall$ $(x, y) \in \Omega_h$ using (3.12) in each time slice and then finding $u^\tau(x, y, t)$ at the net points $(x, y) \in \Omega_h$ using (3.13). The values $u^\tau(x, y, t)$ obtained can be used for the next time step.

Consider (3.12). With respect to its unknowns $u^*(x, y, t)$ this system is a collection of "almost" tridiagonal subsystems. There are only two elements in each subsystem which prevent the matrix from being tridiagonal:

$$
\begin{bmatrix}
b_1 & e_1 & d_1 & & & \\
a_2 & b_2 & e_2 & & & \\
& a_3 & \ddots & \ddots & & \\
& & \ddots & \ddots & \ddots & e_{n-2} \\
& & & a_{n-1} & b_{n-1} & e_{n-1} \\
& & & d_n & a_n & b_b
\end{bmatrix}
\begin{bmatrix}
z_1 \\
z_2 \\
\vdots \\
\\
z_{n-1} \\
z_n
\end{bmatrix}
=
\begin{bmatrix}
g_1 \\
g_2 \\
\vdots \\
\\
g_{n-1} \\
g_n
\end{bmatrix}
\qquad (3.15)
$$

Here z_i are the values $u^*(x, y, t)$, taken in a certain order:

$$z_i = u^*(x_{l+i}, y, t), \qquad i = 1, \ldots, n;$$

the right-hand side contains the values of f and u^τ which are known from the previous time slice; and

$$g_i = f(x_{l+i}, y, t) + \frac{1}{\tau} u^\tau(x_{l+i}, y, t - \tau).$$

System (3.15) contains only unknowns u^* with arguments which are net points of Ω_h for some fixed y; moreover, the set of these net points is such that the points (x_l, y) and (x_{l+n}, y) are irregular in the x-direction, while all intermediate net points are regular. It is obvious that for a nonconvex region Ω there can be several such systems for fixed y.

Introduce the notation $\sigma^- = \delta(x_l, y)$ and $\sigma^+ = \delta(x_{l+n}, y)$. Then the coefficients of (3.15) can be written as

$$b_1 = \frac{1}{\tau} + \frac{3 - \sigma^-}{\sigma^- h^2}, \qquad e_1 = -\frac{4 - 2\sigma^-}{(1 + \sigma^-)h^2}, \qquad d_1 = \frac{1 - \sigma^-}{(2 + \sigma^-)h^2},$$

$$a_i = e_i = -1/h^2, \qquad b_i = 1/\tau + 2/h^2, \qquad i = 2, \ldots, n - 1, \qquad (3.16)$$

$$d_n = \frac{1 + \sigma^+}{(2 + \sigma^+)h^2}, \qquad a_n = -\frac{4 - 2\sigma^+}{(1 + \sigma^+)h^2}, \qquad b_n = \frac{1}{\tau} + \frac{3 - \sigma^+}{\sigma^+ h^2}.$$

Let us prove that given any relation between τ and h, (3.15) is strictly diagonal-dominant (see [103]). There are two possible cases for the first equation: (a) $\sigma^- \in (0, 1]$, (b) $\sigma \in (1, \frac{3}{2})$. Let us consider each in turn.

(a) $\sigma^- \in (0, 1]$. Then

$$|b_1| - |e_1| - |d_1| = \frac{1}{\tau} + \frac{3 - \sigma^-}{\sigma^- h^2} - \frac{4 - 2\sigma^-}{(1 + \sigma^-)h^2} - \frac{1 - \sigma^-}{(2 + \sigma^-)h^2}$$

$$\geq \frac{1}{\tau} + \frac{6 - 5\sigma^- + 2(\sigma^-)^2}{2\sigma^-(1 + \sigma^-)h^2} \geq \frac{1}{\tau} + \frac{3}{4\sigma^- h^2}$$

$$\geq \frac{1}{\tau} + \frac{1}{2\sigma^- h^2}.$$

(b) $\sigma^- \in (1, \frac{3}{2}]$. In this case

$$|b_1| - |e_1| - |d_1| = \frac{1}{\tau} + \frac{3 - \sigma^-}{\sigma^- h^2} - \frac{4 - 2\sigma^-}{(1 + \sigma^-)h^2} + \frac{1 - \sigma^-}{(2 + \sigma^-)h^2}$$

$$\geq \frac{1}{\tau} + \frac{9 - 5\sigma^- + 2(\sigma^-)^2}{3\sigma^-(1 + \sigma^-)h^2} \geq \frac{1}{\tau} + \frac{47}{60\sigma^- h^2}$$

$$\geq \frac{1}{\tau} + \frac{1}{2\sigma^- h^2}.$$

Thus in each case we have the estimate

$$|b_1| - |e_1| - |d_1| \geq \frac{1}{\tau} + \frac{1}{2h^2\sigma^-}. \tag{3.17}$$

The inequality

$$|b_n| - |a_n| - |d_n| \geq \frac{1}{\tau} + \frac{1}{2h^2\sigma^+} \tag{3.18}$$

is proved in a similar way. For all other values of the index i the inequality

$$|b_i| - |a_i| - |e_i| \geq 1/\tau \tag{3.19}$$

holds.

These inequalities are sufficient to prove the regularity of (3.15) (see [103]). If we first exclude the first and the last unknowns this system can then be solved by the factorization method.

The problem of finding the solution of system (3.13) is similar to the above. This system is again a collection of "almost" tridiagonal systems of the (3.15) kind. But this time the z_i are the values of u^τ:

$$z_i = u^\tau(x, y_{m+i}, t), \qquad i = 1, \dots, q$$

and values of u^* we find show up on the right-hand side:

$$g_i = \frac{1}{\tau} u^*(x, y_{m+i}, t).$$

Each subsystem (3.15) has unknown values u^τ with arguments that are points of Ω_h for a certain fixed x. When Ω is nonconvex there can be several such subsystems for one x. For these subsystems (3.17)–(3.19) also hold, the system is again regular and can be solved by the factorization method after some minor preliminaries.

Summing up, we conclude that there is at least one solution of (3.12)–(3.14); we have given an economical method for finding it. To pass from one time slice to the next we need α operations where α is proportional to the number of points of Ω_h.

Let us prove an *a priori* estimate which characterizes the stability of numerical solutions as we pass from one time slice to the next.

Theorem 3.2. *The system*

$$\frac{v^* - v^\tau(x, y, t - \tau)}{\tau} - L_1^h v^* = f_1, \qquad (x, y, t) \in Q_h^\tau,$$

$$v^* = 0, \qquad (x, y, t) \in \Gamma_{h,x} \times \omega_\tau, \tag{3.20}$$

$$\frac{v^\tau - v^*}{\tau} - L_2^h v^\tau = f_2, \qquad (x, y, t) \in Q_h^\tau,$$

$$v^\tau = 0, \qquad (x, y, t) \in \Gamma_{h,y} \times \omega_\tau, \tag{3.21}$$

with the initial condition

$$v^\tau(x, y, 0) = 0, \qquad (x, y) \in \Omega_h, \tag{3.22}$$

obeys the inequality

$$\max_{Q_h^\tau} |v^\tau| \le \sum_{t \in \omega_\tau} \tau \left\{ \max_{\Omega_{h, x}^\tau} |f_1| + \max_{\Omega_{h, y}^\tau} |f_2| \right\}$$

$$+ 2h^2 \left\{ \max_{\Omega_{h, x}^{ir} \times \omega_\tau} |\delta f_1| + \max_{\Omega_{h, y}^{ir} \times \omega_\tau} |\rho f_2| \right\}.$$

PROOF. Divide the functions f_1 and f_2 at the points of the difference net Q_h^τ into two components:

$$f_1 = f_1^r + f_1^{ir}, \qquad f_2 = f_2^r + f_2^{ir}.$$

The support of the net function f_1^r is concentrated at the net points of the region $\Omega_{h, x}^r \times \omega_\tau$, and the support of the function f_1^{ir} is concentrated at the points of the region $\Omega_{h, x}^{ir} \times \omega_\tau$:

$$f_1^r = \begin{cases} f_1 & \text{on } \Omega_{h, x}^r \times \omega_\tau, \\ 0 & \text{on } \Omega_{h, x}^{ir} \times \omega_\tau, \end{cases} \qquad f_1^{ir} = \begin{cases} 0 & \text{on } \Omega_{h, x}^r \times \omega_\tau, \\ f_1 & \text{on } \Omega_{h, x}^{ir} \times \omega_\tau. \end{cases}$$

There are the similar relations for the function f_2:

$$f_2^r = \begin{cases} f_2 & \text{on } \Omega_{h, y}^r \times \omega_\tau, \\ 0 & \text{on } \Omega_{h, y}^{ir} \times \omega_\tau, \end{cases} \qquad f_2^{ir} = \begin{cases} 0 & \text{on } \Omega_{h, y}^r \times \omega_\tau, \\ f_2 & \text{on } \Omega_{h, y}^{ir} \times \omega_\tau. \end{cases}$$

We assume

$$v^\tau = w_1^\tau + w_2^\tau, \qquad v^* = w_1^* + w_2^*,$$

where w_1^τ, w_1^* are the solutions of (3.20)–(3.22) with right-hand sides f_1^r and f_2^r, and w_2^τ, w_2^* are solutions of the same problem but with right-hand sides f_1^{ir} and f_2^{ir}. Let us estimate each of the solutions; we begin with w_1^τ, w_1^*.

Suppose

$$\max_{(x, y) \in \Omega_h} |w_1^\tau(x, y, t - \tau)| \le \sum_{\substack{t' \in \omega_\tau \\ t' \le t - \tau}} \tau \left(\max_{Q_h^\tau} |f_1^r| + \max_{Q_h^\tau} |f_2^r| \right) \tag{3.23}$$

has been proved for some $t \in \bar{\omega}_\tau$. Let us show its validity for any t. To do this consider again the subsystem (3.15) for the two right-hand sides. First assume

$$g_i = \frac{1}{\tau} w_1^\tau(x_{l+i}, y, t)$$

(the right-hand side f_1^r is zero). In this case we have the estimate

$$\max_{(x, y) \in \Omega_h} |w_1^*(x, y, t)| \le \max_{(x, y) \in \Omega_h} |w_1^\tau(x, y, t - \tau)|. \tag{3.24}$$

Indeed, if this inequality is not true then at some net point $(x_0, y_0) \in \Omega_h$ the maximum of $|w_1^*(x, y, t)|$ is attained, which is greater than the maximum of $|w_1^\tau(x, y, t - \tau)|$:

$$|w_1^*(x_0, y_0, t)| > \max_{(x, y) \in \Omega_h} |w_1^\tau(x, y, t - \tau)|. \tag{3.25}$$

Consider the equation corresponding to this net point. If the net point (x_0, y_0) is regular in the x-direction then

$$(2/h^2 + 1/\tau) w_1^*(x_0, y_0, t) = \frac{1}{h^2} w_1^*(x_0 - h, y_0, t)$$

$$+ \frac{1}{h^2} w_1^*(x_0 + h, y_0, t) + \frac{1}{\tau} w_1^\tau(x_0, y_0, t - \tau).$$

From the above and from inequality (3.25) it follows that

$$\left(\frac{2}{h^2} + \frac{1}{\tau}\right) |w_1^*(x_0, y_0, t)| < \left(\frac{2}{h^2} + \frac{1}{\tau}\right) |w_1^*(x_0, y_0, t)|.$$

Since this inequality gives a contradiction, the point (x_0, y_0) cannot be regular in the x-direction.

Now suppose the net point (x_0, y_0) is irregular in the x-direction. Then in accordance with the notation introduced in (3.15), (3.16) we have:

$$\left(\frac{1}{\tau} + \frac{3 - \sigma^\pm}{\sigma^\pm h^2}\right) w_1^*(x_0, y_0, t) = \frac{4 - 2\sigma^\pm}{(1 + \sigma^\pm) h^2} w_1^*(x_0 \pm h, y_0, t)$$

$$- \frac{1 - \sigma^\pm}{(2 + \sigma^\pm) h^2} w_1^*(x_0 \pm 2h, y_0, t)$$

$$+ \frac{1}{\tau} w_1^\tau(x_0, y_0, t - \tau).$$

The sign is chosen depending on where the net point is located. Taking the modulus on both sides and using (3.25) we have

$$\left(\frac{1}{\tau} + \frac{3 - \sigma^\pm}{\sigma^\pm h^2}\right) |w_1^*(x_0, y_0, t)|$$

$$< \left(\frac{4 - 2\sigma^\pm}{(1 + \sigma^\pm) h^2} + \frac{|1 - \sigma^\pm|}{(2 + \sigma^\pm) h^2} + \frac{1}{\tau}\right) |w_1^*(x_0, y_0, t)|.$$

From (3.17), (3.18) we get the contradiction

$$\left(\frac{1}{\tau} + \frac{3 - \sigma^\pm}{\sigma^\pm h^2}\right) |w_1^*(x_0, y_0, t)| \leq \left(\frac{1}{\tau} + \frac{3 - \sigma^\pm}{\sigma^\pm h^2}\right) |w_1^*(x_0, y_0, t)|.$$

Therefore (3.25) is impossible at any point of the region Ω_h and (3.24) is valid for a uniform right-hand side.

Now take

$$g_i = f'_1(x_{l+i}, y, t)$$

(the initial data w'_1 are zero). Let us prove that at the point (x_0, y_0), where the maximum of $|w^*_1(x, y, t)|$ is attained, the estimate

$$|w^*_1(x_0, y_0, t)| \leq \tau |f'_1(x_0, y_0, t)|. \tag{3.26}$$

holds. If the point (x_0, y_0) is regular in the x-direction then

$$(2/h^2 + 1/\tau)w^*_1(x_0, y_0, t) = \frac{1}{h^2} w^*_1(x_0 - h, y_0, t)$$

$$+ \frac{1}{h^2} w^*_1(x_0 + h, y_0, t) + f'_1(x_0, y_0, t).$$

Taking the modulus on both sides in the above and replacing the right-hand side by a greater quantity gives:

$$(2/h^2 + 1/\tau)|w^*_1(x_0, y_0, t)| \leq \frac{2}{h^2}|w^*_1(x_0, y_0, t)| + |f'_1(x_0, y_0, t)|.$$

Hence (3.26) follows.

If the point (x_0, y_0) is irregular in the x-direction then

$$\left(\frac{1}{\tau} + \frac{3 - \sigma^{\pm}}{\sigma^{\pm}h^2}\right)w^*_1(x_0, y_0, t) = \frac{4 - 2\sigma^{\pm}}{(1 + \sigma^{\pm})h^2} w^*_1(x_0 \pm h, y_0, t)$$

$$- \frac{1 - \sigma^{\pm}}{(2 + \sigma^{\pm})h^2} w^*_1(x_0 \pm 2h, y_0, t) + f'_1(x_0, y_0, t).$$

Taking the modulus of both sides and replacing the right-hand side by a larger quantity we have

$$\left(\frac{1}{\tau} + \frac{3 - \sigma^{\pm}}{\sigma^{\pm}h^2}\right)|w^*_1(x_0, y_0, t)|$$

$$\leq \left(\frac{4 - 2\sigma^{\pm}}{(1 + \sigma^{\pm})h^2} + \frac{|1 - \sigma^{\pm}|}{(2 + \sigma^{\pm})h^2}\right)|w^*_1(x_0, y_0, t)| + |f'_1(x_0, y_0, t)|.$$

From (3.17), (3.18) we have

$$\left(\frac{1}{\tau} + \frac{3 - \sigma^{\pm}}{\sigma^{\pm}h^2}\right)|w^*_1(x_0, y_0, t)| \leq \frac{3 - \sigma^{\pm}}{\sigma^{\pm}h^2} |w^*_1(x_0, y_0, t)| + |f'_1(x_0, y_0, t)|$$

which yields (3.26). Considering that (x_0, y_0) gives the maximum of the function $|w^*_1(x, y, t)|$ for fixed t we obtain

$$\max_{(x, y) \in \Omega_h} |w^*_1(x, y, t)| \leq \tau \max_{(x, y) \in \Omega_h} |f'_1(x, y, t)|. \tag{3.27}$$

Now consider the situation where both the right-hand side and the initial values are nonzero. By (3.24), (3.27) the solution w_1^* of

$$\frac{w_1^* - w_1^\tau(x, y, t - \tau)}{\tau} - L_1^h w_1^* = f_1^\tau, \qquad (x, y) \in \Omega_h,$$

$$w_1^* = 0, \qquad (x, y) \in \Gamma_{h, x},$$

obeys the estimate

$$\max_{(x, y) \in \Omega_h} |w_1^*(x, y, t)| \le \max_{(x, y) \in \Omega_h} |w_1^\tau(x, y, t - \tau)| + \tau \max_{(x, y) \in \Omega_h} |f_1^\tau(x, y, t)|.$$

$$(3.28)$$

As in the above the solution w_1^τ of

$$\frac{w_1^\tau - w_1^*}{\tau} - L_2 w_1^\tau = f_2^\tau, \qquad (x, y) \in \Omega_h,$$

$$w_1^\tau = 0, \qquad (x, y) \in \Gamma_{h, x},$$

obeys the inequality

$$\max_{(x, y) \in \Omega_h} |w_1^\tau(x, y, t)| \le \max_{(x, y) \in \Omega_h} |w_1^*(x, y, t)| + \tau \max_{(x, y) \in \Omega_h} |f_2^\tau(x, y, t)|. \qquad (3.29)$$

For a whole step τ (when $w_1^\tau(x, y, t)$ is calculated from $w_1^\tau(x, y, t - \tau)$) from (3.28), (3.29) it follows that

$$\max_{(x, y) \in \Omega_h} |w_1^\tau(x, y, t)| \le \max_{(x, y) \in \Omega_h} |w_1^\tau(x, y, t - \tau)|$$

$$+ \tau \max_{Q_h^\tau} |f_1^\tau| + \tau \max_{Q_h^\tau} |f_2^\tau|.$$

If we use (3.23) for the time slices $t' < t$ then

$$\max_{Q_h^\tau} |w_1^\tau| \le \sum_{\substack{t' \in \omega_\tau \\ t' \le t}} \tau \left(\max_{Q_h^\tau} |f_1^\tau| + \max_{Q_h^\tau} |f_2^\tau| \right). \qquad (3.30)$$

Thus (3.23) is proved for all time slices.

Let us study the system for the function w_2^τ:

$$\frac{w_2^* - w_2^\tau(x, y, t - \tau)}{\tau} - L_1^h w_2^* = f_1^{ir}, \qquad (x, y, t) \in Q_h^\tau,$$

$$w_2^* = 0, \qquad (x, y, t) \in \Gamma_{h, x} \times \omega_\tau, \qquad (3.31)$$

$$\frac{w_2^\tau - w_2^*}{\tau} - L_2^h w_2^\tau = f_2^{ir}, \qquad (x, y, t) \in Q_h^\tau,$$

$$w_2^\tau = 0, \qquad (x, y, t) \in \Gamma_{h, y} \times \omega_\tau, \qquad (3.32)$$

$$w_2^\tau(x, y, 0) = 0, \qquad (x, y) \in \Omega_h. \qquad (3.33)$$

Let us prove the inequality

$$\max\left\{\max_{Q_h^\tau}|w_2^\tau|, \max_{Q_h^\tau}|w_2^*|\right\} \le 2h^2\left(\max_{\Omega_{h,\,x}^{ir}\times\omega_\tau}|\delta f_1^{ir}| + \max_{\Omega_{h,\,y}^{ir}\times\omega_\tau}|\rho f_2^{ir}|\right). \tag{3.34}$$

Suppose that

$$|w_2^\tau(x_0, y_0, t_0)| = \max\left\{\max_{Q_h^\tau}|w_2^\tau|, \max_{Q_h^\tau}|w_2^*|\right\}. \tag{3.35}$$

Assume the point (x_0, y_0) is irregular in the y-direction. Consider the equation corresponding to this point in (3.31). It is of the form

$$\left(\frac{3-\rho}{\rho h^2} + \frac{1}{\tau}\right)w_2^\tau(x_0, y_0, t_0) = \frac{4-2\rho}{(1+\rho)h^2}\,w_2^\tau(x_0, y_0 \pm h, t_0)$$

$$- \frac{1-\rho}{(2+\rho)h^2}\,w_2^\tau(x_0, y_0 \pm 2h, t_0) + \frac{1}{\tau}\,w_2^*(x_0, y_0, t_0)$$

$$+ f_2^{ir}(x_0, y_0, t_0), \qquad \rho = \rho(x_0, y_0).$$

The choice of sign depends on where the point (x_0, y_0) is located and is of no no importance in what follows. In this equality we let us take the modulus on both sides and consider (3.35). We have

$$\left(\frac{3-\rho}{\rho h^2} + \frac{1}{\tau}\right)|w_2^\tau(x_0, y_0, t_0)|$$

$$\le \left(\frac{4-2\rho}{(1+\rho)h^2} + \frac{|1-\rho|}{(2+\rho)h^2} + \frac{1}{\tau}\right)|w_2^\tau(x_0, y_0, t_0)| + |f_2^{ir}(x_0, y_0, t_0)|.$$

Assume $\sigma^- = \rho$ or $\sigma^+ = \rho$ in (3.16); then (3.17), (3.18) can be written as

$$\frac{1}{\tau} + \frac{3-\rho}{\rho h^2} - \frac{4-2\rho}{(1+\rho)h^2} - \frac{|1-\rho|}{(2+\rho)h^2} \ge \frac{1}{\tau} + \frac{1}{2\rho h^2}.$$

Using this on the inequality above we obtain

$$\frac{1}{2\rho h^2}|w_2^\tau(x_0, y_0, t_0)| \le |f_2^{ir}(x_0, y^0, t_0)|,$$

so that from (3.35) inequality (3.34) follows.

If the point (x_0, y_0) is regular in the y-direction then the equation of (3.32) corresponding to the net point (x_0, y_0, t_0) has the form

$$(2/h^2 + 1/\tau)w_2^\tau(x_0, y_0, t_0) = \frac{1}{h^2}\,w_2^\tau(x_0, y_0 + h, t_0)$$

$$+ \frac{1}{h^2}\,w_2^\tau(x_0, y_0 - h, t_0) + \frac{1}{\tau}\,w_2^*(x_0, y_0, t_0).$$

Passing to moduli and noting (3.35) we have

$$(2/h^2 + 1/\tau)|w_2^\tau(x_0, y_0, t_0)| \le (2/h^2 + 1/\tau)|w_2^\tau(x_0, y_0, t_0)|. \tag{3.36}$$

If we suppose that

$$|w_2^\tau(x_0, y_0 + h, t_0)| < |w_2^\tau(x_0, y_0, t_0)|,$$

then (3.36) would be a strict inequality. Since this is impossible then

$$|w_2^\tau(x_0, y_0 + h, t_0)| \geq |w_2^\tau(x_0, y_0, t_0)|.$$

This inequality along with (3.35) results in

$$|w_2^\tau(x_0, y_0 + h, t_0)| = |w_2^\tau(x_0, y_0, t_0)|.$$

Hence the maximal value is attained at the point $(x_0, y_0 + h, t_0)$ as well, and that any conclusion about (x_0, y_0, t_0) is also valid for this point. This means that either the point $(x_0, y_0 + h)$ is irregular in the y-direction and (3.34) is valid, or the point $(x_0, y_0 + h)$ is regular in the y-direction, and

$$|w_2^\tau(x_0, y_0 + h, t_0)| = |w_2^\tau(x_0, y_0 + 2h, t_0)|.$$

If we continue in this way we either obtain (3.34) or obtain the sequence of equalities

$$|w_2^\tau(x_0, y_0, t_0)| = \cdots = |w_2^\tau(x_0, y_0 + kh, t_0)|.$$

But since the region Ω is bounded this sequence of regular points $(x_0, y_0 + kh)$ will terminate in less then $2N$ steps. This terminal net point is irregular in the y-direction. Thus we again have the first case.

The function w_2^* is investigated using a procedure similar to the above. The above is again valid if instead of (3.32) one uses (3.31).

Thus (3.34) is proved. Using $v^\tau = w_1^\tau + w_2^\tau$ we can combine (3.30) and (3.31) to obtain the inequality

$$\max_{(x,y) \in \Omega_h} |v^\tau(x, y, t)| \leq \sum_{\substack{t' \in \omega_\tau \\ t' \leq t}} \tau \left(\max_{Q_h^\tau} |f_1^\tau| + \max_{Q_h^\tau} |f_2^\tau| \right)$$

$$+ 2h^2 \left(\max_{\Omega_{h,x}^{ir} \times \omega_\tau} |\delta f_1^{ir}| + \max_{\Omega_{h,y}^{ir} \times \omega_\tau} |\rho f_2^{ir}| \right),$$

which is valid for all $t \in \omega_\tau$. By adding terms under the summation we have the required inequality. Theorem 3.2 is proved. $\square$

Next we show how to expand the functions u^τ, u^* in powers of τ and h in order to get the solution of (3.12)–(3.14).

Theorem 3.3. *Suppose the coherence conditions of orders 1 and 2 hold for (3.1)–(3.3), and $f \in H^{4+\varepsilon}(\bar{Q})$, $\Gamma \in C^{6+\varepsilon}$ with constant $\varepsilon \in (0, 1)$. Then the solutions u^τ, u^* of (3.12)–(3.14) can be expanded*

$$\begin{aligned} u^\tau &= u + h^2 v_1 + \tau v_2 + (h^4 + \tau^2)\eta^\tau, \\ u^* &= u + h^2 w_1 + \tau w_2 + (h^4 + \tau^2)\eta^*, \end{aligned} \quad \text{on } Q_h^\tau, \qquad (3.37)$$

where the functions v_1, v_2, w_1, w_2 *are continuous on* $\bar{Q}$ *and independent of* τ, h, *and the net functions* η^τ, η^* *are bounded*:

$$\max_{Q_h^\tau} |\eta^\tau| \le c_1, \qquad \max_{Q_h^\tau} |\eta^*| \le c_1, \tag{3.38}$$

PROOF. Choose the functions v_1 and w_1 as the solution of

$$\frac{\partial v_1}{\partial t} = \frac{\partial^2 v_1}{\partial y^2} + \frac{\partial^2 w_1}{\partial x^2} + \frac{1}{12}\left(\frac{\partial^4 u}{\partial x^4} + \frac{\partial^4 u}{\partial y^4}\right) \quad \text{in } Q, \tag{3.39}$$

$$w_1 = v_1 \qquad\qquad\qquad \text{in } Q, \tag{3.40}$$

$$v_1 = 0 \qquad\qquad\qquad \text{in } S, \tag{3.41}$$

$$v_1(x, 0) = 0, \qquad\qquad\qquad x \in \Omega. \tag{3.42}$$

Noting (3.40) we substitute w_1 into equation (3.39) and get

$$\frac{\partial v_1}{\partial t} = \Delta v_1 + \frac{1}{12}\left(\frac{\partial^4 u}{\partial x^4} + \frac{\partial^4 u}{\partial y^4}\right). \tag{3.43}$$

For (3.43), (3.41), (3.42) the assumptions of Theorem 3.1, with $l = 1 + \lambda$, $\lambda \in (0, 1)$, hold. Therefore there is a unique solution $v_1 \in H^{3+\lambda}(\bar{Q})$. Then the function $w_1 \in H^{3+\lambda}(\bar{Q})$ is then defined from (3.40). The independence of the functions v_1 and w_1 on τ, h is obvious. Note that we could have proved additional smoothness for the functions v_1, u if the coherence condition of order 3 held. Thus the derivative $\partial v_1/\partial t$ is bounded while $\partial^2 v_1/\partial t^2$ is not. Neverthless, the higher derivatives increase rather slowly as $t \to 0$. $\square$

Let us prove this fact using some auxiliary results.

Lemma 3.4. *Assume the conditions of Theorem 3.3 hold. Then*

$$t^\alpha \frac{\partial^3 u}{\partial t^3}, t^\alpha \frac{\partial^6 u}{\partial x^6}, t^\alpha \frac{\partial^6 u}{\partial y^6} \in H^\varepsilon(\bar{Q}) \quad \forall \, \alpha \in (\varepsilon, 1).$$

PROOF. Assume

$$z(x, t) = \frac{\partial^2 u}{\partial t^2}(x, t) - \Delta f(x, 0) - \frac{\partial f}{\partial t}(x, 0)$$

and write

$$\frac{\partial z}{\partial t}(x, t) - \Delta z(x, t) = \frac{\partial^2}{\partial t^2}\left(\frac{\partial u}{\partial t}(x, t)\right)$$

$$- \Delta\frac{\partial^2 u}{\partial t^2}(x, t) + \Delta\Delta f(x, 0) + \Delta\frac{\partial f}{\partial t}(x, 0).$$

All derivatives in this equality are continuous in Q. From equation (3.1) it follows that

$$\frac{\partial^3 u}{\partial t^3} = \frac{\partial^2}{\partial t^2}(\Delta u + f) \quad \text{in } Q.$$

So

$$\frac{\partial z}{\partial t} = \Delta z + g \quad \text{in } Q, \tag{3.44}$$

where

$$g(x, t) = \frac{\partial^2 f}{\partial t^2}(x, t) + \Delta \Delta f(x, 0) + \Delta \frac{\partial f}{\partial t}(x, 0).$$

Differentiation of equation (3.1) leads to

$$\frac{\partial^2 u}{\partial t^2} = \Delta \Delta u + \Delta f + \frac{\partial f}{\partial t},$$

from which it follows that

$$z(x, 0) = 0 \quad \forall\, x \in \Omega, \tag{3.45}$$

and from (3.7) we have

$$z = 0 \quad \text{on } S. \tag{3.46}$$

Since the coherence condition of order 0 holds but the condition of order 1 does not hold for (3.44)–(3.46), then $z \in H^{1+\lambda}(\bar{Q})\ \forall\, \lambda \in (0, 1)$ by Theorem 3.1. Next let us consider the new function $w = t^\alpha z$ in (3.44)–(3.46). Then

$$\frac{\partial w}{\partial t} = \Delta w + \alpha t^{\alpha-1} z + t^\alpha g \quad \text{in } Q$$

$$w(x, 0) = 0, \qquad\qquad\qquad x \in \Omega, \tag{3.47}$$

$$w = 0 \qquad\qquad\qquad \text{on } S.$$

From (3.45) and $z \in H^{1+\lambda}(\bar{Q})$ it follows

$$\lim_{t \to 0} t^{-\sigma} z(x, t) = 0 \quad \forall\, \sigma \in \left(0, \frac{1+\lambda}{2}\right), \quad x \in \bar{\Omega}. \tag{3.48}$$

Therefore for $\lambda > 1 - \alpha$ we obtain that $t^{\alpha-1} z \in H^\alpha(\bar{Q})$. Since $\alpha > \varepsilon$ then $t^{\alpha-1} z \in H^\varepsilon(\bar{Q})$. Therefore the right-hand side of (3.47) belongs to $H^\varepsilon(\bar{Q})$. For this problem the coherence conditions of orders 0 and 1 hold (the latter follows from the relation

$$\lim_{t \to 0} (t^\alpha g(x, t) + \alpha t^{\alpha-1} z(x, t)) = 0 \quad \forall\, x \in \bar{\Omega}.$$

Therefore $t^\alpha z \in H^{2+\varepsilon}(\bar{Q})$ from Theorem 3.1. Hence

$$t^\alpha \frac{\partial^2 z}{\partial x^2}, t^\alpha \frac{\partial^2 z}{\partial y^2} \in H^\varepsilon(\bar{Q}), \qquad \alpha t^{\alpha-1} z + t^\alpha \frac{\partial z}{\partial t} \in H^\varepsilon(\bar{Q}).$$

Coming back to (3.48) we conclude that

$$t^{\alpha-1} z \in H^\varepsilon(\bar{Q}),$$

which means that

$$t^\alpha \frac{\partial z}{\partial t} \in H^\varepsilon(\bar{Q}).$$

From the definition of the function z and the smoothness condition on f we conclude that

$$t^\alpha \frac{\partial^3 u}{\partial t^3}, t^\alpha \frac{\partial^4 u}{\partial t^2 \partial x^2}, t^\alpha \frac{\partial^4 u}{\partial t^2 \partial y^2} \in H^\varepsilon(\bar{Q}). \tag{3.49}$$

Now let us take

$$z(x, t) = \frac{\partial^3 u}{\partial t \partial x^2} (x, t) - \frac{\partial^2 f}{\partial x^2} (x, 0)$$

and use a procedure similar to one used above to obtain (3.44)–(3.49). As a result we have

$$t^\alpha \frac{\partial^5 u}{\partial t \partial x^4}, t^\alpha \frac{\partial^5 u}{\partial t \partial x^2 \partial y^2} \in H^\varepsilon(\bar{Q}). \tag{3.50}$$

If we replace x by y then

$$t^\alpha \frac{\partial^5 u}{\partial t \partial y^4} \in H^\varepsilon(\bar{Q}). \tag{3.51}$$

Assuming $z = \partial^4 u/\partial x^4$ and again using this procedure we have

$$t^\alpha \frac{\partial^6 u}{\partial x^6}, t^\alpha \frac{\partial^6 u}{\partial x^4 \partial y^2} \in H^\varepsilon(\bar{Q}). \tag{3.52}$$

This leads to

$$t^\alpha \frac{\partial^6 u}{\partial y^6}, t^\alpha \frac{\partial^6 u}{\partial y^4 \partial x^2} \in H^\varepsilon(\bar{Q}) \tag{3.53}$$

if we replace x by y. Lemma 3.4 is proved. $\qquad\square$

Lemma 3.5. *Assume the conditions of Theorem 3.3 hold. Then the solution of* (3.43), (3.41), (3.42) *obeys*

$$t^\alpha \frac{\partial^2 v_1}{\partial t^2}, t^{\dot\alpha} \frac{\partial^4 v_1}{\partial x^4}, t^\alpha \frac{\partial^4 v_1}{\partial y^4} \in H^\varepsilon(\bar{Q}).$$

PROOF. Assume $z = \partial v_1/\partial t$ and repeat the procedure used in (3.44)–(3.49), taking into account (3.50), (3.51). We obtain

$$t^\alpha \frac{\partial^2 v_1}{\partial t^2}, t^\alpha \frac{\partial^3 v_1}{\partial t\, \partial x^2}, t^\alpha \frac{\partial^3 v_1}{\partial t\, \partial y^2} \in H^\varepsilon(\bar{Q}). \tag{3.54}$$

Take z to be the function $\partial^2 v_1/\partial x^2$. Then, taking into account (3.52) and (3.54), we have

$$t^\alpha \frac{\partial^4 v_1}{\partial x^4}, t^\alpha \frac{\partial^4 v_1}{\partial x^2\, \partial y^2} \in H^\varepsilon(\bar{Q}). \tag{3.55}$$

Replacing x by y leads to

$$t^\alpha \frac{\partial^4 v_1}{\partial y^4} \in H^\varepsilon(\bar{Q}). \tag{3.56}$$

Lemma 3.5 is proved.   □

Let us return to the proof of the theorem. Since $w_1 = v_1$ then (3.54)–(3.56) are also valid for w_1. We choose the functions v_2, w_2 as the solutions of

$$\frac{\partial v_2}{\partial t} = \frac{\partial^2 v_2}{\partial y^2} + \frac{\partial^2 w_2}{\partial x^2} + \frac{1}{2}\frac{\partial^2 u}{\partial t^2} \quad \text{in } Q, \tag{3.57}$$

$$w_2 = v_2 - \frac{\partial^2 u}{\partial y^2} \quad \text{in } Q, \tag{3.58}$$

$$v_2 = 0 \quad \text{on } S, \tag{3.59}$$

$$v_2(x, 0) = 0, \quad x \in \Omega.$$

Eliminating the function w_2 from equation (3.57) using relation (3.58) we obtain an equation for v_2:

$$\frac{\partial v_2}{\partial t} = \Delta v_2 - \frac{\partial^4 u}{\partial x^2\, \partial y^2} + \frac{1}{2}\frac{\partial^2 u}{\partial t^2} \quad \text{in } Q. \tag{3.60}$$

The hypotheses of Theorem 3.1 with constant $l = 1 + \lambda$, $\lambda \in (0, 1)$, hold for (3.59), (3.60). Thus there is a unique solution $v_2 \in H^{3+\lambda}(\bar{Q})$. Using (3.58) and the smoothness of u we can find a function $w_2 \in H^{3+\lambda}(\bar{Q})$. These functions are clearly independent of τ, h. The function v_2 has properties similar to those of the function v_1, as the following analog of Lemma 3.5 shows.

Lemma 3.6. *Assume the conditions of Theorem 3.3 hold. Then*

$$t^\alpha \frac{\partial^2 v_2}{\partial t^2}, t^\alpha \frac{\partial^4 v_2}{\partial x^4}, t^\alpha \frac{\partial^4 v_2}{\partial y^4} \in H^\varepsilon(\bar{Q}).$$

The proof of this result is similar to that of Lemma 3.5, and is omitted. Let us now define the net functions in terms of the known functions

$$\eta^\tau = (u^h - u - h^2 v_1 - \tau v_2)/(h^4 + \tau^2) \quad \text{on } Q_h^\tau \cup (\Gamma_{h,y} \times \omega_\tau),$$

$$\eta^* = (u^* - u - h^2 w_1 - \tau w_2)/(h^4 + \tau^2) \quad \text{on } Q_h^\tau \cup (\Gamma_{h,x} \times \omega_\tau).$$

In order to complete the proof of the theorem we need only establish the validity of (3.38). In (3.12), (3.13) replace the functions u^τ and u^* by their expansions (3.37). We have

$$\frac{u - u(x, y, t - \tau)}{\tau} + \frac{h^2}{\tau}(w_1 - v_1(x, y, t - \tau))$$

$$+ (w_2 - v_2(x, y, t - \tau)) + (h^4 + \tau^2)\frac{\eta^* - \eta^\tau(x, y, t - \tau)}{\tau}$$

$$- L_1^h u - h^2 L_1^h w_1 - \tau L_1^h w_2 - (h^4 + \tau^2)L_1^h \eta^* = f, \quad (3.61)$$

$$\frac{h^2}{\tau}(v_1 - w_1) + (v_2 - w_2) + (h^4 + \tau^2)\frac{\eta^\tau - \eta^*}{\tau}$$

$$- L_2^h u - h^2 L_2^h v_1 - \tau L_2^h v_2 - (h^4 + \tau^2)L_2^h \eta^\tau = 0. \quad (3.62)$$

Let us transform the above using the smoothness hypotheses on the functions u, v_i, w_i. Fix a certain $\alpha \in (\varepsilon, 1)$. Then from Lemmas 3.4, 3.5, and 3.6 the functions ξ_i in the expansions

$$\frac{u - u(x, t - \tau)}{\tau} = \frac{\partial u}{\partial t} - \frac{\tau}{2}\frac{\partial^2 u}{\partial t^2} + \tau^2 \xi_1,$$

$$v_1(x, t - \tau) = v_1 - \tau\frac{\partial v_1}{\partial t} + \tau^2 \xi_2, \quad (3.63)$$

$$v_2(x, t - \tau) = v_2 - \tau\frac{\partial v_2}{\partial t} + \tau^2 \xi_3$$

satisfy the inequalities:

$$|\xi_i(x, t)| \le t^{-\alpha}c_2 \quad \forall (x, t) \in Q_h^\tau, \quad i = 1, 2, 3.$$

Now let us estimate $L_1^h u$. If $x \in \Omega_{h, x}^r$, then

$$L_1^h u = \frac{\partial^2 u}{\partial x^2} + \frac{h^2}{12}\frac{\partial^4 u}{\partial x^4} + h^4 \xi_i \quad (3.64)$$

by Lemma 1.3 of §7.1, where

$$|\xi_4| \le c_3 t^{-\alpha} \quad \forall (x, t) \in \Omega_{h, x}^r \times \omega_\tau.$$

If the knot x is irregular in the x-direction, then from (2.29) of §4.2.3 it follows that

$$L_1^h u = \frac{\partial^2 u}{\partial x^2} + h^2 \xi_5, \quad (3.65)$$

where

$$|\xi_5| \le c_4 \quad \forall (x, t) \in \Omega_{h, x}^{ir} \times \omega_\tau.$$

Both cases (3.64), (3.65) can be described by the same formula

$$L_1^h u = \frac{\partial^2 u}{\partial x^2} + \frac{h^2}{12} \frac{\partial^4 u}{\partial x^4} + h^4 \xi_6, \tag{3.66}$$

where

$$|\xi_6| \le \begin{cases} c_3 t^{-\alpha}, & (x, t) \in \Omega_{h,x}^r \times \omega_\tau, \\ c_4/h^2, & (x, t) \in \Omega_{h,x}^{ir} \times \omega_\tau. \end{cases} \tag{3.67}$$

As in the above we obtain the expressions

$$L_1^h w_1 = \frac{\partial^2 w_1}{\partial x^2} + h^2 \xi_7,$$

$$L_1^h w_2 = \frac{\partial^2 w_2}{\partial x^2} + h^2 \xi_8, \tag{3.68}$$

where

$$|\xi_j| = \begin{cases} c^6 t^{-\alpha}, & (x, t) \in \Omega_{h,x}^r \times \omega_\tau, \\ c_7/h^2, & (x, t) \in \Omega_{h,x}^{ir} \times \omega_\tau, \end{cases} \quad j = 7, 8. \tag{3.69}$$

Replacing x by y we have

$$L_2^h u = \frac{\partial^2 u}{\partial y^2} + \frac{h^2}{12} \frac{\partial^4 u}{\partial y^4} + h^4 \xi_9,$$

$$L_2^h v_1 = \frac{\partial^2 v_1}{\partial y^2} + h^2 \xi_{10},$$

$$L_2^h v_2 = \frac{\partial^2 v_2}{\partial y^2} + h^2 \xi_{11}, \tag{3.70}$$

where

$$|\xi_j| = \begin{cases} c_8 t^{-\alpha}, & (x, t) \in \Omega_{h,y}^r \times \omega_\tau, \\ c_9/h^2, & (x, t) \in \Omega_{h,y}^{ir} \times \omega_\tau, \end{cases} \quad j = 9, 10, 11. \tag{3.71}$$

Transform (3.61) using (3.63)–(3.68). In the equality thus obtained group like powers of τ, h:

$$\left(\frac{\partial u}{\partial t} + w_2 - v_2 - \frac{\partial^2 u}{\partial x^2} \right) + \tau \left(-\frac{1}{2} \frac{\partial^2 u}{\partial t^2} + \frac{\partial v_2}{\partial t} - \frac{\partial^2 w_2}{\partial x^2} \right)$$

$$+ \frac{h^2}{\tau} (w_1 - v_1) + h^2 \left(\frac{\partial v_1}{\partial t} - \frac{1}{12} \frac{\partial^4 u}{\partial x^4} - \frac{\partial^2 w_1}{\partial x^2} \right)$$

$$+ \tau^2 (\xi_1 - \xi_3) - h^4 (\xi_6 + \xi_7) - \tau h^2 (\xi_8 + \xi_2)$$

$$+ (h^4 + \tau^2) \left(\frac{\eta^* - \eta^\tau(x, y, t - \tau)}{\tau} - L_1^h \eta^* \right) = f. \tag{3.72}$$

Simplify this equality using the previous relations. From equations (3.1), (3.58) it follows that

$$\frac{\partial u}{\partial t} + w_2 - v_2 - \frac{\partial^2 u}{\partial x^2} = f.$$

The inequality

$$\tau h^2 \leq (\tau^2 + h^4)/2 \tag{3.73}$$

allows to write the sum of all expressions containing ξ as one term $(h^4 + \tau^2)\xi_{12}$, where

$$|\xi_{12}| \leq \begin{cases} c_{10} t^{-\alpha}, & (x, t) \in \Omega^r_{h,x} \times \omega_\tau, \\ c_{11}/h^2 & (x, t) \in \Omega^{ir}_{h,x} \times \omega_\tau. \end{cases} \tag{3.74}$$

Thus relation (3.72) can be expressed in the form

$$(h^4 + \tau^2)\left(\frac{\eta^* - \eta^\tau(x, y, t - \tau)}{\tau} - L_1^h \eta^*\right)$$

$$= -\tau\left(\frac{\partial v_2}{\partial t} - \frac{1}{2}\frac{\partial^2 u}{\partial t^2} - \frac{\partial^2 w_2}{\partial x^2}\right)$$

$$- h^2\left(\frac{\partial v_1}{\partial t} - \frac{1}{12}\frac{\partial^4 u}{\partial x^4} - \frac{\partial^2 w_1}{\partial x^2}\right) - (h^4 + \tau^2)\xi_{12}. \tag{3.75}$$

Transform (3.62) using formulae (3.70). Grouping like powers of τ, h we have

$$\left(v_2 - w_2 - \frac{\partial^2 u}{\partial y^2}\right) + \frac{h^2}{\tau}(v_1 - w_1) + h^2\left(-\frac{1}{12}\frac{\partial^4 u}{\partial y^4} - \frac{\partial^2 v_1}{\partial y^2}\right)$$

$$- h^4(\xi_9 + \xi_{10}) - \tau\frac{\partial^2 v_2}{\partial y^2} - \tau h^2 \xi_{11}$$

$$+ (h^4 + \tau^2)\left(\frac{\eta^\tau - \eta^*}{\tau} - L_2^h \eta^\tau\right) = 0. \tag{3.76}$$

Simplify the above equality using (3.40), (3.58) and writing the sum of all terms containing ξ_j as $(h^4 + \tau^2)\xi_{13}$, where

$$|\xi_{13}| \leq \begin{cases} c_{12} t^{-\alpha}, & (x, t) \in \Omega^r_{h,y} \times \omega_\tau, \\ c_{13}/h^2, & (x, t) \in \Omega^{ir}_{h,y} \times \omega_\tau. \end{cases} \tag{3.77}$$

From the above (3.76) is transformed to

$$(h^4 + \tau^2)\left(\frac{\eta^\tau - \eta^*}{\tau} - L_2^h \eta^\tau\right) = \tau\frac{\partial^2 v_2}{\partial y^2} + h^2\left(\frac{1}{12}\frac{\partial^4 u}{\partial y^4} + \frac{\partial^2 v_1}{\partial y^2}\right) - (h^4 + \tau^2)\xi_{13}.$$

$$\tag{3.78}$$

Let us impose boundary conditions and initial conditions using the definitions of the functions η^τ and η^*, and the fact that the functions u^τ, u^*, u, v_i, w_i are zero as follows:

$$\eta^* = 0 \quad \text{on } \Gamma_{h,x} \times \omega_\tau, \tag{3.79}$$

$$\eta^\tau = 0 \quad \text{on } \Gamma_{h,y} \times \omega_\tau, \tag{3.80}$$

$$\eta^\tau(x, 0) = 0 \quad \forall\, x \in \Omega_h. \tag{3.81}$$

We have obtained a system (3.75), (3.78)–(3.81) for which the estimate of Theorem 3.2 is valid. This estimate tells us that the solution is unique. We have already proved existence. But this estimate does not give a desirable result since the right-hand sides of (3.75), (3.78) are not of h^4 and τ^2 order. Consider those terms in η^τ and η^* of τ and h^2 order which make contributions of τ^2 and h^4 order to the right-hand side of (3.75), (3.78). These components are the solutions of

$$\frac{z^* - z^\tau(x, y, t - \tau)}{\tau} = -\tau\left(\frac{\partial v_2}{\partial t} - \frac{1}{2}\frac{\partial^2 u}{\partial t^2} - \frac{\partial^2 w_2}{\partial x^2}\right)$$

$$- h^2\left(\frac{\partial v_1}{\partial t} - \frac{1}{12}\frac{\partial^4 u}{\partial x^4} - \frac{\partial^2 w_1}{\partial x^2}\right) \quad \text{on } Q_h^\tau, \tag{3.82}$$

$$z^* = 0 \qquad\qquad\qquad\qquad\qquad\qquad \text{on } \Gamma_{h,x} \times \omega_\tau, \tag{3.83}$$

$$\frac{z^\tau - z^*}{\tau} = \tau\frac{\partial^2 v_2}{\partial y^2} + h^2\left(\frac{1}{12}\frac{\partial^4 u}{\partial y^4} + \frac{\partial^2 v_1}{\partial y^2}\right) \quad \text{on } Q_h^\tau, \tag{3.84}$$

$$z^\tau = 0 \qquad\qquad\qquad\qquad\qquad\qquad \text{on } \Gamma_{h,y} \times \omega_\tau, \tag{3.85}$$

with the initial condition

$$z^\tau(x, 0) = 0, \qquad x \in \Omega_h. \tag{3.86}$$

We wish to prove that the solution of (3.82)–(3.83) has the form

$$z^\tau = 0 \quad \text{on } Q_h^\tau, \tag{3.87}$$

$$z^* = \tau^2\left(-\frac{\partial v_2}{\partial t} + \frac{1}{2}\frac{\partial^2 u}{\partial t^2} + \frac{\partial^2 w_2}{\partial x^2}\right) + \tau h^2\left(-\frac{\partial v_1}{\partial t} + \frac{1}{12}\frac{\partial^4 u}{\partial x^4} + \frac{\partial^2 w_1}{\partial x^2}\right) \quad \text{on } Q_h^\tau. \tag{3.88}$$

Both functions are uniform on the boundary:

$$z^\tau = 0 \quad \text{on } \Gamma_{h,x} \times \omega_\tau, \tag{3.89}$$

$$z^* = 0 \quad \text{on } \Gamma_{h,y} \times \omega_\tau. \tag{3.90}$$

Substituting these functions into equation (3.82) we get the required identity. The initial condition (3.86) holds because the function z^τ is zero everywhere on Q_h^τ. Thus only equation (3.84) has to be checked. Substitute (3.87), (3.88) into this equation. We have

$$-\tau\left(-\frac{\partial v_2}{\partial t} + \frac{1}{2}\frac{\partial^2 u}{\partial t^2} + \frac{\partial^2 w_2}{\partial x^2}\right) - h^2\left(-\frac{\partial v_1}{\partial t} + \frac{1}{12}\frac{\partial^4 u}{\partial x^4} + \frac{\partial^2 w_1}{\partial x^2}\right)$$
$$= \tau\frac{\partial^2 v_2}{\partial y^2} + h^2\left(\frac{1}{12}\frac{\partial^4 u}{\partial y^4} + \frac{\partial^2 v_1}{\partial y^2}\right).$$

This relation is an identity since from the definitions of the functions v_i and w_i we have

$$\frac{\partial v_2}{\partial t} - \frac{1}{2}\frac{\partial^2 u}{\partial t^2} - \frac{\partial^2 w_2}{\partial x^2} = \frac{\partial^2 v_2}{\partial y^2}$$

from (3.60), and

$$\frac{\partial v_1}{\partial t} - \frac{1}{12}\frac{\partial^4 u}{\partial x^4} - \frac{\partial^2 w_1}{\partial x^2} = \frac{1}{12}\frac{\partial^4 u}{\partial y^4} + \frac{\partial^2 v_1}{\partial y^2}$$

from (3.43). Thus the functions z^τ and z^* defined by (3.87)–(3.90) are the solutions of (3.82)–(3.86).

Let us make the following substitution in (3.75), (3.78)–(3.81)

$$\zeta^\tau = \eta^\tau - \frac{1}{h^4 + \tau^2}z^\tau, \qquad \zeta^* = \eta^* - \frac{1}{h^4 + \tau^2}z^*.$$

Then for the function ζ^τ we obtain

$$(h^4 + \tau^2)\left(\frac{\zeta^* - \zeta^\tau(x, y, t - \tau)}{\tau} - L_1^h\zeta^*\right) = -(h^4 + \tau^2)\xi_{12} + L_1^h z^* \quad \text{on } Q_h^\tau,$$
$$\tag{3.91}$$

$$\zeta^* = 0 \qquad \text{on } \Gamma_{h,x} \times \omega_\tau, \tag{3.92}$$

$$(h^4 + \tau^2)\left(\frac{\zeta^\tau - \zeta^*}{\tau} - L_2^h\zeta^\tau\right) = -(h^4 + \tau^2)\xi_{13} \quad \text{on } Q_h^\tau, \tag{3.93}$$

$$\zeta^\tau = 0 \qquad \text{on } \Gamma_{h,y} \times \omega_\tau, \tag{3.94}$$

$$\zeta^\tau(x, 0) = 0, \qquad x \in \Omega_h. \tag{3.95}$$

To justify using Theorem 3.2 we need only estimate the term $L_1^h z^*$ in (3.91). Let $x \in \Omega_{h,x}^\tau$ be a point regular in the x-direction. Then

$$L_1^h z^* = z_{\bar{x}x}^* = \tau^2(g_1)_{\bar{x}x} + \tau h^2(g_2)_{\bar{x}x},$$

where

$$g_1 = -\frac{\partial v_2}{\partial t} + \frac{1}{2}\frac{\partial^2 u}{\partial t^2} + \frac{\partial^2 w_2}{\partial x^2},$$

$$g_2 = -\frac{\partial v_1}{\partial t} + \frac{1}{12}\frac{\partial^4 u}{\partial x^4} + \frac{\partial^2 w_1}{\partial x^2}.$$

From Lemmas 3.4, 3.5, 3.6, and 1.3 of §7.1 we have

$$L_1^h z^* = \tau^2 \xi_{14} + \tau h^2 \xi_{15}, \qquad (3.96)$$

where

$$|\xi_j| \le c_{14} t^{-\alpha} \quad \forall (x, t) \in \Omega_{h,x}^r \times \omega_\tau, \quad j = 14, 15.$$

Let $x \in \Omega_{h,x}^{ir}$ be a point which is irregular in the x-direction. Then

$$L_1^h z^* = \tau^2 L_1^h g_1 + \tau h^2 L_1^h g_2.$$

From the definition (3.9) of the difference operator L_1^h we have

$$L_1^h g_1(x, y, t) = - \frac{3 - \delta}{\delta h^2} g_1(x, y, t)$$

$$+ \frac{4 - 2\delta}{(1 + \delta)h^2} g_1(x \pm h, y, t) - \frac{1 - \delta}{(2 + \delta)h^2} g_1(x \pm 2h, y, t).$$

Since the function g_1 is continuous on $\bar{Q}$ the inequality $|g_1| \le c_{15}$ is valid, from which we have

$$|L_1^h g_1(x, t)| \le \frac{1}{h^2} \left(\frac{3 - \delta}{\delta} + \frac{4 - 2\delta}{1 + \delta} + \frac{|1 - \delta|}{2 + \delta} \right) c_{15}.$$

Noting that $\delta \in (0, \frac{3}{2})$ we have

$$|L_1^h g_1(x, t)| \le \frac{1}{h^2} \left(\frac{3}{\delta} + 4 + \frac{1}{2} \right) c_{15} \le \frac{10 c_{15}}{h^2 \delta}.$$

Hence

$$\tau^2 L_1^h g_1 = \frac{\tau^2}{h^2} \xi_{16},$$

where

$$|\xi_{16}| \le \frac{c_{16}}{\delta} \quad \forall (x, t) \in \Omega_{h,x}^{ir} \times \omega_\tau, \qquad \delta = \delta(x).$$

An estimate of the second term $\tau h^2 L_1^h g_2$ can be found by the similar procedure. We have

$$L_1^h z^* = (\tau^2 + \tau h^2) \xi_{17}/h^2, \qquad (3.97)$$

where

$$|\xi_{17}| \le c_{17}/\delta \quad \forall (x, t) \in \Omega_{h,x}^{ir} \times \omega_\tau.$$

Combining (3.96) and (3.97) we have

$$L_1^h z^* = (\tau^2 + h^4) \xi_{18}, \qquad (3.98)$$

where

$$|\xi_{18}| \leq \begin{cases} 2c_{14}\,t^{-\alpha}, & (x, t) \in \Omega_{h,x}^r \times \omega_\tau, \\ \dfrac{2c_{17}}{h^2\delta}, & (x, t) \in \Omega_{h,x}^{ir} \times \omega_\tau. \end{cases} \tag{3.99}$$

Therefore (3.91) has the form

$$(\tau^2 + h^4)\left(\frac{\zeta^* - \zeta^\tau(x, t - \tau)}{\tau} - L_1^h\zeta^*\right) = (\tau^2 + h^4)(\xi_{18} - \xi_{12}) \quad \text{on } Q_h^\tau. \tag{3.100}$$

Dividing both sides of (3.93), (3.100) by $\tau^2 + h^4$ we arrive at a system which satisfies the hypotheses of Theorem 3.2. According to this we have

$$\max_{Q_h^\tau}|\zeta^*| \leq \sum_{t \in \omega_\tau} \tau\left(\max_{\Omega_{h,x}^r}|\xi_{12}| + \max_{\Omega_{h,x}^r}|\xi_{18}| + \max_{\Omega_{h,x}^r}|\xi_{13}|\right)$$

$$+ 2h^2 \max\left(\max_{\Omega_{h,x}^{ir} \times \omega_\tau}|\delta\xi_{12}| + \max_{\Omega_{h,x}^{ir} \times \omega_\tau}|\delta\xi_{18}| + \max_{\Omega_{h,x}^{ir} \times \omega_\tau}|\rho\xi_{13}|\right).$$

Using (3.74) and (3.99) we obtain

$$\max_{Q_h^\xi}|\zeta^*| \leq \sum_{t \in \omega_\tau} \tau t^{-\alpha}(c_{10} + c_{12} + 2c_{14})$$

$$+ 2\left(\max_{\Omega_{h,x}^{ir} \times \omega_\tau} c_{11}\delta + 2c_{14} + \max_{\Omega_{h,y}^{ir} \times \omega_\tau} c_{13}\rho\right). \tag{3.101}$$

Since $\alpha < 1$ then we have

$$\sum_{t \in \omega_\tau} \tau t^{-\alpha} \leq \int_0^T t^{-\alpha}\,dt = T^{1-\alpha}/(1 - \alpha).$$

We use this inequality along with $\delta(x) \leq \frac{3}{2}$, $\rho(x) \leq \frac{3}{2}$ to simplify (3.101). We obtain

$$\max_{Q_h^\xi}|\zeta^*| \leq c_{18}, \tag{3.102}$$

where

$$c_{18} = \frac{T^{1-\alpha}}{1 - \alpha}(c_{10} + c_{12} + 2c_{17}) + 3c_{11} + 3c_{13} + 2c_{17}.$$

A similar inequality is also obtained for ζ^τ;

$$\max_{Q_h^\xi}|\zeta^\tau| \leq c_{18}.$$

Taking into account

$$\eta^\tau = \zeta^\tau, \qquad \eta^* = \zeta^* + z^*/(\tau^2 + h^4),$$

the equality (3.88), the continuity of all terms in this equality on $\bar{Q}$, and (3.73) we come to (3.38). Theorem 3.3 is proved. $\qquad\square$

We now present a method for increasing the accuracy of our solutions based on the expansions we have obtained. Assume that the conditions of Theorem 3.3 hold. Choose integers $M \geq 2$ and $N \geq 2$. Construct the difference net $\bar{Q}_h^{\tau}$ with mesh-sizes τ and h (in time and space). Find the solution u^{τ} of (3.12)–(3.14). Construct the difference net $\bar{Q}_{h/2}^{\tau/4}$ with mesh-sizes $\tau/4$ and $h/2$ (in time and space). Again find the solution of (3.12)–(3.14), and denote it by $u^{\tau/4}$. We thus obtain two approximate solutions at the points of $\bar{Q}_h^{\tau}$. Consider the linear combination

$$U^H(x, t) = \tfrac{4}{3}u^{\tau/4}(x, t) - \tfrac{1}{3}u^{\tau}(x, t), \qquad (x, t) \in \bar{Q}_h^{\tau}, \qquad (3.103)$$

we will prove that U^H approximates the exact solution (in uniform norm) with an accuracy of order $\tau^2 + h^4$.

Theorem 3.7. *Assume the conditions of Theorem 3.3 hold. Then the corrected solution* (3.103) *obeys the estimate*

$$\max_{\bar{Q}_h^{\tau}} |U^H - u| \leq \tfrac{5}{12}c_1(\tau^2 + h^4). \qquad (3.104)$$

PROOF. We have

$$u^{\tau} = u + h^2 v_1 + \tau v_2 + (\tau^2 + h^4)\eta^{\tau},$$

$$u^{\tau/4} = u + \frac{h^2}{4} v_1 + \frac{\tau}{4} v_2 + \tfrac{1}{16} (\tau^2 + h^4)\eta^{\tau/4}$$

at each point of $\bar{Q}_h^{\tau}$. Since the functions v_1, v_2 are independent of τ and h, then

$$U^H = u + (\tau^2 + h^4)\tfrac{1}{3}(\tfrac{1}{4}\eta^{\tau/4} - \eta^{\tau}).$$

Using (3.38) we obtain

$$|U^H - u| \leq \tfrac{1}{3}(\tau^2 + h^4)\tfrac{5}{4}c_1 \quad \text{on } \bar{Q}_h^{\tau}.$$

Theorem 3.7 is proved. $\qquad\square$

Let us consider a locally-one-dimensional scheme (see [113]) which differs from the above in that it uses a simpler approximation of operators $\partial^2/\partial x^2$, $\partial^2/\partial y^2$ near the boundary (this approximation being given in §4.2.3 for $n = 1$). For this scheme the main ideas of the proof remains the same, but the calculations become much simpler. We present the main result without proof.

Theorem 3.8. *Assume the conditions of Theorem 3.3 hold. Then we have*

$$\max_{\bar{Q}_h^{\tau}} |U^H - u| \leq c_{19}(\tau^2 + h^3) \qquad (3.105)$$

for the corrected solution (3.103) *which uses the solutions of the locally-one-dimensional scheme.*

5.4. The Equation of Motion

In this section we consider the initial-value problem for the equation of motion

$$\frac{\partial \phi}{\partial t} + \sum_{\alpha=1}^{p} v_\alpha(t, x) \frac{\partial \phi}{\partial x_\alpha} = f(t, x), \qquad (t, x) \in T \times R^p,$$

$$\phi(0, x) = g(x), \qquad x = (x_1, \ldots, x_p) \in R^p, \qquad (4.1)$$

where $T = [0, T_0]$. The coefficients v_α are assumed to satisfy the continuity equation

$$\sum_{\alpha=1}^{p} \frac{\partial v_\alpha}{\partial x_\alpha} = 0 \quad \text{in } T \times R^p. \qquad (4.2)$$

The equation of motion has a very important property which allows one to convert its domain from a half-space to a bounded region. This is because how the solution depends on the initial data and the forcing function is controlled by the characteristics of the differential equation (see [27]) so that we can solve the problem in the region of interest to us without solving it on the whole space. Let us prove this. Let $\varkappa_\sigma^{x_0}(x) \in C^\infty(R^p)$ be on Urysohn function, which is equal to 1 inside the sphere $B_\sigma(x_0)$ with center at $x_0 \in R^p$ and radius $\sigma > 0$, and equal to zero outside the sphere $B_{2\sigma}(x_0)$. If the velocities v_α are finite everywhere in R^p then for any region $\Omega \in R^p$ of interest to us, at time t one can always choose $x_0 \in \Omega$ and sufficiently large σ such that the solution of (4.1), (4.2) coincides on Ω with the solution of

$$\frac{\partial \phi_1}{\partial t} + \sum_{\alpha=1}^{p} v_\alpha \frac{\partial \phi_1}{\partial x_\alpha} = f\varkappa_\sigma^{x_0} \quad \text{in } T \times R^p,$$

$$\phi_1(0, x) = g(x)\varkappa_\sigma^{x_0}, \quad x \in R^p.$$

We will use the usual notation $C^k(Q)$, where k is an integer, to denote the class of continuous functions in Q which have continuous derivatives up to order k in the region Q. In order to emphasize the boundedness of the support of the function φ of $C^k(Q)$ we will write $\varphi \in \tilde{C}^k(Q)$. If the function φ belongs to this class it follows that it is bounded, along with all its partial derivatives up to order k.

We will study the system with initial data and forcing function having local support:

$$v_\alpha \in C^{2r}(T \times R^p), \qquad f \in \tilde{C}^{2r}(T \times R^p), \qquad g \in \tilde{C}^{2r}(R^p), \qquad (4.3)$$

where r is a natural number. Then (4.1), (4.2) has a unique solution (see [27]) and the solution ϕ belongs to $\tilde{C}^{2r}(T \times R^p)$.

Assume now that the support of the function ϕ is contained in the parallelepiped $Q = \{(t, x_1, \ldots, x_p), 0 \le t \le T_0, 0 \le x_\alpha \le 1, \alpha = 1, \ldots, p\}$. This can always be achieved using a linear transformation in the spatial variables. Then ϕ will be equal to zero on the side S of this parallelepiped:

$$x_\alpha = 0, \qquad x_\alpha = 1, \qquad \alpha = 1, .., p.$$

Construct uniform difference nets in the space variable:

$$\bar{\Omega}_h = \{x; x_\alpha = i_\alpha h, i_\alpha = 0, 1, \ldots, N\},$$

$$\Omega_h = \{x; x_\alpha = i_\alpha h, i_\alpha = 1, \ldots, N - 1\}$$

(4.4)

and in the time variable

$$\bar{\omega}_\tau = \{t_j = j\tau, j = 0, 1, \ldots, M\},$$

$$\omega_\tau = \{t_j = j\tau, j = 1, \ldots, M\},$$

where $\tau = T_0/M$, $h = 1/N$. We will assume that the mesh-sizes are related by

$$\tau = c_0 h, \tag{4.5}$$

where c_0 is a constant independent of τ, h.

Denote the Cartesian product of the nets $\bar{\omega}_\tau \times \bar{\Omega}_h$ by $\bar{Q}_h^\tau$ and the Cartesian product of the nets $\omega_\tau \times \Omega_h$ by Q_h^τ. Assume $S_h^\tau = \bar{Q}_h^\tau \cap S$.

Using an implicit form of the splitting-up method we replace (4.1)–(4.3) by the approximate system

$$(I + \tau\Lambda_1)\cdots(I + \tau\Lambda p)\phi_\tau(t, x) - \phi_\tau(t - \tau, x) = \tau f(t, x), \qquad (t, x) \in Q_h^\tau,$$

(4.6)

$$\phi_\tau(t, x) = 0, \qquad (t, x) \in S_h^\tau,$$

$$\phi_\tau(0, x) = g(x), \qquad x \in \Omega_h.$$

(4.7)

Here I is the identity operator,

$$\Lambda_\alpha \varphi(t, x) = \frac{1}{4h} \{(v_\alpha(t, x_1, \ldots, x_\alpha + h, \ldots, x_p)$$

$$+ v_\alpha(t, x_1, \ldots, x_p))\varphi(t, x_1, \ldots, x_\alpha + h, \ldots, x_p)$$

$$- (v_\alpha(t, x_1, \ldots, x_p) + v_\alpha(t, x_1, \ldots, x_\alpha - h, \ldots, x_p))$$

$$\times \varphi(t, x_1, \ldots, x_\alpha - h, \ldots, x_p)\}, \qquad (t, x) \in \omega_\tau \times \Omega_h.$$

Let us define the scalar product on Ω_h

$$(a, b) = \sum_{x \in \Omega_h} a(x)b(x)h^p,$$

together with the norm $\|a\| = (a, a)^{1/2}$.

Lemma 4.1. *We have*

$$(\Lambda_\alpha \varphi, \varphi) = 0, \qquad \alpha = 1, \ldots, p; \quad t \in \omega_\tau, \tag{4.8}$$

for any net function φ defined on $\bar{\Omega}_h$ which is zero on $\bar{\Omega}_h \backslash \Omega_h$.

PROOF. The equality $(\Lambda_\alpha \varphi, \varphi) = -(\varphi, \Lambda_\alpha \varphi)$, which is equivalent to (4.8), is the difference analog of the formula for integrating by parts, and can be proved directly. $\square$

The following lemma follows immediately

Lemma 4.2. *The inequalities*

$$\|\varphi\| \le \|(I + \tau\Lambda_\alpha)\varphi\|, \qquad \alpha = 1, 2, \ldots, p,$$

are valid for any net function φ defined on $\overline{\Omega}_h$ and equal to zero on $\overline{\Omega}_h \backslash \Omega_h$, for any $\tau, t \in T$.

A proof of this lemma is given in [86] in slightly different notation.

Let us prove two more auxiliary results for a special form of the approximation error.

Lemma 4.3. *Assume that the function φ belongs to $\tilde{C}^q(T \times R^p)$, $q \ge 2$, and its support is inside the parallelepiped Q. Then*

$$\Lambda_\alpha \varphi = \frac{\partial \varphi}{\partial x_\alpha} v_\alpha + \frac{\varphi}{2} \frac{\partial v_\alpha}{\partial x_\alpha} + \sum_{i=1}^{q-2} \tau^i \varphi_i + \tau^{q-1} \varphi_{q-1,\tau} \quad on \ Q_h^\tau.$$

Here $\varphi_i \in \tilde{C}^{q-i-1}(T \times R^p)$ for $i = 1, \ldots, q-2$ and is independent of τ, and the support of φ_i is contained in Q. The net function $\varphi_{q-1,\tau}$ is bounded:

$$\max_{Q_h^\tau} |\varphi_{q-1,\tau}| \le c_2.$$

If ψ is an arbitrary net function defined on $\overline{\Omega}_h$, then

$$\max_{\Omega_h} |\Lambda_\alpha \psi| \le \frac{c_3}{\tau} \max_{\overline{\Omega}_h} |\psi|.$$

PROOF. The last result follows from the definition of the operator Λ_α. The constant c_3 is equal to

$$\frac{c_0}{2} \max_{\overline{Q}} |v_\alpha|.$$

In order to prove the validity of the expansion in τ we rewrite the operator Λ_α in the form

$$\Lambda_\alpha \varphi(t, x) = \frac{h}{4} \{v_\alpha \varphi|_{(t, x_1, \ldots, x_\alpha + h, \ldots, x_p)} - v_\alpha \varphi|_{(t, x_1, \ldots, x_\alpha - h, \ldots, x_p)} + v_\alpha(t, x)$$

$$\times \{\varphi(t, x_1, \ldots, x_\alpha + h, \ldots, x_p) - \varphi(t, x_1, \ldots, x_\alpha - h, \ldots, x_p)\}\}$$

and apply Lemma 1.1 of §7.1 twice. Then we obtain the expansion in τ, with

$$\varphi_i = \begin{cases} 0, & \text{if } i \text{ is odd, in } T \times R^p, \\ \dfrac{c_0^{-i}}{2(i+1)!}\left\{\dfrac{\partial^{i+1}(v_\alpha \varphi)}{\partial x_\alpha^{i+1}} + v_\alpha \dfrac{\partial^{i+1}\varphi}{\partial x_\alpha^{i+1}}\right\}, & \text{if } i \text{ is even,} \end{cases}$$

$$|\varphi_{q-1,\tau}| \leq \frac{c_0^{-q+1}}{2q!}\left(\max_{T \times R^p}\left|\frac{\partial^q(v_\alpha \varphi)}{\partial x_\alpha^q}\right| + \max_{T \times R^p}\left|v_\alpha \frac{\partial^q \varphi}{\partial x_\alpha^q}\right|\right) \quad \text{in } Q_h^\tau. \qquad \square$$

Lemma 4.4. *Assume condition* (4.3) *holds. Then the solution of* (4.1), (4.2) *obeys the relations*

$$(I + \tau\Lambda_1)\cdots(I + \tau\Lambda_p)\phi(t, x) - \phi(t - \tau, x)$$

$$= \tau\left\{f + \sum_{i=1}^{r-1}\tau^i f_i + \tau^r f_{r,\tau}\right\}\Bigg|_{(t,x)}, \quad \phi(t, x) \in Q_h^\tau.$$

Here for $i = 1, \ldots, r - 1$ the functions $f_i \in \tilde{C}^{r-i}(T \times R^p)$ are independent of τ and their supports are concentrated inside Q. The function $f_{r,\tau}$ is bounded:

$$\max_{Q_h^\tau}|f_{r,\tau}| \leq c_4.$$

PROOF. Using Lemma 4.3 successively we obtain

$$(I + \tau\Lambda_1)\cdots(I + \tau\Lambda_p)\phi$$

$$= \phi + \tau\sum_{\alpha=1}^{p}\left(v_\alpha \frac{\partial\phi}{\partial x_\alpha} + \frac{\phi}{2}\frac{\partial v_\alpha}{\partial x_\alpha}\right) + \sum_{i=2}^{r}\tau^i F_i + \tau^{r+1}F_{r+1,\tau} \quad \text{in } Q_h^\tau.$$

Here, for $i = 2, \ldots, r$, the functions $F_i \in \tilde{C}^{r+1-i}(T \times R^p)$ are independent of τ, their supports are concentrated in Q, and the net function $F_{r+1,\tau}$ is bounded:

$$\max_{Q_h^\tau}|F_{r+1,\tau}| \leq c_5.$$

When (4.2) is taken into account the coefficient of τ is found to be equal to

$$\sum_{\alpha=1}^{p} v_\alpha \frac{\partial\phi}{\partial x_\alpha}.$$

Apply the Taylor formula to $\phi(t - \tau, x)$. We have

$$(I + \tau\Lambda_1)\cdots(I + r\Lambda_p)\phi(t, x) - \phi(t - \tau, x)$$

$$= \tau\left\{\frac{\partial\phi}{\partial t} + \sum_{\alpha=1}^{p}\frac{\partial\phi}{\partial x_\alpha}v_\alpha + \sum_{i=2}^{r}\tau^{i-1}\bar{F}_i + \tau^r\bar{F}_{r+1,\tau}\right\}\Bigg|_{(t,x)}, \quad (t, x) \in Q_h^\tau.$$

Here for $i = 2, \ldots, r$

$$\bar{F}_i = F_i + (-1)^{i-1}\frac{1}{i!}\frac{\partial^i\phi}{\partial t^i} \quad \text{on } T \times R^p,$$

$$|\bar{F}_{r+1,\tau}(t, x)| \leq |F_{r+1,\tau}(t, x)| + \frac{1}{(r+1)!}\max_{T \times R^p}\left|\frac{\partial^{r+1}\phi}{\partial t^{r+1}}\right|, \quad (t, x) \in Q_h^\tau.$$

Since the coefficient of τ is equal to f the lemma is proved. $\qquad \square$

Lemma 4.5. *The difference scheme* (4.6), (4.7) *satisfies the a priori estimate*

$$\|\phi_\tau(t, x)\| \le \|g(x)\| + T_0 \max_{t \in \omega_\tau}\|f(t, x)\| \quad \forall\, t \in \omega_\tau.$$

The proof is based on Lemma 4.2 and follows by induction on $t_j \times \omega_\tau$ (see, for example, [87]).

Theorem 4.6. *Assume* (4.3) *holds for* (4.1), (4.2), *and condition* (4.5) *holds for* (4.6), (4.7); *then the solutions of these systems satisfy the relation*

$$\phi_\tau = \phi + \sum_{i=1}^{r-1} \tau^i \phi_i + \tau^r \phi_{r,\tau} \quad \text{on } Q_h^\tau.$$

Here the functions $\phi_i \in \tilde{C}^{2r-2i}(T \times R^p)$ *are independent of* τ, h *for* $i = 1, \ldots,$ $r - 1$, *and their supports are concentrated in* Q. *The net function* $\phi_{r,\tau}$ *is bounded:*

$$\max_{t \in \bar{\omega}_\tau}\|\phi_{r,\tau}\| \le c_6. \tag{4.9}$$

PROOF. Taking into account Lemma 4.2 and formulae (4.6), (4.7) for the difference solution we obtain

$$(I + \tau\Lambda_1) \cdots (I + \tau\Lambda_p)(\phi_\tau - \phi)|_{(t, x)} - (\phi_\tau - \phi)|_{(t-\tau, x)}$$

$$= -\tau\left(\sum_{i=1}^{r-1} \tau^i f_i + \tau^r f_{r,r}\right)\Bigg|_{(t, x)}, \qquad (t, x) \in Q_h^\tau,$$

$$(\phi_\tau - \phi)|_{(0, x)} = 0, \qquad x \in \Omega_h,$$

$$(\phi_\tau - \phi)|_{(t, x)} = 0, \qquad (t, x) \in S_h^\tau. \tag{4.10}$$

Now choose ϕ_1 as the solution of

$$\frac{\partial\phi_1}{\partial t} + \sum_{\alpha=1}^{p} v_\alpha \frac{\partial\phi_1}{\partial x_\alpha} = -f_1 \quad \text{on } T \times R^p,$$

$$\phi_1(0, x) = 0, \qquad x \in R^p.$$

Since $f_1 \in \tilde{C}^{2r-2}(T \times R^p)$ and is independent of τ the function $\phi_1 \in \tilde{C}^{2r-2}(T \times R^p)$ is also independent of τ. Moreover, since f_1 is nonzero only along the characteristics which are inside Q in fact ϕ_1 can also be nonzero only on these curves. Therefore the support of ϕ_1 is concentrated inside Q. Therefore Lemma 4.4 can be again applied; this results in

$$(I + \tau\Lambda_1) \cdots (I + \tau\Lambda_p)\phi_1(t, x) - \phi_1(t - \tau, x)$$

$$= -\tau\left(f_1 + \sum_{i=1}^{r-2} \tau^i \bar{f}_i + \tau^{r-1}\bar{f}_{r,\tau}\right)\Bigg|_{(t, x)}, \qquad (t, x) \in Q_h^\tau,$$

$$\phi_1(0, x) = 0, \qquad x \in \Omega_h,$$

$$\phi_1(t, x) = 0, \qquad (t, x) \in S_h^\tau. \tag{4.11}$$

Multiply (4.11) by τ and subtract the result from (4.10). We have

$$(I + \tau\Lambda_1)\cdots(I + \tau\Lambda_p)(\phi_\tau - \phi - \tau\phi_1)|_{(t,\,x)}$$

$$- (\phi_\tau - \phi - \tau\phi_1)|_{(t-\tau,\,x)} = -\tau\left(\sum_{i=2}^{r-1}\tau^i\tilde{f}_i + \tau^r\tilde{f}_{r,\,\tau}\right)\bigg|_{(t,\,x)}, \qquad (t,\,x) \in Q_h^\tau,$$

$$(\phi_\tau - \phi - \tau\phi_1)|_{(0,\,x)} = 0, \qquad x \in \Omega_h,$$

$$(\phi_\tau - \phi - \tau\phi_1)|_{(t,\,x)} = 0, \qquad (t,\,x) \in S_h^\tau. \qquad (4.12)$$

The remaining $r - 2$ functions ϕ_i can be chosen so that after the $(r - 1)$th step the following relations are obtained

$$(I + \tau\Lambda_1)\cdots(I + \tau\Lambda_p)\left(\phi_\tau - \phi - \sum_{i=1}^{r-1}\tau^i\phi_i\right)\bigg|_{(t,\,x)}$$

$$- \left(\phi_r - \phi - \sum_{i=1}^{r-1}\tau^i\phi_i\right)\bigg|_{(t-\tau,\,x)} = -\tau^{r+1}\hat{f}_{r,\,\tau}(t,\,x), \qquad (t,\,x) \in Q_h^\tau,$$

$$\left(\phi_\tau - \phi - \sum_{i=1}^{r-1}\tau^i\phi_i\right)\bigg|_{(0,\,x)} = 0, \qquad x \in \Omega_h,$$

$$\left(\phi_\tau - \phi - \sum_{i=1}^{r-1}\tau^i\phi_i\right)\bigg|_{(t,\,x)} = 0, \qquad (t,\,x) \in S_h^\tau.$$

Here

$$\max_{Q_h^\tau}|\hat{f}_{r,\,\tau}| \le c_7,$$

hence

$$\max_{t\in\omega_\tau}\|\hat{f}_{r,\,\tau}\| \le c_7.$$

Applying the *a priori* estimate from Lemma 4.5 to the net function

$$\tau^r\phi_{r,\,\tau} = \phi_\tau - \phi - \sum_{i=1}^{r-1}\tau^i\phi_i$$

we finally obtain the result of the theorem. □

We will now formulate a method for increasing the accuracy of the difference solution on the basis of this theorem.

Assume (4.3) holds for (4.1), (4.2). Fix the constant c_0 and construct for integers $M_1 < M_2 < \cdots < M_r$ temporal difference nets $\omega_{\tau_1}, \omega_{\tau_2}, \ldots, \omega_{\tau_r}$ and the correspond spatial nets $\Omega_{h_1}, \Omega_{h_2}, \ldots, \Omega_{h_r}$, Take

$$\tau_i = c_0 h_i, \qquad i = 1, \ldots, r.$$

Solve (4.6), (4.7) on each net Q_h^τ and obtain r solutions $\phi_{\tau_1}, \ldots, \phi_{\tau_r}$. Let us consider the case when $Q_{h_1}^{\tau_1} \subset Q_{h_i}^{\tau_i} \ \forall \ i = 1, \ldots, r$. Then at each point one can form a linear combination of all solutions

$$\bar{\phi} = \sum_{i=1}^{r} \gamma_i \phi_i \quad \text{on } Q_{h_1}^{\tau_1}. \tag{4.13}$$

Theorem 4.7. *Assume* (4.3) *holds for* (4.1), (4.2). *Then if for the* M_i *we have*

$$c_9 \geq M_{i+1}/M_i \geq 1 + c_8 \quad \forall \ i = 1, \ldots, r - 1, \tag{4.14}$$

and the γ_i *satisfy*

$$\sum_{i=1}^{r} \gamma_i = 1, \tag{4.15}$$

$$\sum_{i=1}^{r} \gamma_i M_i^{-l} = 0, \qquad l = 1, \ldots, r - 1,$$

then the solution constructed from (4.13) *obeys the estimate*

$$|\phi - \bar{\phi}| \leq \left(\sum_{i=1}^{r} |\gamma_i| \tau_i^r \right) b, \quad \text{on } Q_{h_1}^{\tau_1}$$

where the γ_i *are bounded. The net function* b *is bounded in the root-mean-square norm*

$$\max_{t \in \bar{\omega}_{\tau_1}} \left(\sum_{x \in \Omega_{h_1}} |b(t, x)|^2 h_1^p \right)^{1/2} \leq c_{10}$$

and the constant c_{10} *is independent of* τ *and* h.

PROOF. Choosing the γ_i to solve (4.15), together with Theorem 4.6, guarantees

$$|\phi - \bar{\phi}| \leq \sum_{i=1}^{r} |\gamma_i| \tau_i^r |\phi_{r, \tau_i}| \quad \text{on } Q_{h_1}^{\tau_1},$$

where the $|\gamma_i|$ are bounded form Lemma 2.3 of §7.2 because of condition (4.14). It is this condition that gives

$$h_1/h_r \leq c_9^{r-1}.$$

Therefore

$$\left(\sum_{x \in \Omega_{h_i}} |\phi_{r, \tau_i}(t, x)|^2 h_i^p \right)^{1/2} \geq \left(\sum_{x \in \Omega_{h_1}} |\phi_{r, \tau_i}(t, x)|^2 h_i^p \right)^{1/2}$$

$$\geq c_9^{(1-r)p/2} \left(\sum_{x \in \Omega_{h_1}} |\phi_{r, \tau_i}(t, x)|^2 h_1^p \right)^{1/2}.$$

Using (4.9) we arrive at

$$\left(\sum_{x \in \Omega_{h_1}} |\phi_{r, \tau_i}(t, x)|^2 h_1^p \right)^{1/2} \leq c_6 c_9^{(r-1)p/2}.$$

Now, in order to complete the proof of the theorem we need only use the above inequality to estimate the difference $\phi - \bar{\phi}$. $\square$

Remark. Note that there is also an expansion (when $r \geq p/2 + 1$) with uniformly bounded remainder, namely:

$$\phi_\tau = \phi + \sum_{i=1}^{S_0} \tau^i \phi_i + \tau^{r-p/2} \phi_{r-p/2,\tau} \quad \text{on } \bar{Q}_h^\tau,$$

$$S_0 = [r - p/2 - \tfrac{1}{2}],$$

which follows from Theorem 4.6. Here ϕ_i, $i = 1, \ldots, S_0$ are analogous to the functions in Theorem 4.6, and the net function $\phi_{r-p/2,\tau}$ is bounded in the uniform metric:

$$\max_{\bar{Q}_h^\tau} |\phi_{r-p/2,\tau}| \leq \max_{\bar{Q}_h^\tau} |\phi_{r,\tau}| h^{p/2} + \sum_{i=S_0+1}^{r} h^{i-S_0} \max_{\bar{Q}_h^\tau} |\phi_i|.$$

The functions ϕ_i are all uniformly bounded by the constant c_{11} because they are continuous on $\bar{Q}$, and the uniform estimate of $\phi_{r,\tau}$ follows from the inequality

$$\max_{x \in \Omega_h} |\phi_{r,\tau}(t, x)| \leq h^{-p/2} \|\phi_{r,\tau}\|$$

noting that

$$\|\phi_{r,\tau}\| \leq c_6 \quad \forall t \in \bar{\omega}_\tau.$$

Combining these inequalities we have

$$\max_{\bar{Q}_h^\tau} |\phi_{r-p/2,\tau}| \leq c_6 + \sum_{i=S_0+1}^{r} c_{11} \leq c_6 + (r-1)c_{11}.$$

Thus this analog of Theorem 4.7. is valid in the uniform metric.

Theorem 4.8. *Assume the conditions* (4.3) *hold for* (4.1), (4.2) *at some* $r \geq p/2 + 1$. *If for* M_i *we obtain*

$$M_{i+1}/M_i \geq 1 + c_{12}, \quad i = 1, \ldots, S_1 - 1, \, S_1 = [r - p/2 - 1/2],$$

and the γ_i *satisfy*

$$\sum_{i=1}^{S_1} \gamma_i = 1,$$

$$\sum_{i=1}^{S_1} \gamma_i M_i^{-l} = 0, \quad l = 1, \ldots, S_1 - 1,$$

then the γ_i *are bounded and the solution* $\bar{\phi}$ *constructed using* (4.13) *obeys the estimate*

$$\max_{\bar{Q}_{h1}^\tau} |\phi - \bar{\phi}| \leq \sum_{i=1}^{S_1} |\gamma_i| \tau_i^{r-p/2} c_{11}.$$

CHAPTER 6
Extrapolation for Algebraic Problems and Integral Equations

The method of extrapolation can be applied to problems other than the numerical solution of differential equations. This method, with some modifications, can also be applied to other problems of numerical mathematics.

Extrapolation with respect to small parameters is of special interest. When one deals with differential problems equations the parameter is the mesh size of the difference scheme approximating the original problem. In other types of problems other parameters may appear which have to be taken to a limit in the process of improving the accuracy of the solution. For example, we encounter this in the regularization method, when we solve a nonsingular system of algebraic equations.

Differential or integral equations with continuous or discrete spectra of eigenvalues with a condensation point at $\lambda = 0$ can be the objects of regularization. By first reducing them to problems of linear algebra we obtain two small parameters: the mesh size, and a regularization parameter. Both of these parameters must go to zero in order to attain a high degree of accuracy. The theory of regularization was developed by A. N. Tikhonov, M. M. Lavrentiev, V. K. Ivanov, J.-L. Lions, V. A. Morozov, and others.

At the beginning of the present chapter we focus our attention on the extrapolation of solutions of a system of linear algebraic equations with respect to a regularization parameter. In more complex situations (for example, in the case of regularized differential equations) one should also use the methods described in the previous chapters of this book.

At the end of the chapter we will consider some model types of integral equations. We have chosen three typical equations involving an unknown function of one argument: the Volterra equations of the first and second kind, and the Fredholm equation of the second kind. The Fredholm equation of the first kind is omitted. This class of improperly-posed problems, which is of interest and importance from the point of view of applications, can be extrapolated with respect to the mesh-size as well as with respect to a regularization parameter. The methods of this chapter apply in principle. But in order to justify their use with the Fredholm equation of the first kind it is necessary to use a large body of auxilary results and special methodology which we have not used earlier. We thus decided that the Fredholm equation of the first kind need not be considered.

6.1. Regularization of a Singular System of Linear Algebraic Equations

Let C^n be an n-dimensional unitary space. The scalar product and the norm in C^n are defined by †

$$(u, v) = \sum_{i=1}^{n} u_i \bar{v}_i, \qquad \|u\| = (u, u)^{1/2}.$$

Consider a system of linear algebaic equations

$$Ax = f \tag{1.1}$$

with singular complex matrix A of order $n \times n$, and vectors $x, f \in C^n$.

Let us introduce the notation

$$U = \left\{ u \in C^n : \|Au - f\| = \inf_{v \in C^n} \|Av - f\| \right\}.$$

The element $u^f \in U$ will be referred to a *normal pseudosolution* of system (1.1), if

$$\|u^f\| = \min_{u \in U} \|u\|. \tag{1.2}$$

Under these conditions the vector u^f is defined uniquely (see [134]). We will exhibit an algorithm for the construction of a normal pseudosolution. Multiply system (1.1) on the left by the adjoint of A:

$$A^*Ax = A^*f. \tag{1.3}$$

Now find a unitary matrix P and a diagonal matrix $\Lambda = \mathrm{diag}(\lambda_1, \ldots, \lambda_n)$ such that

$$A^*A = P\Lambda P^*. \tag{1.4}$$

Let us pass to new variables $y = P^*x$ in (1.3)

$$\Lambda y = P^*A^*f. \tag{1.5}$$

Assume $F = P^*A^*f$; then a normal pseudosolution is constructed from

$$u^f = Py, \tag{1.6}$$

† A subscript on a vector denotes a component.

where the components of the vector y are defined by

$$y_i = \begin{cases} F_i/\lambda_i & \text{if } \lambda_i \neq 0, \\ 0 & \text{if } \lambda_i = 0. \end{cases} \tag{1.7}$$

The problem of minimizing the norm on U can be approximated by the problem of finding the minimum in the entire space C^n (see [134, 129]).

Consider the following regularization method: find in C^n the point which is the minimum of the functional

$$\|u\|^2 + \frac{1}{\varepsilon} \|Au - f\|^2, \tag{1.8}$$

where ε is a certain positive parameter. This problem has a unique solution u^ε, where

$$\|u^f - u^\varepsilon\| \to 0 \quad \text{at } \varepsilon \to 0.$$

From the differentiability of (1.8) there follows

$$B^\varepsilon u^\varepsilon = H, \tag{1.9}$$

where $B^\varepsilon = A^*A + \varepsilon I$, $H = A^*f$ (I is unit matrix).

Let matrices P and Λ be as in (1.4). Then by substituting $y^\varepsilon = P^*u^\varepsilon$ (1.9) results in a system of equations with a diagonal matrix

$$(\Lambda + \varepsilon I)y^\varepsilon = F. \tag{1.10}$$

Recall that $F = P^*A^*f$, $\lambda_i \geq 0$ for all $i = 1, \ldots, n$. From the last condition it follows that (1.10) is solvable and the unique solution is given by

$$y_i^\varepsilon = F_i/(\lambda_i + \varepsilon), \qquad i = 1, \ldots, n. \tag{1.11}$$

Our aim is to approximate the solution u^ε by a sum of solutions with other regularization parameters. Note that in (1.11) the function

$$\gamma(\varepsilon) = 1/(\lambda + \varepsilon) \tag{1.12}$$

occurs. When $\lambda > 0$ the function $\gamma(\varepsilon)$ is infinitely differentiable at any nonnegative ε and the derivatives are easily estimated:

$$|\gamma^k(\varepsilon)| \leq k!/\lambda^{k+1}. \tag{1.13}$$

Let $\varepsilon_1 > \cdots > \varepsilon_{l+1} \geq 0$ be a decreasing sequence of values of ε. Application of the Lagrange formula yields

$$\gamma(\varepsilon_{l+1}) = \sum_{i=1}^{l} \alpha_i \gamma(\varepsilon_i) + Q(\varepsilon_{l+1}), \tag{1.14}$$

where

$$\alpha_i = \prod_{\substack{j=1 \\ j \neq i}}^{l} (\varepsilon_{l+1} - \varepsilon_j)/(\varepsilon_i - \varepsilon_j). \tag{1.15}$$

From [14] we can estimate the remainder

$$|Q(\varepsilon_{l+1})| \geq \lambda^{-l-1} \prod_{j=1}^{l} \varepsilon_j.$$

Thus if $\lambda_i \neq 0$ from (1.11) we have

$$y_i^{\varepsilon_{l+1}} = \sum_{j=1}^{l} \alpha_j y_i^{\varepsilon_j} + q_i(\varepsilon_{l+1}), \tag{1.16}$$

where

$$|q_i(\varepsilon_{l+1})| \leq F_i \lambda^{-l-1} \prod_{j=1}^{l} \varepsilon_j.$$

If $\lambda_i = 0$ the the function $\gamma(\varepsilon)$ is not smooth at zero and the above is not valid. However, in this case from the orthogonality of vector $H = A^*f$ to the null space of A^*A it follows that $F_i = 0$. We denote the minimum overall λ_i not equal to zero by λ_0.

From (1.16) (for nonzero λ_i) and $F_i = 0$ (for zero λ_i) there follows

$$y^{\varepsilon_{l+1}} = \sum_{j=1}^{l} \alpha_j y^{\varepsilon_j} + q(\varepsilon_{l+1}), \tag{1.17}$$

where

$$\|q\| \leq \lambda_0^{-l-1} \left(\prod_{j=1}^{l} \varepsilon_j \right) \|F\|.$$

Using (1.17) we will prove the following result.

Lemma 1.1. *The solution of the regularization problem* (1.9) *obeys the relation*

$$u^{\varepsilon_{l+1}} = \sum_{j=1}^{l} \alpha_j u^{\varepsilon_j} + r(\varepsilon_{l+1}), \tag{1.18}$$

where

$$\|r\| \leq c_1 \prod_{j=1}^{l} \varepsilon_j \tag{1.19}$$

and c_1 does not depend on ε_j.

PROOF. Multiplying (1.17) from the left by the matrix P we obtain (1.18), where $r = Pq$. To prove (1.19), we use the norm-preserving property of unitary matrices twice:

$$\|H\| = \|PF\| = \|F\|, \qquad \|r\| = \|Pq\| = \|q\|.$$

One of the most important questions is that of the stability of the solution to (1.9) when the right-hand side varies. We need to approximate the right-hand side in (1.2) for two reasons. First, an error will appear in calculating A^*f in practice even if the vector f is exactly known. Second, (1.9) is being solved approximately, i.e., we find a vector $\breve{u}$, so that $B^\varepsilon \breve{u} - H$ is sufficiently small (but not zero). Thus we are actually finding the exact solution of the system

$$B^\varepsilon v^\varepsilon = H^\delta \tag{1.20}$$

with an unknown vector H^δ, which must satisfy the inequality

$$\|H^\delta - H\| \le \delta, \tag{1.21}$$

where $\delta > 0$ is small. It is obvious that $B^\varepsilon(v^\varepsilon - u^\varepsilon) = H^\delta - H$. Since the minimal eigenvalue of the matrix B^ε is equal to ε from (1.21) it immediately follows that

$$\|v^\varepsilon - u^\varepsilon\| \le \delta/\varepsilon. \tag{1.22}$$

On the basis of these estimates we arrive at the following method for improving the solutions of regularization problems. Let $\varepsilon_1 > \varepsilon_2 > \cdots > \varepsilon_k > 0$ be a sequence of regularization parameters, for which we have solved

$$B^{\varepsilon_i} v^{\varepsilon_i} = H^{\delta_i} \tag{1.23}$$

with positive definitive matrices $B^{\varepsilon_i} = (A^*A + \varepsilon_i I)$, and with $H^{\delta_i} \in C^n$, where

$$\|H^{\delta_i} - H\| \le \delta_i. \tag{1.24}$$

Then the corrected solution is constructed by the formula

$$w^k = \sum_{j=1}^{k} \alpha_j v^{\varepsilon_j} \tag{1.25}$$

with weights

$$\alpha_j = \prod_{\substack{i=1 \\ i \ne j}}^{k} - \varepsilon_i/(\varepsilon_j - \varepsilon_i). \qquad \square \tag{}$$

Theorem 1.2. *Let u^f be a normal pseudosolution of system* (1.1) *and let w^k be as in* (1.25). *Then for difference $w^k - u^f$ obeys the estimate*

$$\|w^k - u^f\| \le \sum_{j=1}^{k} |\alpha_j|\delta_j/\varepsilon_j + c_2 \prod_{j=1}^{k} \varepsilon_j, \tag{1.26}$$

where c_2 is a constant independent of ε_j.

PROOF. Since $u^f = u^\varepsilon$ at $\varepsilon = 0$, an application of Lemma 1.1 for $l = k$ and $\varepsilon_{k+1} = 0$ yields

$$u^f - \sum_{j=1}^{k} \alpha_j u^{\varepsilon_j} = r(0),$$

where

$$\|r(0)\| \le c_2 \prod_{j=1}^{l} \varepsilon_j.$$

From

$$\|u^{\varepsilon_j} - v^{\varepsilon_j}\| \le \delta_j/\varepsilon_j.$$

we have

$$\|u^f - w^k\| \le \left\| u^f - \sum_{j=1}^{k} \alpha_j u^{\varepsilon_j} \right\| + \left\| \sum_{j=1}^{k} \alpha_j (u^{\varepsilon_j} - v^{\varepsilon_j}) \right\|$$

$$\le \|r(0)\| + \sum_{j=1}^{k} |\alpha_j| \, \|u^{\varepsilon_j} - v^{\varepsilon_j}\|.$$

From this the theorem follows.   □

Further, let us note the following. Generally speaking, the coefficients α_j will depend on the choice of ε_j. However, one can choose ε_j so that $|\alpha_j|$ remains bounded when all the ε_j tend to zero. One way is to subdivide the interval $[0, \varepsilon_1]$ uniformly at the points $\varepsilon_2, \ldots, \varepsilon_k$. In this case the $|\alpha_j|$ are independent of the ε_j. A more general choice for which this conclusion is valid is

$$\varepsilon_j/\varepsilon_{j+1} \ge c_3 > 1, \qquad j = 1, \ldots, k-1, \qquad (1.27)$$

with c_3 independent of ε_j. Then from Lemma 2.3 of §7.2 it follows that

$$|\alpha_j| \le \left(\frac{c_3}{c_3 - 1} \right)^k$$

for all $j = 1, \ldots, k$. The behavior of δ_j when ε_j decreases, depends on the method used to solve (1.23). However, an analysis shows that if an iterative method is used which seeks to minimize the remainder then the values δ_j vary gradually. The number of iterations necessary, however, is large. We will show how to overcome this difficulty.

In order to avoid a large amount of computation when the regularization parameters are small we propose the following algorithm, which allows one to use all the smoothness properties of the solution with respect to ε.

Choose parameters $\varepsilon_1 > \varepsilon_2 > \cdots > \varepsilon_k > 0$. Consider the problem

$$(A^*A + \varepsilon_1 I)u^{\varepsilon_1} = A^*f. \qquad (1.28)$$

Let its approximate solution be v^{ε_1}, and let the norm of the remainder equal δ_1. Consider then

$$(A^*A + \varepsilon_2 I)u^{\varepsilon_2} = A^*f.$$

Take the vector v^{ε_1} as an initial approximation to the solution of this problem. In this case we should continue the iteration until we obtain an approximate solution v^{ε_2} with the norm of the remainder of order $\varepsilon_2 \delta_1/\varepsilon_1$. The contributions of errors from v^{ε_1} and v^{ε_2} in the final solution are then of the same order.

For the third problem

$$(A^*A + \varepsilon_3 I)u^{\varepsilon_3} = A^*f$$

we use the lemma and define the initial approximation w^3

$$w^3 = \frac{\varepsilon_3 - \varepsilon_2}{\varepsilon_1 - \varepsilon_2} v^{\varepsilon_1} + \frac{\varepsilon_3 - \varepsilon_1}{\varepsilon_2 - \varepsilon_1} v^{\varepsilon_2}.$$

The iterative process should be continued until the remainder was norm of order $\varepsilon_3 \delta_1/\varepsilon_1$.

Thus, the ith problem

$$(A^*A + \varepsilon_i I)u^{\varepsilon_i} = A^*f$$

is solved approximately until the remainder has norm of order $\varepsilon_i \delta_1/\varepsilon_1$. The vector

$$w^i = \sum_{j=1}^{i-1} \beta_j v^{\varepsilon_j}$$

is chosen as the initial approximation in this process, where

$$\beta_j = \prod_{\substack{l=1 \\ l \neq j}}^{i-1} (\varepsilon_i - \varepsilon_l)/(\varepsilon_j - \varepsilon_l)$$

and the v^{ε_j} have been found earlier.

After we have found the solution of the kth problem the final solution is constructed by (1.25). Theorem 1.2 then holds. ☐

Remark. The above remain valid in a real case.

EXAMPLE. From Table 6.1 let us take a real matrix A and a real vector f as the data of (1.1).

The results obtained are in good agreement with Theorem 1.2. For this matrix the eigenvalues are known: $-2, -1, 0, 0, 0, 1, 2, 3, 4$. From the last row of Table 6.1 it is seen that the system is incompatible. Solve (1.9) for

Table 6.1. The Matrix A and Right-Hand Side f.

−0.954	−0.028	−0.014	0.102	0.078	0.130	0.030	−0.330	−0.054	0.0	1.0
−0.028	1.704	−0.148	0.564	−0.204	−0.340	−0.540	−0.06	−0.228	0.0	2.0
−0.014	−0.148	1.926	0.282	−0.102	−0.170	−0.270	−0.03	−0.114	0.0	3.0
0.102	−0.564	0.282	1.974	0.486	0.810	1.110	−0.21	0.402	0.0	4.0
0.078	−0.204	−0.102	0.486	0.054	0.090	−0.210	−0.690	−0.222	0.0	5.0
0.130	−0.340	−0.170	0.810	0.090	0.150	−0.350	−1.150	−0.370	0.0	6.0
0.030	−0.540	−0.270	1.110	0.210	−0.350	0.150	−0.650	−0.470	0.0	7.0
−0.330	−0.06	−0.03	−0.21	−0.690	−1.150	−0.650	0.150	0.170	0.0	8.0
−0.054	−0.228	−0.114	0.402	−0.222	−0.370	−0.470	0.170	3.84	0.0	9.0
0.0	0.0	0.0	0.0	0.0	0.0	0.0	0.0	0.0	0.0	10.0

Table 6.2

Number of comp.	u^f	$u^f - u^{\varepsilon_1}$	$u^f - u^{\varepsilon_2}$	$u^f - u^\varepsilon$	$u^f - w^3$
1	-0.9045	-4.9×10^{-3}	-3.3×10^{-3}	-1.6×10^{-3}	-7.8×10^{-8}
2	-1.5090	-7.6×10^{-3}	-5.1×10^{-3}	-2.5×10^{-3}	-1.1×10^{-7}
3	0.2455	-1.3×10^{-3}	-8.9×10^{-4}	-4.5×10^{-4}	-5.2×10^{-8}
4	4.6468	1.4×10^{-2}	9.5×10^{-3}	4.7×10^{-3}	1.7×10^{-7}
5	-2.4135	-1.1×10^{-2}	-7.9×10^{-3}	-3.9×10^{-3}	-1.7×10^{-7}
6	-4.0225	-1.9×10^{-2}	-1.3×10^{-2}	-6.6×10^{-3}	-2.8×10^{-7}
7	-1.5225	4.9×10^{-3}	3.3×10^{-2}	1.7×10^{-3}	2.6×10^{-7}
8	-2.2275	4.1×10^{-3}	2.8×10^{-3}	1.4×10^{-3}	2.6×10^{-7}
9	1.2205	-2.6×10^{-3}	-1.8×10^{-3}	-9.0×10^{-4}	-5.6×10^{-8}
10	0.0	0.0	0.0	0.0	0.0

$\varepsilon_1 = 0.01$, $\varepsilon_2 = 0.5 \times 10^{-2}$, $\varepsilon_3 = \frac{1}{3} \times 10^{-2}$. Then from the three solutions u^{ε_1}, u^{ε_2}, u^{ε_3} obtained construct a linear corrector w^3 by (1.25). Table 6.2 gives the normal pseudosolution u^f, and the errors $u^f - u^{\varepsilon_i}$ and $u^f - w^3$ taking into account the fact that the effect of round-off errors does not exceed the last significant figure of the results.

6.2. Regularization of a System with a Selfadjoint Matrix

Let us consider a system of linear algebraic equations

$$Ax = f \tag{2.1}$$

with a singular self adjoint matrix A of $n \times n$ order, and with $x, f \in E^n$.

Let us write the matrix A in the form of a product

$$A = P\Lambda P^*, \tag{2.2}$$

where P is a unitary matrix, and $\Lambda = \text{diag}(\lambda_1, \ldots, \lambda_n)$ is a diagonal matrix with real entries λ_i.

In (2.1) we pass to new variables $y = P^*x$:

$$\Lambda y = P^*f.$$

Assume $F = P^*f$; then the normal pseudosolution of (2.1) is defined by

$$x^f = Py, \tag{2.3}$$

where the components of the vector $y \in E^n$ are given by

$$y_j = \begin{cases} F_j/\lambda_j, & \text{if } \lambda_j \neq 0, \\ 0, & \text{if } \lambda_j = 0. \end{cases} \tag{2.4}$$

Instead of (2.1) we will solve the following system with a small real parameter ε:

$$(A + i\varepsilon I)x^\varepsilon = f, \qquad (2.5)$$

where $i = \sqrt{-1}$ and I is the unit matrix. All eigenvalues of the matrix $A + i\varepsilon I$ exceed $|\varepsilon|$ in modulus and lie in half-plane $\operatorname{Im} \lambda > 0$ if $\varepsilon > 0$ and in $\operatorname{Im} \lambda < 0$ if $\varepsilon < 0$.

Remark. In the case of a real symmetric matrix A and a real vector f, the complex system (2.5) is equivalent to a system of linear algebraic equations over the field of real numbers, namely:

$$\begin{bmatrix} \varepsilon I & -A \\ A & \varepsilon I \end{bmatrix} \begin{bmatrix} v^\varepsilon \\ w^\varepsilon \end{bmatrix} = \begin{bmatrix} 0 \\ f \end{bmatrix}, \qquad (2.6)$$

the solutions being related by $x^\varepsilon = v^\varepsilon - iw^\varepsilon$. The matrix of (2.6) is positive definite for $\varepsilon > 0$ and negative definite for $\varepsilon < 0$ in $2n$-dimensional real Euclidean space with scalar product

$$(a, b) = \sum_{i=1}^{2n} a_i b_i.$$

By changing variables $z^\varepsilon = P^* x^\varepsilon$, (2.5) can be brought to the form

$$(\Lambda + i\varepsilon I)z^\varepsilon = P^* f.$$

We will seek a normal pseudosolution in the form of the sum of several regularized solutions with various regularization parameters.

Let $k > 1$ be an integer. From (2.4) it is not difficult to see that the behavior of the interpolants is connected with the behavior of the function

$$\gamma_\lambda(\varepsilon) = \frac{1}{\lambda + i\varepsilon}$$

for real λ and ε. When $\lambda \neq 0$, the function $\gamma_\lambda(\varepsilon)$ is infinitely differentiable with respect to ε and its derivatives are easily estimated in modulus:

$$|\gamma_\lambda^{(k)}| \leq k! |\lambda|^{-k-1}.$$

Let us expand the function $\gamma_\lambda(\varepsilon)$ using the Taylor formula, restricting ourselves to the first k terms. We have

$$\gamma_\lambda(\varepsilon) = \sum_{j=0}^{k-1} \varepsilon^j \lambda^{-j-1}(-i)^j + \xi(\varepsilon), \qquad (2.7)$$

where $|\xi(\varepsilon)| \leq |\varepsilon|^k |\lambda|^{-k-1}$. Now, let ε_j, $j = 1, \ldots, k$ be a set of nonzero parameters. Construct a linear combination

$$\sum_{j=1}^{k} \alpha_j \gamma_\lambda(\varepsilon_j) \qquad (2.8)$$

and choose the weights α_j so as to approximate $\gamma_\lambda(0)$ at $\varepsilon_j \to 0$ as accurately as possible. Summing up the expansions (2.7) and setting the first k coefficients of the powers of λ equal to zero, we have

$$\sum_{j=1}^k \alpha_j = 1,$$

$$\sum_{j=1}^k \alpha_j \varepsilon_j^l = 0, \qquad l = 1, \ldots, k-1. \tag{2.9}$$

The weights

$$\alpha_j = \prod_{\substack{l=1 \\ l \neq j}}^k - \varepsilon_l/(\varepsilon_j - \varepsilon_l) \tag{2.10}$$

are the solutions of this system. These weights allow us to approximate $\gamma_\lambda(0)$ by (2.8) with the following accuracy (see [14]):

$$\left| \gamma_\lambda(0) - \sum_{j=1}^k \alpha_j \gamma_\lambda(\varepsilon_j) \right| \leq |\lambda|^{-k-1} \prod_{j=1}^k \varepsilon_j.$$

Thus, if $\lambda_l \neq 0$ we have

$$y_l = z_l^0 = \sum_{j=1}^k \alpha_j z_l^{\varepsilon_j} + q_l \prod_{j=1}^k \varepsilon_j, \tag{2.11}$$

where

$$|q_l| \leq F_l \lambda_l^{-k-1}.$$

If $\lambda_l = 0$ and (2.1) is not consistent, then it is necessary to require that

$$\sum_{j=1}^k \alpha_j/\varepsilon_j = 0. \tag{2.12}$$

This yields

$$y_l = 0 = \sum_{j=1}^k \alpha_j z_l^{\varepsilon_j}.$$

It is possible to realize (2.12) in two ways. In the first we replace (2.9) by the following system:

$$\sum_{j=1}^k \alpha_j \varepsilon_j^{-1} = 0,$$

$$\sum_{j=1}^k \alpha_j = 1,$$

$$\sum_{j=1}^k \alpha_j \varepsilon_j^l = 0, \qquad l = 1, 2, \ldots, k-2. \tag{2.13}$$

Its determinant can be calculated explicitly and equals

$$\left(\prod_{1 \le j < i \le k} (\varepsilon_i - \varepsilon_j) \right) \bigg/ \prod_{i=1}^{k} \varepsilon_i.$$

This is zero only if $\varepsilon_i = \varepsilon_j$ for $i \ne j$. The disadvantage of this approach is that our level of accuracy suffers because, despite the presence of k parameters, the last equation in (2.9), generally speaking, fails to hold and the extrapolation thus actually uses only $k - 1$ parameters.

The second method imposes an additional restriction on (2.12) at the expense of our choice of the ε_j.

Lemma 2.1. *Let ε_j be pairwise distinct, nonzero and satisfy the condition*

$$\sum_{j=1}^{k} \varepsilon_j^{-1} = 0. \tag{2.14}$$

Then the solutions α_j of (2.9) obey the relation

$$\sum_{j=1}^{k} \alpha_j \varepsilon_j^{-1} = 0.$$

PROOF. Consider the system

$$\begin{bmatrix} 1 & \varepsilon_1 & \varepsilon_1^2 & \cdots & \varepsilon_1^{k-1} \\ 1 & \varepsilon_2 & \varepsilon_2^2 & \cdots & \varepsilon_2^{k-1} \\ \cdots & \cdots & \cdots & \cdots & \cdots \\ 1 & \varepsilon_k & \varepsilon_k^2 & \cdots & \varepsilon_k^{k-1} \end{bmatrix} \begin{bmatrix} b_0 \\ b_1 \\ \cdots \\ b_{k-1} \end{bmatrix} = \begin{bmatrix} \varepsilon_1^{-1} \\ \varepsilon_2^{-1} \\ \cdots \\ \varepsilon_k^{-1} \end{bmatrix}. \tag{2.15}$$

Since the determinant of the system matrix is equal to the Vandermonde determinant $V(\varepsilon_1, \varepsilon_2, \ldots, \varepsilon_k)$, (2.15) is uniquely solvable because all the ε_i are pairwise distinct. We will find b_0 by Kramer's rule:

$$b_0 = \det \begin{bmatrix} \varepsilon_1^{-1} & \varepsilon_1 & \varepsilon_1^2 & \cdots & \varepsilon_1^{k-1} \\ \varepsilon_2^{-1} & \varepsilon_2 & \varepsilon_2^2 & \cdots & \varepsilon_2^{k-1} \\ \cdots & \cdots & \cdots & \cdots & \cdots \\ \varepsilon_k^{-1} & \varepsilon_k & \varepsilon_k^2 & \cdots & \varepsilon_k^{k-1} \end{bmatrix} : V(\varepsilon_1, \ldots, \varepsilon_k).$$

Lemma 2.7 of §7.2 gives an explicit expression for the last determinant. Thus, because of (2.14) we have

$$b_0 = \sum_{j=1}^{k} \varepsilon_j^{-1} V(\varepsilon_1, \ldots, \varepsilon_{j-1}, \varepsilon_{j+1}, \ldots, \varepsilon_k) / V(\varepsilon_1, \ldots, \varepsilon_k).$$

The rest of the $b_j, j = 1, \ldots, k - 1$ are found similarly and

$$\sum_{i=1}^{k-1} \varepsilon_j^i b_i = \varepsilon_j^{-1}, \qquad j = 1, \ldots, n. \tag{2.16}$$

Let us sum up the equations of (2.9) with weights b_i :

$$\sum_{l=1}^{k-1} b_l \sum_{j=1}^{k} \alpha_j \varepsilon_j^l = 0.$$

Changing the order of summation and using (2.16) we have

$$\sum_{j=1}^{k} \alpha_j \sum_{l=1}^{k-1} b_l \varepsilon_j^l = \sum_{j=1}^{k} \alpha_j \varepsilon_j^{-1} = 0,$$

Q.E.D.   □

These results justify the following method for finding the normal pseudo-solution. Let $k \geq 2$ be an integer and let $\varepsilon_j, j = 1, \ldots, k$ be a sequence of real parameters satisfying (2.14). Let k solutions x^{ε_j} of the regularized systems (2.5) with parameters ε_j having been found. From the x^{ε_j} we construct a linear combination

$$\bar{x} = \sum_{j=1}^{k} \alpha_j x^{\varepsilon_j}, \tag{2.17}$$

where α_j are the solutions of (2.9). The following theorem then holds.

Theorem 2.2. *The vector $\bar{x}$ defined by (2.17) approximates the normal pseudo-solution of problem (2.1) with the following relative accuracy:*

$$\|x^f - \bar{x}\| / \|\bar{x}\| \leq \mu^{-k} \sum_{j=1}^{k} |\varepsilon_j|, \tag{2.18}$$

where μ is the eigenvalue of matrix A of minimum modulus (but not zero), and x^f is a normal pseudosolution of (2.1).

PROOF. Let us expand the vectors x^f and $\bar{x}$ in the basis of eignevectors of the matrix A. Using (2.2)–(2.4) we rewrite the square of the numerator of the left-hand side in (2.18):

$$\|x^f - \bar{x}\|^2 = \sum_{l=1}^{n} F_l^2 \left| 1/\lambda_l - \sum_{j=1}^{k} \alpha_j / (\lambda_l + i\varepsilon_j) \right|^2.$$

Here the terms corresponding to $\lambda_l = 0$ are dropped. Using (2.11) we estimate each term on the right-hand side of the last relation:

$$F_l^2 \left| 1/\lambda_l - \sum_{j=1}^{k} \alpha_j / (\lambda_l + i\varepsilon_j) \right|^2 \leq F_l^2 \lambda_l^{-2k-2} \left(\prod_{j=1}^{k} \varepsilon_j \right)^2, \qquad \lambda_l \neq 0.$$

Thus we have an inequality

$$\|x^f - \bar{x}\|^2 \leq \mu^{-2k} \left(\prod_{j=1}^{k} \varepsilon_j \right)^2 \sum_{l=1}^{k} F_l^2 \lambda_l^{-2} = \mu^{-2k} \left(\prod_{j=1}^{k} \varepsilon_j \right)^2 \|\bar{x}\|^2$$

from which the results of the theorem follow immediately.   □

Since we seek several solutions of (2.5) corresponding to different values of ε, it is worthwhile to choose the initial approximation for a new problem using those already available.

For instance, suppose we have solved (2.5) with parameter ε_1 (to reduce computing time the parameter ε_1 is chosen from the k available parameters so that it has its maximum modulus). Then we can construct an initial approximation $x^{\varepsilon_2,0}$ for the solution of (2.5) with parameter ε_2 by

$$x^{\varepsilon_2,0} = \frac{\varepsilon_1}{\varepsilon_2} x^{\varepsilon_1}. \tag{2.19}$$

Such a choice provides a necessary coefficient on the component corresponding to the null space of the matrix A. Indeed, the same component is included in x^{ε_2} with weight $1/(i\varepsilon_2)$ and in x^{ε_1} with weight $1/(i\varepsilon_1)$

If we know two solutions x^{ε_1} and x^{ε_2} the initial approximation is constructed from

$$x^{\varepsilon_3,0} = \beta_1 x^{\varepsilon_1} + \beta_2 x^{\varepsilon_2}$$

with the weights β_i found by solving the system

$$\beta_1/\varepsilon_1 + \beta_2/\varepsilon_2 = 1/\varepsilon_3,$$

$$\beta_1 + \beta_2 = 1.$$

From (2.11) it is seen that this gives us an approximation of x^{ε_3} from the initial approximation $x^{\varepsilon_3,0}$ with accuracy $O(\bar{\varepsilon})$, where

$$\bar{\varepsilon} = \max_{1 \le i \le k} |\varepsilon_i|.$$

If it is necessary to find x^{ε_4}, then the initial approximation is constructed from the formula

$$x^{\varepsilon_4,0} = \gamma_1 x^{\varepsilon_1} + \gamma_2 x^{\varepsilon_2} + \gamma_3 x^{\varepsilon_3},$$

where the weights are determined from

$$\gamma_1/\varepsilon_1 + \gamma_2/\varepsilon_2 + \gamma_3/\varepsilon_3 = 1/\varepsilon_4,$$

$$\gamma_1 + \gamma_2 + \gamma_3 = 1,$$

$$\varepsilon_1 \gamma_1 + \varepsilon_2 \gamma_2 + \varepsilon_2 \gamma_3 = \varepsilon_4;$$

in this case

$$\|x^{\varepsilon_4,0} - x^{\varepsilon_4}\| = O(\bar{\varepsilon}^2).$$

To illustrate these results, we present a numerical example, using the matrix A and the vector f given in Table 6.1. We solve (2.5) with the values $\varepsilon_1 = 0.01$, $\varepsilon_2 = 0.5 \times 10^{-2}$ and $\varepsilon_3 = -\frac{1}{3} \times 10^{-2}$. Such a choice insures that condition (2.14) in Lemma 2.1 holds. The normal pseudosolution, the error of the regularized solutions x^{ε_i} and the error of the extrapolated solution $\bar{x}$ are given in the following table. The real parts of the corresponding vectors are presented in Table 6.3, and the imaginary parts in Table 6.4. Comparing the results of our extrapolation with those obtained in §6.1 we see that the method of extrapolation which uses the symmetry of the matrix A is more effective.

Table 6.3. Real Parts.

Number of component	$Re(x^f)$	$Re(x^f - x^{\varepsilon_1})$ $\varepsilon_1 = 0.01$	$Re(x^f - x^{\varepsilon_2})$ $\varepsilon_2 = \varepsilon_1/2$	$Re(x^f - x^{\varepsilon_3})$ $\varepsilon_3 = -\varepsilon_1/3$	$Re(x^f - \bar{x})$
1	-0.9045	-4.9×10^{-5}	-1.2×10^{-5}	-5.5×10^{-6}	6.5×10^{-9}
2	-1.5090	-7.7×10^{-5}	-1.9×10^{-5}	-8.5×10^{-6}	1.1×10^{-9}
3	0.2455	-1.3×10^{-5}	-3.3×10^{-6}	-1.5×10^{-6}	4.2×10^{-9}
4	4.6468	1.4×10^{-4}	-3.6×10^{-5}	1.6×10^{-5}	1.7×10^{-9}
5	-2.4135	-1.2×10^{-4}	-2.9×10^{-5}	-1.3×10^{-5}	-7.0×10^{-10}
6	-4.0225	-2.0×10^{-4}	-4.9×10^{-5}	-2.2×10^{-5}	1.9×10^{-9}
7	-1.5225	5.1×10^{-5}	1.2×10^{-5}	5.6×10^{-6}	2.1×10^{-9}
8	-2.2276	4.3×10^{-5}	1.0×10^{-5}	4.7×10^{-6}	1.9×10^{-9}
9	1.2205	-2.7×10^{-5}	-6.8×10^{-6}	-3.0×10^{-6}	5.3×10^{-10}
10	0.0	0.0	0.0	0.0	0.0

Table 6.4. Imaginary Parts.

Number of component	$Im(x^f)$	$Im(x^f - x^{\varepsilon_1})$	$Im(x^f - x^{\varepsilon_2})$	$Im(x^f - x^{\varepsilon_3})$	$Im(x^f - \bar{x})$
1	0.0	-2.87×10^1	-5.75×10^1	8.63×10^1	10^{-8}
2	0.0	-5.75×10^1	-1.15×10^2	1.72×10^2	10^{-8}
3	0.0	-2.88×10^1	-5.75×10^1	8.63×10^1	10^{-9}
4	0.0	8.64×10^1	1.72×10^2	-2.59×10^2	10^{-7}
5	0.0	1.43×10^2	2.87×10^2	-4.30×10^2	10^{-8}
6	0.0	6.00	1.2×10^1	-1.8×10^1	10^{-7}
7	0.0	-1.43×10^2	-2.87×10^2	4.3×10^2	10^{-7}
8	0.0	1.44×10^2	2.88×10^2	-4.3×10^2	10^{-7}
9	0.0	-2.87×10^1	-5.75×10^1	8.6×10^1	10^{-8}
10	0.0	10^3	2.0×10^3	-3.0×10^3	10^{-12}

6.3. Extrapolation of Solutions Containing Boundary-Layer Functions

In a number of problems of mathematical physics the introduction of a small parameter leads to interesting and useful results. One of the most widely-used applications of this is in the addition of higher derivatives with a small weight in order to change the type of boundary conditions, or the type of differential equation present. In this case extrapolation methods can be of great help when constructing economical numerical algorithms. We will consider two examples in which one can succeed in using an extrapolation algorithm to exploit certain special properties of the solutions.

EXAMPLE 1. Replace the problem

$$-\Delta u = f \quad \text{in } \Omega,$$
$$u = g \quad \text{on } \Gamma, \tag{3.1}$$

by the following:

$$-\Delta u_\varepsilon = f \quad \text{in } \Omega,$$
$$u_\varepsilon + \varepsilon \frac{\partial u_\varepsilon}{\partial n} = g \quad \text{on } \Gamma. \tag{3.2}$$

Here Ω is a two-dimensional bounded region with smooth boundary Γ, $\partial u/\partial n$ is the outward normal derivative on Γ, and $\varepsilon > 0$ is a small parameter.

A similar change can be applied to pass from one kind of boundary condition to another (see [93]). This simplifies applications of the finite element method (see [125, 9, 90]).

Before we begin to investigate the asymptotic expansion for u_ε let us prove a comparison theorem for (3.2).

Theorem 3.1. *Let u_1 be the solution of*

$$-\Delta u_1 = 0 \quad \text{in } \Omega,$$
$$u_1 + \varepsilon \frac{\partial u_1}{\partial n} = g_1 \quad \text{on } \Gamma,$$

and u_2 the solution of

$$-\Delta u_2 = 0 \quad \text{in } \Omega,$$
$$u_2 + \varepsilon \frac{\partial u_2}{\partial n} = g_2 \quad \text{on } \Gamma,$$

where $\Gamma \in C^{2+\lambda}$, $g_1, g_2 \in C^{1+\lambda}(\Gamma)$, $\lambda \in (0, 1)$.
Then from

$$|g_1| < g_2 \quad \text{on } \Gamma$$

it follows that

$$|u_1| < u_2 \quad \text{on } \overline{\Omega}. \tag{3.3}$$

PROOF. Consider the equation satisfied by the difference $v = u_2 - u_1$; we have

$$-\Delta v = 0 \qquad \text{in } \Omega,$$

$$v + \varepsilon \frac{\partial v}{\partial n} = g_2 - g_1 \quad \text{on } \Gamma.$$

Suppose that the function v has negative values in $\bar{\Omega}$. Then because it is continuous it attains a negative minimum at a certain point x_0. Let us show that $x_0 \in \Omega$. Indeed, if $x_0 \in \Gamma$ then

$$\varepsilon \frac{\partial v}{\partial n}(x_0) = g_2(x_0) - g_1(x_0) - v(x_0) > 0;$$

the derivative $\partial v/\partial n$ (by continuity) preserves its sign in the intersection of some neighborhood of x_0 with Ω. Therefore in the direction of the inner normal to Γ at the point x_0 the function v is strictly decreasing, which contradicts the fact that the point x_0 is a minimum. Thus $x_0 \in \Omega$. From the maximum principle it follows that $v \equiv \text{const on } \bar{\Omega}$.

Consider an arbitrary point $x \in \Gamma$. Since $v \equiv \text{const}$, $\partial v/\partial n(x) = 0$. Therefore the boundary condition is transformed into

$$v(x) = g_2(x) - g_1(x).$$

By assumption $v(x) < 0$ (true everywhere on $\bar{\Omega}$), yet according to the condition of the theorem $g_2(x) - g_1(x) \geq 0$. We came to a contradiction if we assume that the function v is negative at least one point $x \in \bar{\Omega}$. Thus

$$v = u_2 - u_1 > 0 \quad \text{on } \bar{\Omega}.$$

Assuming $v = u_2 + u_1$ we have

$$u_2 + u_1 > 0 \quad \text{on } \bar{\Omega}.$$

combining the last two inequalities we have (3.3). Theorem 3.1 is proved. $\square$

Let us exhibit the conditions which guarantee the existence of an asymptotic expansion of the function u_ε.

Theorem 3.2. *Suppose the following smoothness conditions*

$$\Gamma \in C^{l+2}, \qquad g \in C^{l+2}(\Gamma), \qquad f \in C^l(\bar{\Omega}) \tag{3.4}$$

are satisfied by the solution of (3.1), where l is nonintegral. Then the solution of (3.2) has an expansion

$$u_\varepsilon = u + \sum_{k=1}^{s} \varepsilon^k v_k + \varepsilon^{s+1} w_\varepsilon \quad \text{on } \bar{\Omega}, \tag{3.5}$$

where $s = [l]$, the functions v_k are independent of ε, and

$$\max_{\bar{\Omega}} |w_\varepsilon| \leq c_1. \tag{3.6}$$

PROOF. Assuming $v_0 = u$ let us define the functions v_i via the sequence of equations

$$-\Delta v_i = 0 \qquad \text{in } \Omega, \qquad i = 1, \ldots, S. \tag{3.7}$$
$$v_i = -\frac{\partial v_{i-1}}{\partial n} \quad \text{on } \Gamma,$$

By Theorem 1.2 of §4.1 the solution u of (3.1) belongs to $C^{l+2}(\overline{\Omega})$, and the functions $v_i \in C^{l+2-i}(\overline{\Omega})$ are independent of ε. Introduce the function

$$w_\varepsilon = \varepsilon^{-s-1}\left(u_\varepsilon - \sum_{i=0}^{s} \varepsilon^i v_i\right) \quad \text{on } \overline{\Omega}.$$

From (3.2) it follows that

$$-\Delta w_\varepsilon = 0 \qquad \text{in } \Omega,$$
$$w_\varepsilon + \varepsilon\frac{\partial w_\varepsilon}{\partial n} = -\frac{\partial v_s}{\partial n} \quad \text{on } \Gamma. \tag{3.8}$$

Since $v_s \in C^{l-s+2}(\overline{\Omega})$ then $\partial v_s/\partial n \in C^{l-s+1}(\Gamma)$. Therefore there is a unique solution of (3.8) which belongs to $C^{l-s+2}(\overline{\Omega})$ (see [73]). Estimates on the solution and its derivatives will generally depend on ε. Let us show, however, that the function w_ε can be estimated independently of ε. Apply Theorem 3.1 to

$$w_\varepsilon \quad \text{and} \quad w = \max_\Gamma\left|\frac{\partial v_s}{\partial n}\right|.$$

It is obvious that

$$-\Delta w = 0 \qquad \text{in } \Omega,$$
$$w + \varepsilon\frac{\partial w}{\partial n} = \max_\Gamma\left|\frac{\partial v_s}{\partial n}\right| \quad \text{on } \Gamma.$$

Consequently

$$|w_\varepsilon| \le w \quad \text{on } \overline{\Omega}.$$

Assuming

$$c_1 = \max_\Gamma\left|\frac{\partial v_s}{\partial n}\right|,$$

we come to (3.6). Theorem 3.2. is proved.   □

Since we wish to find u rather than u_ε, then using (3.5) the usual linear extrapolation with respect to ε can be applied.

Suppose the conditions of Theorem 3.2 hold. Let us find solutions $u_{\varepsilon/k}$ of (3.2) with parameters ε/k, $k = 1, \ldots, s + 1$. Form the linear corrector

$$U^E = \sum_{k=1}^{s+1} \gamma_k u_{\varepsilon/k} \quad \text{on } \overline{\Omega}, \tag{3.9}$$

where

$$\gamma_k = \frac{(-1)^{s-k+1} k^{s+1}}{k!\,(s - k + 1)!}.$$

Theorem 3.3. *Assume the conditions of Theorem 3.2 hold. Then the corrector (3.9) obeys the estimate*

$$|U^E - u| \le c_2 \varepsilon^{s+1} \quad on \ \overline{\Omega} \tag{3.10}$$

with constant c_2 independent of ε.

PROOF. Using (3.5) for $k = 1, \ldots, s + 1$ let us form the corrector U^E and apply Lemma 2.1 of §7.2. We have

$$U^E = u + \varepsilon^{s+1} \sum_{k=1}^{s+1} \gamma_k k^{-s-1} w_{\varepsilon/k}.$$

Subtract u from both sides of the equality, take the modulus on both sides, and use (3.6). Assuming

$$c_2 = c_1 \sum_{k=1}^{s+1} |\gamma_k| k^{-s-1},$$

we come to (3.10). Theorem 3.3 is proved. $\qquad\square$

This result allows us to calculate u with great accuracy by using several approximate solutions $u_{\varepsilon/k}$ at comparatively large values of ε/k. This increases the economy of the numerical algorithms considerably. The most beneficial effect is attained if we align the extrapolation in ε with the extrapolation in mesh-size.

EXAMPLE 2. For simplicity let us consider a one-dimensional case. We will study the following model system, which appears in numerical calculation of the motion of a medium with low viscosity:

$$-\varepsilon^2 y_\varepsilon'' + a y_\varepsilon = f \quad \text{on } (0, 1),$$
$$y_\varepsilon(0) = y_0, \qquad y_\varepsilon(1) = y_1. \tag{3.11}$$

Here $\varepsilon > 0$ is the small parameter, and $a(x) > 0$ and $f(x)$ are sufficiently smooth functions. According to the results of [132] the solution of (3.11) can be written as a sum

$$y_\varepsilon = u + b e^{-d/\varepsilon} + \varepsilon w_\varepsilon \quad \text{on } [0, 1], \tag{3.12}$$

where u is the solution of the limiting problem, for (3.11):

$$au = f \quad \text{on } (0, 1). \tag{3.13}$$

The functions $b(x)$ and $d(x) \geq 0$ are continuous on $[0, 1]$ and are independent of ε. The remainder w_ε obeys the estimate

$$|w_\varepsilon| \leq c_3 \tag{3.14}$$

with constant c_3 independent of ε.

In (3.11) it is desirable to find an approximation of y_{ε_0} for certain sufficiently small ε_0, rather than a limiting solution u (as was the case in Example 1). Let us try to construct this approximation using several solutions y_{ε_i}, for values of ε_i substantially greater than ε_0.

Consider the case when the solution of (3.13) is known. Find two more solutions y_{ε_1}, y_{ε_2} of (3.11) with parameters ε_1 and $\varepsilon_2 = \varepsilon_1/2$ greater than ε_0. Omit the remainder on the right-hand side of (3.12) and pass to the approximate formulae

$$y_{\varepsilon_1} \approx u + be^{-d/\varepsilon_1}, \qquad y_{\varepsilon_2} \approx u + be^{-2d/\varepsilon_1}$$

Hence the unknowns $b(x)$, $d(x)$, $x \in (0, 1)$ satisfy the approximate system of equations

$$be^{-d/\varepsilon_1} \approx y_{\varepsilon_1} - u, \qquad be^{-ed/\varepsilon_1} \approx y_{\varepsilon_2} - u. \tag{3.15}$$

Square the first equation and divide it by the second one. We have

$$b \approx (y_{\varepsilon_1} - u)^2/(y_{\varepsilon_2} - u).$$

Dividing the second equation of (3.15) by the first one we have

$$e^{-d/\varepsilon_1} \approx (y_{\varepsilon_2} - u)/(y_{\varepsilon_1} - u).$$

Recall that

$$y_{\varepsilon_0} \approx u + be^{-d/\varepsilon_0}.$$

Therefore

$$y_{\varepsilon_0} \approx u + \frac{(y_{\varepsilon_1} - u)^2}{y_{\varepsilon_2} - u} \left(\frac{y_{\varepsilon_2} - u}{y_{\varepsilon_1} - u}\right)^{\varepsilon_1/\varepsilon_0}.$$

Since all terms on the right-hand side of this formula are known we can take this expression to be our approximation to y_{ε_0}. Let

$$Y_{\varepsilon_0} = u + \frac{(y_{\varepsilon_1} - u)^2}{y_{\varepsilon_2} - u} \left(\frac{y_{\varepsilon_2} - u}{y_{\varepsilon_1} - u}\right)^{\varepsilon_1/\varepsilon_0}. \tag{3.16}$$

This formula should be used only for those $x \in [0, 1]$ for which the boundary-layer function differs from zero by a term of at least order ε_0:

$$|b(x)| \exp(-d(x)/\varepsilon_0) \geq c_4 \varepsilon_0 \tag{3.17}$$

(with an arbitrary positive constant c_4 independent of ε). If $b(0) \neq 0$ then near zero there is an interval $[0, \delta_1]$ on which condition (3.17) is satisfied. If $b(1) \neq 0$ then there is such an interval near $1 : [\delta_2, 1]$. We denote the union of these intervals by ω.

Let us now determine the accuracy of the extrapolated solution Y_{ε_0}, and stability of (3.16) relative to the errors which arise in computing u, y_{ε_1}, y_{ε_2}. Let $x \in \omega$ and consider the errors instead of the approximations $\tilde{u}$, $\tilde{y}_{\varepsilon_1}$, $\tilde{y}_{\varepsilon_2}$:

$$\alpha = \tilde{u}(x) - u(x), \qquad \alpha_{\varepsilon_1} = \tilde{y}_{\varepsilon_1}(x) - y_{\varepsilon_1}(x), \qquad \alpha_{\varepsilon_2}(x) = \tilde{y}_{\varepsilon_2}(x) - y_{\varepsilon_2}(x). \tag{3.18}$$

Using (3.16) we get the extrapolated value

$$\tilde{Y}_{\varepsilon_0} = \tilde{u} + (\tilde{y}_{\varepsilon_1} - \tilde{u}) \left(\frac{\tilde{y}_{\varepsilon_2} - \tilde{u}}{\tilde{y}_{\varepsilon_1} - \tilde{u}} \right)^{\varepsilon_1/\varepsilon_0 - 1} \tag{3.19}$$

Let us investigate the accuracy of this approximation with small ε_1 but with ratio

$$\varepsilon_1/\varepsilon_0 = k > 2 \tag{3.20}$$

held fixed.

Theorem 3.4. *Assume that at some point $x \in \omega$ the expansion (3.12) together with the estimate (3.14) is valid; assume further that the errors (3.18) are of $O(\varepsilon_1^l)$ order:*

$$\max(|\alpha|, |\alpha_{\varepsilon_1}|, |\alpha_{\varepsilon_2}|) \leq c_5 \varepsilon_1^l, \quad \text{where } l > 1/k. \tag{3.21}$$

Then for sufficiently small ε_1 we have

$$|y_{\varepsilon_0}(x) - \tilde{Y}_{\varepsilon_0}(x)| \leq c_6(\varepsilon_1 + |\alpha| + |\alpha_{\varepsilon_1}| + |\alpha_{\varepsilon_2}|). \tag{3.22}$$

PROOF. From (3.12), (3.18) it follows that

$$\tilde{y}_{\varepsilon_1} - \tilde{u} = b \exp(-d/\varepsilon_1) + \varepsilon_1 w_{\varepsilon_1} + \alpha_{\varepsilon_1} - \alpha,$$

$$\tilde{y}_{\varepsilon_2} - \tilde{u} = b \exp(-2d/\varepsilon_1) + \varepsilon_1 w_{\varepsilon_2}/2 + \alpha_{\varepsilon_2} - \alpha.$$

write

$$\beta_1 = \varepsilon_1 w_{\varepsilon_1} + \alpha_{\varepsilon_1} - \alpha, \qquad \beta_2 = \varepsilon_1 w_{\varepsilon_2}/2 + \alpha_{\varepsilon_2} - \alpha, \qquad \beta_3 = \alpha - \varepsilon_0 w_{\varepsilon_0}$$

and introduce the function

$$U(t_1, t_2, t_3) = u + \varepsilon_0 w_{\varepsilon_0} + t_3 + (b \exp(-d/\varepsilon_1) + t_1)$$
$$\times \{(b \exp(-2d/\varepsilon_1) + t_2)/(b \exp(-d/\varepsilon_1) + t_1)\}^{k-1}.$$

Note that $U(0, 0, 0) = y_{\varepsilon_0}$ and $U(\beta_1, \beta_2, \beta_3) = \tilde{Y}_{\varepsilon_0}$. We will prove that for sufficiently small t_i the function U has bounded derivatives

$$\frac{\partial U}{\partial t_1}(t_1, t_2, t_3) = (2 - k)\{(b \exp(-2d/\varepsilon_1) + t_2)/(b \exp(-d/\varepsilon_1) + t_1)\}^{k-1},$$

$$\frac{\partial U}{\partial t_2}(t_1, t_2, t_3) = (k - 1)\{(b \exp(-2d/\varepsilon_1) + t_2)/(b \exp(-d/\varepsilon_1) + t_1)\}^{k-2},$$

$$\frac{\partial U}{\partial t_3}(t_1, t_2, t_3) = 1.$$

First let us investigate the denominator of

$$A = (b \exp(-2d/\varepsilon_1) + t_2)/(b \exp(-d/\varepsilon_1) + t_1).$$

From (3.17) it follows that

$$\exp(-d/\varepsilon_1) \geq (c_4 \varepsilon_0/|b|)^{\varepsilon_0/\varepsilon_1} = (c_4 \varepsilon_1/|b|k)^{1/k}.$$

Therefore

$$|b| \exp(-d/\varepsilon_1) \geq c_7 \varepsilon_1^{1/k}. \tag{3.23}$$

From (3.14) and (3.21) we have

$$|t_1| \leq |\beta_1| \leq c_3 \varepsilon_1 + 2\varepsilon_1^l c_5. \tag{3.24}$$

If $l > 1$ then from (3.24) and $\varepsilon_1 \leq 1$ we have $|t_1| \leq (c_3 + 2c_5)\varepsilon_1$. If $l \leq 1$ then from (3.24) it follows that $|t_1| \leq (c_3 + 2c_5)\varepsilon_1^l$. In both cases the $|t_1|$ term is a higher-order term than the right-hand side of (3.23), Therefore

$$|b| \exp(-d/\varepsilon_1) \geq 2|t_1| \quad \forall \, \varepsilon_1 \leq \min\{1, \{c_7/(c_3 + 2c_5)2\}^{k/(kl-1)},$$
$$\{c_7/(c_3 + 2c_5)2\}^{k/(k-1)}\}.$$

Hence

$$|b \exp(-d/\varepsilon_1) + t_1| \geq \tfrac{1}{2}|b| \exp(-d/\varepsilon_1).$$

From these inequalities and (3.23) we have

$$|A| \leq \{|b| \exp(-2d/\varepsilon_1) + |t_2|\}/\{\tfrac{1}{2}|b| \exp(-d/\varepsilon_1)\}$$
$$\leq 2 \exp(-d/\varepsilon_1) + 2\varepsilon_1^{-1/k}|t_2|/c_7.$$

Equations (3.14) and (3.21) yield

$$|t_2| \leq c_3 \varepsilon_1/2 + 2c_5 \varepsilon_1^l.$$

Combining these inequalities and taking into account that $\varepsilon_1 < 1$, $d \geq 0$ $1/k < l$,

we have

$$|A| \leq 2 + (c_3 + 4c_5)/c_7 \equiv c_8.$$

Therefore

$$\left| \frac{\partial u}{\partial t_i}(t_1, t_2, t_3) \right| \leq c_9 \quad \forall i = 1, 2, 3, \quad \forall t_i \in [0, \beta_i], \qquad (3.25)$$

where

$$c_9 \equiv \max\{1, (k - 2)c_8^{k-1}, (k - 1)c_8^{k-2}\}.$$

According to the Lagrange theorem

$$U(\beta_1, \beta_2, \beta_3) = U(0, 0, 0) + \beta_1 \frac{\partial U}{\partial t_1}(\xi_1, \beta_2, \beta_3)$$

$$+ \beta_2 \frac{\partial U}{\partial t_2}(0, \xi_2, \beta_3) + \beta_3 \frac{\partial U}{\partial t_3}(0, 0, \xi_3),$$

where

$$\xi_i \in [0, \beta_i].$$

From this relation and (3.25) we have

$$|U(\beta_1, \beta_2, \beta_3) - U(0, 0, 0)| \leq c_9(|\beta_1| + |\beta_2| + |\beta_3|).$$

Because the left-hand side of this inequality is equal to $|y_{\varepsilon_0}(x) - \tilde{Y}_{\varepsilon_0}(x)|$ using (3.14) on the right-hand side we get (3.22). The theorem is proved.   □

Let us present an example illustrating the effectiveness of this extrapolation algorithm. Let $a(x) = 1$, $f(x) = 10(2 - e^x)$, $y_0 = y_1 = 0$ in (3.11). Then the solution is

$$y_\varepsilon(x) = 20 - de^x + (20 - d)(\coth \gamma \sinh \gamma x - \cosh \gamma x)$$

$$+ de(20 - de)\frac{\sinh \gamma x}{\sinh \gamma}, \quad \text{where } d = 10/(1 - \varepsilon^2), \quad \gamma = 1/\varepsilon.$$

Table 6.5

x	y_{ε_1}	y_{ε_2}	y_{ε_0}	Y_{ε_0}	u
0	0	0	0	0	10
0.01	0.3406	0.7074	6.2201	6.2356	9.8995
0.02	1.5905	3.0860	8.4437	8.4554	9.7980
0.03	2.2582	4.1553	9.1966	9.2033	9.6955
0.04	2.8515	5.0838	9.4077	9.4114	9.5919
0.05	3.3774	5.7914	9.4189	9.4210	9.4873
0.06	3.8421	6.3506	9.3558	9.3573	9.3816
0.07	4.2514	6.7882	9.2647	9.2660	9.2749
0.08	4.6104	7.1261	9.1627	9.1638	9.1671
0.09	4.9238	7.3820	9.0559	9.0570	9.0583
0.10	5.1958	7.5706	8.9468	8.9478	8.9483

The values of the solutions y_ε found from this formula for $\varepsilon_1 = 0.1$, $\varepsilon_2 = 0.05$ and $\varepsilon_0 = 0.01$ are given in Table 6.5. We also give the solution $u = f/a$ of the singular problem, and Y_{ε_0}.

We see from the table that the extrapolations are in good agreement with the exact solutions near the point $x = 0$ in spite of the significant difference between the sought-after solution y_{ε_0} and the values y_{ε_1}, y_{ε_2}, u used in the extrapolation.

6.4. The Fredholm Equation of the Second Kind

One-dimensional integral equations with sufficiently smooth stable solutions in the absence of boundary conditions are the simplest objects we can use to illustrate the general theorems of Chapter 1. For the one-dimensional Fredholm equation of the second kind Richardson extrapolation is uneconomical in comparison with higher-order quadratures; therefore the example in this section is primarily of pedagogic interest. But when dealing with the equations in higher dimensions application of the simplest type of quadrature, followed by Richardson extrapolation, appears to be more efficient. This can be vividly illustrated when trying to find eigenvalues of an operator with a symmetric kernel. In this case the simplest form of quadrature gives the symmetric matrices for a discrete eigenvalue problem, as a rule, while higher-order quadratures do not.

Let us consider the Fredholm equation of the second kind

$$\varphi(x) = \int_0^1 K(x, t)\varphi(t)\, dt + f(x) \quad \text{on } [0, 1]. \tag{4.1}$$

Assume that the right-hand side f and the kernel K are sufficiently smooth functions:

$$f \in C^m[0, 1], \quad K \in C^m([0, 1] \times [0, 1]). \tag{4.2}$$

where m is an integer, $m \geq 1$. In order that (4.1) have a unique solution we require that the kernel be sufficiently small:

$$\varkappa = \max_{[0,1] \times [0,1]} |K| < 1. \tag{4.3}$$

Theorem 4.1. *The equation* (4.1), *under the conditions* (4.2), (4.3) *has a unique solution* $\varphi \in C^m[0, 1]$.

PROOF. From [65] by the continuity of f, K and by (4.3) it follows, that (4.1) has a unique continuous solution φ. To determine whether its derivatives $\varphi^{(k)}$, $k \leq m$ are continuous we differentiate (4.1) with respect to x, k times. From (4.2) and from the fact that integral $\int_0^1 K(x, t)\varphi(t)\, dt$ is continuously differentiable with respect to x it follows that the addends terms on the right-hand side of the relation thus obtained are continuous.

Thus Condition A of §1.2 with $M_k(\Omega) = P_k(\overline{\Omega}) = C^k[0, 1]$, with no require-ment on the boundary values is satisfied. To find an approximate solution of (4.1) we introduce the difference net $\breve{\omega}_h = \{x_{i+1/2} = (i + 1/2)h;\ i = 0, \ldots,$ $N - 1\}$ with integer $N \geq 2$ and mesh-size $h = 1/N$. Replace the integral in (4.1) by its approximation using the trapezoidal rule. As a result we have

$$\varphi^h(x) = \sum_{t \in \breve{\omega}_h} K(x, t)\varphi^h(t)h + f(x) \quad \forall\, x \in \breve{\omega}_h. \tag{4.4}$$

The number of equations here will equal the number of unknowns $\varphi^h(x)$, $x \in \breve{\omega}_h$. Assuming that there is at least one solution of this system we derive an *a priori* estimate. Let the component of the vector φ^h having maximum modulus be the kth component:

$$\max_{\breve{\omega}_h} |\varphi^h| = |\varphi^h(x_{k+1/2})|.$$

Then from (4.4) and (4.3) it follows that

$$|\varphi^h(x_{k+1/2})| \leq \sum_{t \in \breve{\omega}_h} |K(x_{k+1/2}, t)|\, |\varphi^h(t)|h$$

$$+ |f(x_{k+1/2})| \leq \varkappa|\varphi^h(x_{k+1/2})| + \max_{\breve{\omega}_h}|f|.$$

Hence we have

$$\max_{\breve{\omega}_h}|\varphi^h| \leq \frac{1}{1 - \varkappa} \max_{\breve{\omega}_h} |f|. \tag{4.5}$$

This guarantees that (4.4) is stable and has a unique solution. Thus Condition B' of §1.2 holds, where

$$\|u\|_{\overline{\Omega}_h} = \|u\|_{\breve{\Omega}_h} = \max_{\breve{\omega}_h}|u|.$$

Now, to verify Condition D of §1.2 we introduce the operators L, L_h defined by

$$Lu(x) = u(x) - \int_0^1 K(x, t)u(t)\, dt, \qquad x \in [0, 1],$$

$$L_h u(x) = u(x) - \sum_{t \in \breve{\omega}_h} K(x, t)u(t)h, \qquad x \in \breve{\omega}_h.$$

Then for any function $u \in C^k[0, 1]$, $k \leq m$ from Theorem 2.7 of §2.2 we have

$$L_h u = Lu + \sum_{j=1}^{[(k-1)/2]} h^{2j}g_j + \sigma^h \quad \text{on } \breve{\omega}_h, \tag{4.6}$$

where the g_j are independent of h, $g_j \in C^{k-2j}[0, 1]$, and $\max_{\breve{\omega}_h}|\sigma^h| = O(h^k)$. Because Conditions A, B', D are fulfilled, Theorem 2.2 and the remark from §1.2, insure that we can write

$$\varphi^h = \varphi + \sum_{j=1}^{s} h^{2j}v_j + \eta^h \quad \text{on } \breve{\omega}_h, \tag{4.7}$$

where $s = [(m - 1)/2]$, the v_j are independent of h, and $\max_{\breve{\omega}_h}|\eta^h| = O(h^m)$.

From (4.7) we can use the correction method of §1.3 for expansions in even powers of h. To do this we build difference nets $\ddot{\omega}_{h_i}$ with mesh-sizes h_i, equal to $h, h/3, \ldots, h/(2s + 1)$ and solve the system of linear equations (4.4) on each net. All solutions $\varphi^{h_i}, i = 1, \ldots, s + 1$, thus obtained are defined on the net $\ddot{\omega}_h$. We choose the μ_i as the solutions of the system

$$\sum_{i=1}^{s+1} \mu_i = 1,$$

$$\sum_{i=1}^{s+1} \mu_i h_i^{2k} = 0, \qquad k = 1, \ldots, s. \tag{4.8}$$

Form a linear combination with these weights:

$$\phi(x) = \sum_{k=1}^{s+1} \mu_k \varphi^{h_k}(x) \quad \text{on } \ddot{\omega}_h. \tag{4.9}$$

Then from Theorem 3.2 of §1.3 we have the estimate

$$\max_{\ddot{\omega}_h} |\phi - \varphi| \le c_3 h^m, \tag{4.10}$$

which completes the justification of our extrapolation. ☐

Note that the division of the mesh-size of the initial net $\ddot{\omega}_h$ into only odd parts is important in this case. For example, the usual divisions of the mesh-size $h, h/2, h/3, \ldots$ or $h, h/2, h/4, \ldots$ result in no overlap of the nets which consist of the knots not located at the endpoints of the interval.

On the basis of Lemma 4.2 of Chapter 3 the results of this section can be used for the trapezoidal rule [3].

The inequality (4.5), in addition to its role in justifying (4.7), plays an important role in the estimation of the effect of the error when the right-hand side is given approximately or as a set of data points. It is also effective in controlling the accuracy of the sequence of approximations when iterative methods are used to solve system (4.4).

6.5. The Volterra Equation of the Second Kind

For equations of this kind, if we apply the trapezoidal rule, and then use Richardson extrapolation a simple uniform algorithm results. The application of quadrature rules with higher powers [68] increases the overall complexity of the algorithm because it is necessary to compute several values values by different methods.

Consider the equation

$$\varphi(x) = \int_0^x K(x, t)\varphi(t) \, dt + f(x), \qquad x \in [0, 1]. \tag{5.1}$$

In order to prove that the problem has a unique solution which is sufficiently smooth it is sufficient that the right-hand side and the kernel are smooth.

Theorem 5.1. *Let*

$$f \in C^m[0, 1], \qquad K \in C^m(\bar{Q}), \tag{5.2}$$

where the integer $m \geq 2$, and $\bar{Q}$ is a triangle: $0 \leq t \leq x \leq 1$. Then equation (5.1) has a unique solution which belongs to $C^m[0, 1]$.

PROOF. From the continuity of f and K from [68] it follows that equation (5.1) has the unique continuous solution. The continuity of derivatives of the solution is determined by differentiation of both sides of equation (5.1) as many times as is necessary. □

This theorem means that Condition A of §1.2 is satisfied with the classes $M_k(\Omega) = P_k(\bar{\Omega}) = C^k[0, 1]$. The requirements on the boundary values are naturally omitted.

We will compute the approximate solution on the net $\breve{\omega}_h = \{x_i = ih, i = 0, 1, \ldots, N\}$ with mesh-size $h = 1/N$, and $N \geq 2$, N an integer. We use the trapezoidal rule. We have

$$\varphi^h(t) = \sum_{\substack{x \in \bar{\omega}_h \\ x \leq t}} \{K(t, x)\varphi^h(x)\}_{\bar{x}} h + f(t), \qquad t \in \bar{\omega}_h \tag{5.3}$$

or, what is the same,

$$\varphi^h(0) = f(0),$$

$$\varphi^h(x_i) = \sum_{j=0}^{i-1} \{K(x_i, x_j)\varphi^h(x_j) + K(x_i, x_{j+1})\varphi^h(x_{j+1})\}h/2 + f(x_i), \tag{5.4}$$

$$i = 1, 2, \ldots, N.$$

The system has a triangular matrix. For the unique solvability it is enough that its diagonal elements differ from zero. This can be insured by the choice of a sufficiently small h:

$$h\varkappa/2 \leq q < 1, \quad \text{where } \varkappa = \max_{t \in [0, 1]} |K(t, t)|. \tag{5.5}$$

Lemma 5.2. *If (5.5) is satisfied then the estimate*

$$\|\varphi^h\|_{C,h} \leq c_1 \|f\|_{C,h}, \quad \text{where } c_1 = \frac{1}{1 - q} \exp\left(\frac{\varkappa}{2(1 - q)}\right), \tag{5.6}$$

holds for the solution of the difference problem (5.3).

PROOF. First we derive a somewhat different inequality

$$|\varphi^h(x_i)| \leq \alpha(1 + \beta)^i, \qquad i = 0, 1, \ldots, N, \tag{5.7}$$

where

$$\alpha = \frac{1}{1 - q} \max_{\bar{\omega}_h} |f|, \qquad \beta = \frac{\varkappa h}{2(1 - q)}.$$

For $i = 0$ this inequality evidently follows from the first equation of system (5.4). Assume that it is satisfied for $l = 0, 1, \ldots, i$, we will prove that it is also satisfied for $i + 1$. Consider equation (5.4) with index $i + 1$. Let us take the modulus of both sides

$$|\varphi^h(x_{i+1})| \le \sum_{j=0}^{i} \varkappa(|\varphi^h(x_j)| + |\varphi^h(x_{j+1})|)h/2 + \max_{\bar{\omega}_h} |f|.$$

From this, using (5.5), we have

$$|\varphi^h(x_{i+1})| \le \beta \sum_{j=0}^{i} |\varphi^h(x_j)| + \alpha$$

$$\le \alpha + \alpha\beta \sum_{j=0}^{i} (1 + \beta)^j = \alpha(1 + \beta)^{i+1}.$$

Thus inequality (5.7) is proved. From (5.7) and $1 + x \le e^x \; \forall\, x \ge 0$ (5.6) follows:

$$|\varphi^h(x_i)| \le \alpha(1 + \beta)^i \le \alpha e^{\beta i} \le \alpha \exp\left(\frac{\varkappa}{2(1 - q)}\right) \quad \forall\, i = 0, \ldots, N.$$

The lemma is proved.   □

We have thus shown that Condition B′ of §1.2 is satisfied for (5.3). Here

$$\|u\|_{\bar{\Omega}_h} = \|u\|_{\bar{\Omega}_h} = \|u\|_{C, h}.$$

Let us introduce the operators

$$L\varphi(t) = \varphi(t) - \int_0^t K(t, x)\varphi(x)\, dx, \qquad t \in [0, 1],$$

$$L_h\varphi(t) = \varphi(t) - \sum_{\substack{x \in \bar{\omega}_h \\ x \le t}} \{K(t, x)\varphi(x)\}_{\bar{x}} h, \qquad t \in \bar{\omega}_h.$$

Using Theorem 2.8 of §2.2 for $\varphi \in C^k[0, 1]$, $k \le m$, we have

$$L_h\varphi = L\varphi + \sum_{j=1}^{[(k-1)/2]} h^{2j} g_j + \sigma^h \quad \text{on } \bar{\omega}_h,$$

where g_j is independent of h, $g_j \in C^{k-2j}[0, 1]$, $\|\sigma^h\|_{C, h} = O(h^k)$. This expansion means that Condition D is satisfied for the operators L, L_h. Thus from Theorem 2.2 and the remark from §1.2 we have

$$\varphi^h = \varphi + \sum_{j=1}^{s} h^{2j} v_j + \eta^h \quad \text{on } \bar{\omega}_h, \tag{5.8}$$

in which $s = [(m - 1)/2]$, the v_j do not depend on h, and $\|\eta^h\|_{C, h} = O(h^m)$.

On the basis of this expansion we use the correction method from §1.3 for even powers of h. To do this we will build difference nets $\bar{\omega}_{h_i}$ with mesh-sizes $h_i = h, h/2, \ldots, h/(s+1)$ and solve the system of linear equation (5.3) on each net. All solutions thus obtained are determined on the net $\bar{\omega}_h$. We choose μ_i as the solution of (4.8) and form the linear combination

$$\phi = \sum_{k=1}^{s+1} \mu_k \varphi^{h_k} \quad \text{on } \bar{\omega}_h.$$

Then from Theorem 3.2 of §1.3 the following estimate for the difference between the exact and extrapolated solution

$$\|\phi - \varphi\|_{C,h} \leq c_4 h^m$$

holds, with the order m dependent on the degree of smoothness of the kernel and of the right-hand side.

Note that (5.6) allows to estimate the influence of round-off errors, as well as the contribution of the error introduced when the right-hand side is given by data points is given approximately.

6.6. The Volterra Equation of the First Kind

In contradistinction to the two previously studied types of integral equations when we discretize the Volterra integral equation of the First kind using the familiar quadrature rules we get an algebraic system of equations, whose sensitivity to errors in the right-hand side depends strongly on the value of the mesh-size. As the mesh-size decreases the influence of the error of the right-hand side function increases. Therefore, the problem of achieving a high order of accuracy at a reasonable mesh-size is particularly acute. In [16, 83] it is shown that a stable solution can be guaranteed only when using quadrature rules employing the upper, lower, and central rectangles, or trapezoids. The two latter rules are of second-order accuracy and hence are preferable to the two former ones. We will focus on the central-rectangles rule, and show that it is more than convenient than the trapezoidal rule for Richardson extrapolation.

We consider the Volterra equation of the first kind

$$\int_0^x K(x, t)\varphi(t) \, dt = f(x), \qquad x \in [0, 1]. \tag{6.1}$$

We assume the following

$$f \in C^{m+1}[0, 1], \qquad K \in C^{m+1}(\bar{Q}) \tag{6.2}$$

are satisfied, where the integer $m \geq 1$, and $\bar{Q}$ is a triangle $0 \leq t \leq x \leq 1$. The degrees of smoothness of the kernel and right-hand side function in this case are greater than in the two previous sections. Assume also

$$f(0) = 0, \qquad \min_{x \in [0,1]} |K(x, x)| = c_1 \neq 0. \tag{6.3}$$

Theorem 6.1. *When conditions* (6.2), (6.3) *are satisfied* (6.1) *will have a unique solution* $\varphi \in C^m[0, 1]$.

PROOF. Consider the Volterra equation of the second kind

$$\varphi(x) + \int_0^x \left\{ \frac{\partial K}{\partial x}(x, t)/K(x, x) \right\} \varphi(t)\, dt = f'(x)/K(x, x), \qquad x \in [0, 1]. \tag{6.4}$$

The conditions of Theorem 5.1 are satisfied, and hence there is a unique solution $\varphi \in C^m[0, 1]$. We show that it satisfies equation (6.1). Multiply both parts of equation (6.4) by $K(x, x)$ and integrate with respect to x. Keeping in mind that

$$K(x, x)\varphi(x) + \int_0^x \frac{\partial K}{\partial x}(x, t)\varphi(t)\, dt$$

is the derivative of $\int_0^x K(x, t)\varphi(t)\, dt$, we have

$$\int_0^x K(x, t)\varphi(t)\, dt = f(x) - f(0).$$

Since $f(0) = 0$ we come to equation (6.1). There can be no other continuous solutions of this equation. Let, for example, φ_1 be another continuous solution of equation (6.1). Substitute it into (6.1), differentiate with respect to x and divide by $K(x, x)$. Then we get (6.1), which has no other solutions besides φ. This contradiction proves there are no other solutions besides φ. $\qquad \square$

From this theorem Condition A of §1.2 for equation (6.1) with $M_k(\Omega) = C^{k+1}[0, 1]$, $P_k(\bar{\Omega}) = C^k[0, 1]$ follows (boundary conditions being absent, and $D = \varnothing$).

We will now search for an approximate solution of equation (6.1) on the net $\ddot{\omega}_h$, by constructing algebraic equations at the points $\bar{\omega}_h \backslash \{0\}$ according to the central-rectangles rule:

$$\sum_{\substack{t \in \ddot{\omega}_h \\ t \leq x}} K(x, t)\varphi^h(t)h = f(x), \qquad x \in \bar{\omega}_h \backslash \{0\} \tag{6.5}$$

or in indexed form

$$\sum_{j=0}^{i-1} K(x_i, x_{j+1/2})\varphi^h(x_{j+1/2})h = f(x_i), \qquad i = 1, \ldots, N. \tag{6.6}$$

Equations (6.6) form a system of linear algebraic equations with respect to φ^h with a triangular matrix. From (6.3) and the continuity of $K(x, t)$ it follows that for sufficiently small h the diagonal elements of this matrix are strictly positive, for example

$$|hK(x_i, x_{i-1/2})| \geq hc_1/2. \tag{6.7}$$

Then system (6.6) has a unique solution. We will now derive an *a priori* estimate for it. Take the difference of the two equations (6.6) with indices $i + 1, i$ and rewrite it:

$$K(x_{i+1}, x_{i+1/2})\varphi^h(x_{i+1/2})$$

$$= -\sum_{j=0}^{i} \{K(x_{i+1}, x_{j+1/2}) - K(x_i, x_{j+1/2})\}\varphi^h(x_{j+1/2}) + \{f(x_{i+1})$$

$$- f(x_i)\}/h.$$

Using (6.2), (6.7) we have

$$|\varphi^h(x_{i+1/2})| \leq \frac{2}{c_1}\left(c_2 \sum_{j=0}^{i-1} |\varphi^h(x_{j+1/2})|h + \|f\|_{1,h}\right), \tag{6.8}$$

where

$$c_2 = \max_{\bar{Q}}\left|\frac{\partial K}{\partial x}\right|, \qquad \|f\|_{1,h} = \max_{\bar{\omega}_h}|f_{\bar{x}}|.$$

Using the method of mathematical induction as in Lemma 5.2 we have

$$|\varphi^h(x_{i+1/2})| \leq \|f\|_{1,h}2(1 + 2hc_2/c_1)/c_1$$

$$\leq \exp(2c_2/c_1)2\|f\|_{1,h}/c_1 \quad \forall i = 0, 1, \ldots, N-1.$$

Assuming $\|\varphi^h\|_{\bar{\omega}_h} = \max_{\bar{\omega}_h}|\varphi^h|$ we write this estimate in the form

$$\|\varphi^h\|_{\bar{\omega}_h} \leq c_3\|f\|_{1,h}. \tag{6.9}$$

This proves that Condition B′ of §1.2 is satisfied.

Now we have to verify Condition D of §1.2. We define the operators

$$Lu(x) = \int_0^x K(x, t)u(t)\, dt, \qquad x \in [0, 1],$$

$$L_h u(x) = \sum_{\substack{t \in \bar{\omega}_h \\ t \leq x}} K(x, t)u(t)h, \qquad x \in \bar{\omega}_h.$$

For an arbitrary function $u \in C^k[0, 1]$, $k \leq m$, from Theorem 2.7 of §1.2 it follows that

$$L_h u = Lu + \sum_{j=1}^{[(k-1)/2]} h^{2j}g_j + \sigma^h \quad \text{on } \bar{\omega}_h,$$

where the functions $g_j \in C^{k-2j}[0, 1]$ are independent of h, and the remainder obeys the estimate

$$\|\sigma^h\|_{1,h} \le c_4 h^k.$$

This expansion is adequate to verify Condition D of §1.2 for the operators L_h, L. Therefore, from Theorem 2.2 and the remark at the end of §1.2 the solution of system (6.5) has the expansion

$$\varphi^h = \varphi + \sum_{j=1}^{s} h^{2j} v_j + \eta^h \quad \text{on } \bar{\omega}_h, \tag{6.10}$$

where $s = [(m-1)/2]$, $v_j \in C^{m-2j}[0, 1]$ and are independent of h, and $\|\eta^h\|_{c,h} = O(h^m)$. The approximate solution can be corrected as in the previous section. We choose h and construct nets $\bar{\omega}_{h_i}$ with mesh-sizes $h_i = h, h/2, \ldots, h/(s+1)$. On each of these we solve (6.5), or (6.6), which is the same. Using these solutions we calculate the extrapolated solution

$$\phi = \sum_{k=1}^{s+1} \mu_k \varphi^{h_k} \quad \text{on } \bar{\omega}_h$$

with weights μ_k defined from system (4.8). For this solution ϕ, from Theorem 3.2 of §1.3 we have the estimate

$$\|\phi - \varphi\|_{c,h} \le c_5 h^m,$$

where the power m is determined by the smoothness of the kernel K and the right-hand side f (see (6.2)).

When the value of the right-hand side is given approximately or is given tabularly a nonsystematic error δf appears. Despite the small value of this error $\|\delta f\|_{c,h} = \varepsilon$, in the norm $\|\delta f\|_{1,h}$ it will generally behave like $2\varepsilon/h$, and will increase as $h \to 0$. At small h the contribution of this error to the solution $\tilde{\varphi}^h$ becomes dominant. This is clearly seen by numerical experiment. The combined influence of this error and approximation error on $\tilde{\varphi}^h$ is determined by the inequality

$$\|\tilde{\varphi}^h - \varphi\|_{c,h} \le 2\varepsilon/h + c_6 h^2.$$

The minimum value of the right-hand side is attained at $h = (\varepsilon/c_6)^{1/3}$. In this case it equals $3c_6^{1/3} \varepsilon^{2/3} = O(\varepsilon^{2/3})$. Thus for a difference scheme of the second order an accuracy of $O(\varepsilon^{2/3})$ is, in general, limiting.

When we extrapolate the error δf will enter into $\tilde{\phi}$ via each solution $\tilde{\varphi}^{h_k}$ with weights μ_k. As a result we have the estimate

$$\|\tilde{\phi} - \varphi\|_{c,h} \le \frac{2\varepsilon}{h} \sum_{k=1}^{s+1} |\mu_k| + c_5 h^m.$$

Now the minimal value is attained at

$$h = \left(\frac{2\varepsilon}{mc_5} \sum_{k=1}^{s+1} |\mu_k| \right)^{1/(m+1)}$$

In this case the total error is $O(\varepsilon^{m/(m+1)})$. Hence for small values of ε and for a sufficiently smooth right-hand side and kernel the extrapolated solution gives a higher order of accuracy.

If we use the trapezoidal rule instead of the central-rectangle rule the solution will have an oscillatory error. In particular [16, 61] give the following expansion for the approximate solution:

$$\varphi^h(x) = \varphi(x) + h^2 a(x) + (-1)^{x/h} h^2 b(x) + O(h^3) \quad \forall\, x \in \bar{\omega}_h.$$

Here a, and b are smooth functions on $[0, 1]$, independent of h. The oscillating term $(-1)^{x/h} h^2 b(x)$ somewhat complicates the extrapolation algorithm, and therefore we have used the central-rectangle rule, which has no such effect.

CHAPTER 7
Appendix

In this chapter we present some simple results often used in the previous chapters.

7.1. Expansion of Difference Relations in the Mesh-Size

The following results are a simple consequence of the Taylor formula.

Lemma 1.1. *If the points* x, $x \pm h$ *belong to the interval* $[0, 1]$, *then the function* $v \in C^m[0, 1]$ *has the expansions:*

$$v_{\hat{x}}(x) = \{v(x + h/2) - v(x - h/2)\}/h$$

$$= \sum_{i=0}^{r} \frac{h^{2i}}{(2i + 1)! \, 4^i} v^{(2i+1)}(x) + 2^{-m+1} h^{m-1} \varkappa_1, \tag{1.1}$$

where $r = [m/2] - 1$;

$$v_{\hat{x}}(x) = \{v(x + h/2) + v(x - h/2)\}/2$$

$$= \sum_{i=0}^{q} \frac{h^{2i}}{(2i)! \, 4^i} v^{(2i)}(x) + 2^{-m} h^m \varkappa_2, \tag{1.2}$$

where $q = [(m - 1)/2]$;

$$v_x(x) = \{v(x + h) - v(x)\}/h = \sum_{i=0}^{m-2} \frac{h^i}{(i + 1)!} v^{(i+1)}(x) + h^{m-1} \varkappa_3, \tag{1.3}$$

and

$$v_{\bar{x}}(x) = \{v(x) - v(x - h)\}/h = \sum_{i=0}^{m-2} \frac{(-h)^i}{(i+1)!} v^{(i+1)}(x) + h^{m-1}\varkappa_4. \quad (1.4)$$

All the $\varkappa_i$ are bounded independent of x and h by

$$\frac{1}{m!} \max_{[0,\,1]} |v^{(m)}|.$$

Using these expansions we can prove the following result.

Lemma 1.2. *Assume that the points x, $x \pm h$ belong to the interval $[0, 1]$, and that $p \in C^m[0, 1]$ and $v \in C^{m+1}[0, 1]$. Then we have*

$$(pv_{\mathring{x}})_{\mathring{x}}|_x = \{p(x + h/2)(v(x + h) - v(x)) - p(x - h/2)(v(x) - v(x - h))\}/h^2$$

$$= \sum_{s=0}^{q} h^{2s} \sum_{0 \le i+j \le s} \frac{(p(x)v^{(2i+1)}(x))^{(2j+1)}}{(2i+1)!\,(2j+1)!\,4^{i+j}} + h^{m-1}\varkappa_5 \quad (1.5)$$

where $q = [(m-1)/2]$,

$$|\varkappa_5| \le c_1 \max_{0 \le s \le m} |p^{(s)}| \max_{0 \le s \le m+1} |v^{(s)}| \quad (1.6)$$

with constant c_1 independent of x, h, p, q.

PROOF. It follows from (1.1) that

$$v_{\mathring{x}}(x - h/2) = \{v(x) - v(x - h)\}/h$$

$$= \sum_{k=0}^{q} \frac{h^{2k}}{4^k(2k+1)!} v^{(2k+1)}(x - h/2) + h^m \varkappa^-,$$

$$v_{\mathring{x}}(x + h/2) = \sum_{k=0}^{q} \frac{h^{2k}}{4^k(2k+1)!} v^{(2k+1)}(x + h/2) + h^m \varkappa^+,$$

where

$$|\varkappa^{\pm}| \le \frac{2^{-m}}{(m+1)!} \max_{[0,\,1]} |v^{(m+1)}|. \quad (1.7)$$

Multiply these divided differences by the corresponding values of the function p and subtract them from one another, divide the result by h and apply Lemma 1.1 to each of the addends $pv^{(2k+1)}$. We have

$$p(x + h/2)v_{\mathring{x}}(x + h/2) - p(x - h/2)v_{\mathring{x}}(x - h/2)$$

$$= \sum_{k=0}^{q} \frac{h^{2k}}{4^k(2k+1)!} \{p(x + h/2)v^{(2k+1)}(x + h/2)$$

$$- p(x - h/2)v^{(2k+1)}(x - h/2)\}/h + h^{m-1}(p(x + h/2)\varkappa^+ - p(x - h/2)\varkappa^-)$$

$$= \sum_{k=0}^{q} \frac{h^{2k}}{4^k(2k+1)!} \left\{ \sum_{j=0}^{q-k} \frac{h^{2j}}{4^j(2j+1)!} (p(x)v^{(2k+1)}(x))^{(2j+1)} + \mu_k h^{m-2k-1} \right\}$$

$$+ h^{m-1}(p(x + h/2)\varkappa^+ - p(x - h/2)\varkappa^-), \quad (1.8)$$

where

$$|\mu_k| \leq \frac{2^{2k-m-1}}{(m+1-2k)!} \max_{[0,1]} |(pv^{(2k+1)})^{(m+1-2k)}|$$

$$\leq \frac{1}{(m+1-2k)!} \max_{\substack{0 \leq s \leq m \\ [0,1]}} |p^{(s)}| \max_{\substack{0 \leq s \leq m+1 \\ [0,1]}} |v^{(s)}|. \tag{1.9}$$

The last inequality follows from the formula for repeated differentiation of a product. If we collect similar terms in h in (1.8) we get (1.5); (1.6) follows from inequalities (1.7) and (1.9). $\qquad\qquad\square$

Lemma 1.3. *Let* $v \in C^{l+\alpha}[x - h, x + h]$ *with integer* $l \geq 2$, *and* $\alpha \in (0, 1)$. *Then we have*

$$v_{\hat{x}\hat{x}}(x) = \{v(x - h) - 2v(x) + v(x + h)\}/h^2$$

$$= 2\sum_{i=0}^{s} h^{2i} \frac{1}{(2i+2)!} v^{(2i+2)}(x) + h^{l-2+\alpha}\rho(x),$$

where $s = [l/2] - 1$, *and*

$$|\rho(x)| \leq \frac{2}{l!} \|v\|_{C^{l+\alpha}[x-h,\,x+h]}.$$

PROOF. It follows from the Taylor formula that

$$v(x \pm h) = \sum_{i=0}^{l-1} \frac{(\pm h)^i}{i!} v^{(i)}(x) + \frac{(\pm h)^l}{l!} v^{(l)}(\xi^{\pm}). \tag{1.10}$$

Since the derivative $v^{(l)}$ is continuous in the sense of Hölder with exponent α, then

$$|v^{(l)}(\xi^{\pm}) - v^{(l)}(x)|/|\xi^{\pm} - x|^{\alpha} \leq c_3 = \|v\|_{C^{l+\alpha}[x-h,\,x+h]}.$$

From $|\xi^{\pm} - x| \leq h$ we have

$$|v^{(l)}(\xi^{\pm}) - v^{(l)}(x)| \leq c_3 h^{\alpha}.$$

Then, from (1.10) we have the expansion

$$v(x \pm h) = \sum_{i=0}^{l} \frac{(\pm h)^i}{i!} v^{(i)}(x) + h^{l+\alpha}\rho^{\pm}(x), \tag{1.11}$$

where $|\rho^{\pm}(x)| \leq c_3/l!$.

We use expansion (1.11) to compute $v_{\hat{x}\hat{x}}(x)$.
We have

$$v_{\hat{x}\hat{x}}(x) = 2\sum_{i=0}^{s} \frac{h^{2i}}{(2i+2)!} v^{(2i+2)}(x) + h^{l-2+\alpha}(\rho^+(x) - \rho^-(x)).$$

Hence

$$c_2 = 2c_3/l! = \frac{2}{l!} \|v\|_{C^{l+\alpha}[x-h,\,x+h]}.$$

Lemma 1.3 is proved. $\square$

7.2. On the Solution of Some Special Systems of Equations

Let us consider the following system of equations in γ_i:

$$\sum_{i=1}^{s+1} \gamma_i = 1,$$

$$\sum_{i=1}^{s+1} \gamma_i \mu_i^l = 0, \qquad l = 1, \ldots, s. \tag{2.1}$$

It is known (see [69]) that the Vandermonde determinant

$$V(v_1, v_2, \ldots, v_{s+1}) = \det \begin{bmatrix} 1 & 1 & \cdots & 1 \\ v_1 & v_2 & \cdots & v_{s+1} \\ v_1^2 & v_2^2 & \cdots & v_{s+1}^2 \\ \cdots\cdots\cdots\cdots\cdots\cdots \\ v_1^s & v_2^s & \cdots & v_{s+1}^s \end{bmatrix}$$

can be calculated by the formula

$$V(v_1, v_2, \ldots, v_{s+1}) = \prod_{i \le j} (v_j - v_i). \tag{2.2}$$

It is thus seen that system (2.1) is solvable only when all v_i are distinct. In this case we may apply Kramer's rule to (2.1):

$$\gamma_i = \frac{V(\mu_1, \ldots, \mu_{i-1}, 0, \mu_{i+1}, \ldots, \mu_{s+1})}{V(\mu_1, \ldots, \mu_{i-1}, \mu_i, \mu_{i+1}, \ldots, \mu_{s+1})}. \tag{2.3}$$

In particular, the following lemma is valid.

Lemma 2.1. *If the system* (2.1) *has coefficients* $\mu_i = 1/i$, *then*

$$\gamma_k = \frac{(-1)^{s-k+1} k^{s+1}}{k! (s - k + 1)!}, \qquad k = 1, \ldots, s + 1.$$

PROOF. Replace the numerator and the denominator of (2.3) by their values according to formula (2.2), and cancel coinciding multipliers. We get

$$\gamma_k = \prod_{i=1}^{k-1} \frac{-\mu_i}{\mu_k - \mu_i} \prod_{i=k+1}^{s+1} \frac{\mu_i}{\mu_i - \mu_k}. \tag{2.4}$$

Replacing μ_i by $1/i$ we come to the result of the lemma. $\square$

The following lemma is proved in a similar way.

Lemma 2.2. *If the system* (2.1) *has coefficients* $\mu_i = 1/i^2$, *then*

$$\gamma_k = \frac{2(-1)^{s-k+1}k^{2s+2}}{(s+k+1)!\,(s-k+1)!}, \qquad k = 1,\ldots,s+1.$$

Lemma 2.3. *If the coefficients of the system of equations* (2.1) *satisfy the condition*

$$\mu_i/\mu_{i+1} \geq 1 + c, \qquad i = 1,\ldots,s,$$

with positive constant c, then the solution of the system obeys the estimate

$$|\gamma_i| \leq \left(\frac{1+c}{c}\right)^s, \qquad i = 1,\ldots,s+1.$$

PROOF. Transform formula (2.4) to the form

$$|\gamma_k| = \prod_{i=1}^{k-1} \frac{1}{\mu_i/\mu_k - 1} \prod_{i=k+1}^{s+1} \left(1 + \frac{1}{\mu_k/\mu_i - 1}\right).$$

In order to use the inequality from the hypothesis of the lemma, we rewrite it in the form

$$\mu_i/\mu_k \geq (1+c)^{k-i} \geq 1 + c \quad \text{for } i < k.$$

The estimate of the previous equality leads to

$$|\gamma_k| \leq \left(\frac{1}{c}\right)^{k-1}\left(1 + \frac{1}{c}\right)^{s-k+1} \leq \left(\frac{1+c}{c}\right)^s.$$

The lemma is proved.   □

Lemma 2.4. *If the coefficients in the system of equations*

$$\sum_{i=1}^{s+1} \beta_i = 1,$$

$$\sum_{i=1}^{s+1} \beta_i v_i^{2l} = 0, \qquad l = 1,\ldots,s,$$

satisfy the condition

$$v_i/v_{i+1} \geq 1 + d, \qquad i = 1,\ldots,s; d > 0, \tag{2.5}$$

then there is a unique solution which satisfies the inequality

$$|\beta_k| \leq \left(\frac{1 + 2d + d^2}{2d + d^2}\right)^s, \qquad k = 1,\ldots,s.$$

The proof is based on the previous lemma with $v_i^2 = \mu_i$. Then inequality (2.5) takes the form

$$\mu_i/\mu_{i+1} \geq (1 + d)^2 = 1 + 2d + d^2$$

and in the condition of Lemma 2.3 the constant c can be taken to be $2d + d^2$.

Consider the following system of equations in the α_i:

$$\sum_{i=1}^{s} \alpha_i \mu_i^{l-1} = \mu_l^s \qquad l = 1, \ldots, s + 1; \qquad (2.6)$$

the determinant here will equal the Vandermonde determinant $V(\mu_1, \ldots, \mu_{s+1})$, since the first can be obtained by transposing the second (see [69]). Therefore, if all the μ_i are distinct then (2.6) has the unique solution. We are interested in the coefficient

$$\alpha_1 = \frac{\det \begin{bmatrix} \mu_1^{s+1} & \mu_1 & \cdots & \mu_1^s \\ \mu_2^{s+1} & \mu_2 & \cdots & \mu_2^s \\ \cdots\cdots\cdots\cdots\cdots\cdots\cdots \\ \mu_{s+1}^{s+1} & \mu_{s+1} & \cdots & \mu_{s+1}^s \end{bmatrix}}{V(\mu_1, \mu_2, \ldots, \mu_{s+1})}$$

From each row of the determinant in the numerator we put factor out a μ_i', and then move the first column to the extreme right. This results in

$$\alpha_1 = (-1)^s \mu_1 \mu_2 \cdots \mu_{s+1} V(\mu_1, \mu_2, \ldots, \mu_{s+1})/V(\mu_1, \mu_2, \ldots, \mu_{s+1})$$

$$= (-1)^s \mu_1 \cdots \mu_{s+1}. \qquad (2.7)$$

Lemma 2.5. *Let the μ_i be distinct for $i = 1, \ldots, s$ and let γ_i solve (2.1). Then*

$$\sum_{i=1}^{s+1} \gamma_i \mu_i^{s+1} = (-1)^s \mu_1 \mu_2 \cdots \mu_{s+1}.$$

PROOF. Use (2.6) to rewrite the left-hand side:

$$\sum_{i=1}^{s+1} \gamma_i \mu_i^{s+1} = \sum_{i=1}^{s+1} \gamma_i \sum_{j=1}^{s+1} \alpha_j \mu_i^{j-1}$$

Change the order of summation:

$$\sum_{i=1}^{s+1} \gamma_i \mu_i^{s+1} = \sum_{j=1}^{s+1} \alpha_j \sum_{i=1}^{s+1} \gamma_i \mu_i^{j-1}.$$

Using (2.1) we have

$$\sum_{i=1}^{s+1} \gamma_i \mu_i^{s+1} = \alpha_1 = (-1)^s \mu_1 \mu_2 \cdots \mu_{s+1}.$$

Lemma 2.5 is proved. □

We will formulate a similar result for even powers without proof since it follows from Lemma 2.5.

Lemma 2.6. *If the assumptions of Lemma 2.4 are satisfied we have*

$$\sum_{i=1}^{s+1} \beta_i v_i^{2s+2} = (-1)^s v_1^2 v_2^2 \cdots v_{s+1}^2.$$

Lemma 2.7. *Let $\varepsilon_i \neq 0$ for all $i = 1, \ldots, k$. Then*

$$W = \det \begin{bmatrix} \varepsilon_1^{-1} & \varepsilon_2^{-1} & \cdots & \varepsilon_k^{-1} \\ \varepsilon_1 & \varepsilon_2 & \cdots & \varepsilon_k \\ \varepsilon_1^2 & \varepsilon_2^2 & \cdots & \varepsilon_k^2 \\ \cdots\cdots\cdots\cdots\cdots\cdots\cdots \\ \varepsilon_1^{k-1} & \varepsilon_2^{k-1} & \cdots & \varepsilon_k^{k-1} \end{bmatrix} = \sum_{i=1}^{k} \varepsilon_i^{-1} V(\varepsilon_1, \varepsilon_2, \ldots, \varepsilon_k).$$

PROOF. Multiply column number i of W by ε_i; then we have

$$\varepsilon_1 \ldots \varepsilon_k W = \det \begin{bmatrix} 1 & 1 & \cdots & 1 \\ \varepsilon_1^2 & \varepsilon_2^2 & \cdots & \varepsilon_k^2 \\ \varepsilon_1^3 & \varepsilon_2^3 & \cdots & \varepsilon_k^3 \\ \cdots\cdots\cdots\cdots\cdots \\ \varepsilon_1^k & \varepsilon_2^k & \cdots & \varepsilon_k^k \end{bmatrix} \tag{2.8}$$

Consider the determinant

$$V(t, \varepsilon_1, \ldots, \varepsilon_k) = \det \begin{bmatrix} 1 & 1 & 1 & \cdots & 1 \\ t & \varepsilon_1 & \varepsilon_2 & \cdots & \varepsilon_k \\ t^2 & \varepsilon_1^2 & \varepsilon_2^2 & \cdots & \varepsilon_k^2 \\ \cdots\cdots\cdots\cdots\cdots\cdots\cdots \\ t^k & \varepsilon_1^k & \varepsilon_2^k & \cdots & \varepsilon_k^k \end{bmatrix} \tag{2.9}$$

It is seen from the expansion of this determinant in its first column that (2.8) is equal to (-1) times the coefficient of t in this expansion. The determinant (2.9) is equal to

$$\left\{ \prod_{k \geq j > i \geq 1} (\varepsilon_j - \varepsilon_i) \right\} \prod_{l=1}^{k} (\varepsilon_l - t) = V(\varepsilon_1, \varepsilon_2, \ldots, \varepsilon_k)(-1)^k \prod_{l=1}^{k} (t - \varepsilon_l).$$

The coefficient of t in the polynomial $\prod(t - \varepsilon_l)$ is equal to

$$\sum_{l=1}^{k} \frac{\varepsilon_1 \cdots \varepsilon_k}{\varepsilon_l} (-1)^{k+1}.$$

From this we see that the coefficient of the polynomial (2.9) in t is equal to

$$-V(\varepsilon_1, \varepsilon_2, \ldots, \varepsilon_k) \varepsilon_1 \cdots \varepsilon_k \sum_{l=1}^{k} \varepsilon_l^{-1}.$$

Equating this with $-\varepsilon_1 \cdots \varepsilon_k W$ we have the required result. Lemma 2.7 is proved.   □

Now we present a result concerning the Neville algorithm [53].

Lemma 2.8. *Let $T_j^{(0)}$ be a given set of numbers for $j = 1, \ldots, s + 1$, and let the sequence $T_j^{(i)}$ be determined by the Neville algorithm*

$$T_j^{(i)} = (\mu_{i+j} T_j^{(i-1)} - \mu_j T_{j+1}^{(i-1)})/(\mu_{i+j} - \mu_j), \qquad j = 1, \ldots, s - i + 1;$$

$$i = 1, 2, \ldots, s. \tag{2.10}$$

Then we have

$$T_1^{(s)} = \sum_{k=1}^{s+1} \gamma_k T_k^{(0)}, \tag{2.11}$$

where γ_k is the solution of (2.1).

PROOF. Consider this statement for $s = 1$. From (2.1) it follows that

$$\gamma_1 = \mu_2/(\mu_2 - \mu_1), \qquad \gamma_2 = -\mu_1/(\mu_2 - \mu_1)$$

and therefore (2.10) and (2.11) coincide, since

$$T_1^{(1)} = \gamma_1 T_1^{(0)} + \gamma_2 T_2^{(0)}.$$

Now assume that (2.11) has been proved for $s - 1$; we will prove it for s. If (2.11) holds for $s - 1$ then we have

$$T_1^{(s-1)} = \sum_{k=1}^{s} \beta_k T_k^{(0)} \quad \text{and} \quad T_2^{(s-1)} = \sum_{k=1}^{s} \delta_k T_{k+1}^{(0)},$$

where the coefficients β_k and δ_k are the solutions of

$$\sum_{k=1}^{s} \beta_k = 1, \qquad \sum_{k=1}^{s} \mu_k^l \beta_k = 0, \qquad l = 1, 2, \ldots, s - 1;$$

and

$$\sum_{k=1}^{s} \delta_k = 1, \qquad \sum_{k=1}^{s} \mu_{k+1}^l \delta_k = 0, \qquad l = 1, 2, \ldots, s - 1. \tag{2.12}$$

We calculate

$$T_1^{(s)} = (\mu_{s+1} T_1^{(s-1)} - \mu_1 T_2^{(s-1)})/(\mu_{s+1} - \mu_1) = T_1^{(0)} \mu_{s+1} \beta_1/(\mu_{s+1} - \mu_1)$$

$$+ \sum_{k=2}^{s} T_k^{(0)}(\mu_{s+1}\beta_k - \mu_1 \delta_{k-1})/(\mu_{s+1} - \mu_1) - T_{s+1}^{(0)} \mu_1 \delta_s/(\mu_{s+1} - \mu_1)$$

$$= \sum_{k=1}^{s+1} \rho_k T_k^{(0)}, \tag{2.13}$$

where

$$\rho_1 = \mu_{s+1} \beta_1/(\mu_{s+1} - \mu_1), \qquad \rho_{s+1} = -\mu_1 \delta_s/(\mu_{s+1} - \mu_1),$$

$$\rho_k = (\mu_{s+1} \beta_k - \mu_1 \delta_{k-1})/(\mu_{s+1} - \mu_1), \qquad k = 2, 3, \ldots, s.$$

Let us show that these coefficients satisfy (2.1). Using equation (2.12) we have

$$\sum_{k=1}^{s+1} \rho_k = \left(\sum_{k=1}^{s} \beta_k\right)\mu_{s+1}/(\mu_{s+1} - \mu_1) - \left(\sum_{k=1}^{s} \delta_k\right)\mu_1/(\mu_{s+1} - \mu_1)$$

$$= (\mu_{s+1} - \mu_1)/(\mu_{s+1} - \mu_1) = 1;$$

(2.14)

$$\sum_{k=1}^{s+1} \mu_k^l \rho_k = \left\{\mu_{s+1}\beta_1\mu_1^l + \sum_{k=2}^{s} (\mu_{s+1}\beta_k - \mu_1\delta_{k-1})\mu_k^l - \mu_1\delta_s\mu_1^l\right\}/(\mu_{s+1} - \mu_1)$$

$$= \left\{\mu_{s+1}\sum_{k=1}^{s} \beta_k\mu_k^l - \mu_1\sum_{k=1}^{s} \delta_k\mu_{k+1}^l\right\}/(\mu_{s+1} - \mu_1).$$

If $l = 1, 2, \ldots, s - 1$ then both sums in brackets (in 2.14) are zero, due to (2.12). If $l = s$ then from Lemma 2.5 we have

$$\sum_{k=1}^{s} \beta_k\mu_k^l = (-1)^{s-1}\mu_1\mu_2\cdots\mu_s,$$

$$\sum_{k=1}^{s} \delta_k\mu_{k+1}^l = (-1)^{s-1}\mu_2\mu_3\cdots\mu_{s+1}.$$

Using these equalities in (2.14) yields

$$\sum_{k=1}^{s+1} \mu_k^s \rho_k = 0.$$

Thus the ρ_k satisfy (2.1). But the solution of (2.1) is unique (for the μ_i distinct). Therefore $\rho_k = \gamma_k$ and the right-hand sides of (2.11) and (2.13) coincide. $\square$

7.3. Some Results on the Lagrange Interpolation Polynomials

Let the function $f(x)$ be continuous on the interval $[0, 1]$ and assume its values at the s uniformly spaced points $x_i = x_1 + (i - 1)h$ of $[0, 1]$ are known. The function

$$L_s(x) = \sum_{i=1}^{s} f(x_i)\prod_{\substack{j \neq i \\ j=1}}^{s} \frac{x - x_j}{x_i - x_j}$$

(3.1)

is referred to as the Lagrange interpolation polynomial. If $f \in C^s[0, 1]$ then the relation

$$f(x) - L_s(x) = \frac{1}{s!} f^{(s)}(\xi)\omega_s(x)$$

(3.2)

holds for $x \in [x_1, x_s]$ (see [14]), where $\xi \in [0, 1]$ and

$$\omega_s(x) = \prod_{j=1}^{s} (x - x_j). \tag{3.3}$$

We will estimate these quantities in terms of powers of h. The maximum value of the modulus of ω_2 in the interval $[x_1, x_2]$ is attained at the midpoint of the interval, where it is equal to $h^2/4$. Therefore

$$|\omega_2(x)| \le h^2/4 \quad \forall\, x \in [x_1, x_2].$$

Let us prove the estimate

$$|\omega_l(x)| \le \frac{h^l}{4}(l - 1)! \quad \forall\, x \in [x_1, x_l]. \tag{3.4}$$

For $l = 2$ this inequality is obvious. Assuming that estimate (3.4) has been proved for l we will prove it for $l + 1$. Two uses are possible. When $x \in [x_1, x_l]$

$$|\omega_{l+1}(x)| \le |\omega_l(x)| |x - x_{l+1}| \le \frac{h^l}{4}(l - 1)! \, lh = \frac{h^{l+1}}{4} l!$$

and (3.4) is proved. When $x \in [x_l, x_{l+1}]$ the inequalities

$$|x - x_l| |x - x_{l+1}| \le \frac{h^2}{4}, \qquad |x - x_i| \le (l - i + 1)h$$

obviously hold. Therefore

$$|\omega_{l+1}(x)| \le h^{l-1} l(l - 1) \cdots 2 \frac{h^2}{4} = \frac{h^{l+1}}{4} l!$$

Combining (3.2)–(3.4) we have

$$|f(x) - L_s(x)| \le \frac{h^s}{4s} \|f\|_{C^s[0, 1]} \quad \forall\, x \in [x_1, x_s]. \tag{3.5}$$

If the function $f(x)$ is not sufficiently smooth the error is of a lower order.

Lemma 3.1. Let $s > k > 0$ and $f \in C^k[0, 1]$, then we have

$$|f(x) - L_s(x)| \le \frac{h^k}{k} 2^{s-k-2} \|f\|_{C^k[0, 1]} \quad \forall\, x \in [x_1, x_s]. \tag{3.6}$$

PROOF. With the reference to [14] let us denote by $f(x_1, \ldots, x_n)$ the divided difference of order n. Then $L_s(x)$ can be written as a telescoping series of interpolation polynomials:

$$L_s(x) = L_k(x) + (L_{k+1}(x) - L_k(x)) + \cdots + (L_s(x) - L_{s-1}(x)).$$

Since

$$|L_{j+1}(x) - L_j(x)| = f(x_1, \ldots, x_j)\omega_j(x),$$

then

$$|f(x) - L_s(x)| \le |f(x) - L_k(x)| + |L_{k+1}(x) - L_k(x)| + \cdots + |L_s(x)$$

$$- L_{s-1}(x)| \le \frac{h^k}{4k} \|f\|_{C^k[0,\,1]} + |f(x_1, \ldots, x_{k+1})\omega_k(x)|$$

$$+ \cdots + |f(x_1, \ldots, x_s)\omega_{s-1}(x)|. \tag{3.7}$$

Using mathematical induction we will prove that for an arbitrary function $f \in C^k[0, 1]$ and any $l = k + 1, \ldots, s$ the inequality

$$|f(x_1, x_2, \ldots, x_l)| \le \frac{h^{k+1-l}}{(l-1)!} 2^{l-k-1} \|f\|_{C^k[0,\,1]} \tag{3.8}$$

is satisfied. Let $l = k + 1$. It is known from [14] that

$$f(x_1, x_2, \ldots, x_{k+1}) = f(\theta)/k!, \qquad \theta \in [x_1, x_{k+1}],$$

and so (3.8) is evidently true. Let us suppose now that (3.8) is valid for a certain integer $l > k + 1$. Then the inequality

$$|f(x_2, x_3, \ldots, x_{l+1})| \le \frac{h^{k+1-l}}{(l-1)!} 2^{l-k-1} \|f\|_{C^k[0,\,1]}$$

is a consequence of (3.8) for the function $g(x) = f(x + h)$:

$$|f(x_2, x_3, \ldots, x_{l+1})| = |g(x_1, \ldots, x_l)| \le \frac{h^{k+1-l}}{(l-1)!} 2^{l-k-1} \|g\|_{C^k[0,\,1]}.$$

According to the definition of the divided difference

$$f(x_1, \ldots, x_{l+1}) = \{f(x_2, \ldots, x_{l+1}) - f(x_1, \ldots, x_l)\}/(x_{l+1} - x_1)$$

from which

$$|f(x_1, \ldots, x_{l+1})| \le \frac{h^{k+1-l}}{(l-1)!} 2^{l-k-1} \{\|f\|_{C^k[0,\,1]} + \|f\|_{C^k[0,\,1]}\}/(lh)$$

$$\le \frac{h^{k-l}}{l!} 2^{l-k} \|f\|_{C^k[0,\,1]}.$$

Estimate (3.8) is proved.
Using inequality (3.8) let us extend the estimate (3.7) for the difference $f(x) - L_s(x)$:

$$|f(x) - L_s(x)| \le \frac{h^k}{4k} \|f\|_{C^k[0,\,1]} + \frac{h^k}{4k} \|f\|_{C^k[0,\,1]}$$

$$+ \cdots + \frac{h^{k+1-s}}{4(s-1)} 2^{s-k-1} \|f\|_{C^k[0,\,1]}$$

$$\le \frac{h^k}{4} \|f\|_{C^k[0,\,1]} \left(\frac{2}{k} + \frac{2}{k+1} + \cdots + 2^{s-k-1}/(s-1) \right).$$

Since $s > k \geq 1$

$$|f(x) - L_s(x)| \leq \frac{h^k}{4} \|f\|_{C^k[0,\,1]} 2^{s-k}/k.$$

Lemma 3.1 is proved.   □

We now find an estimate for the polynomials

$$\alpha_i(x) = \prod_{\substack{j \neq i \\ j=1}}^{s} \frac{x - x_j}{x_i - x_j}, \tag{3.9}$$

which characterizes the value of the error of the interpolation formula when the values $f(x_i)$ are not known exactly.

Lemma 3.2. *The inequality*

$$\sum_{i=1}^{s} |\alpha_i(x)| \leq 2^{s-1} \tag{3.10}$$

holds for $x \in [x_1, x_s]$.

PROOF. When the net points are equidistant (see [14]) we have the formula

$$|\alpha_i| = C_{s-1}^{i-1} t(t-1) \cdots (t-i+2)(t-i) \cdots (t-s+1)/(s-1)!,$$

where $t = (x - x_1)/h$, and therefore $t \in [0, s-1]$. It is easy to see that the numerator of the fraction does not exceed the constant $(s-1)!$. Therefore

$$\sum_{i=1}^{s} |\alpha_i| \leq \sum_{i=1}^{s} C_{s-1}^{i-1} = 2^{s-1}.$$ □

List of Notation

1. Regions and Boundaries

Ω open bounded region   159
Γ boundary of Ω 159
Ω' subregion of Ω, i.e., $\Omega' \subset \Omega$ 160
$\overline{\Omega}$ closure of Ω, i.e., $\overline{\Omega} = \Omega \cup \Gamma$ 160
Q open cylinder $\Omega \times (0, T)$ 246
S lateral surface of Q 246
$\overline{Q}$ closure of Q 246
S_1 sector   215
∂S_1 boundary of S_1 224

2. Spaces and Norms

Spaces

$M_k(\Omega)$ 16	R^p 159	$\mathring{W}^l_2(\Omega)$ 160
$N_k(D)$ 16	$C^{l+\alpha}(\overline{\Omega})$ 160	C^k 160
$P_k(\overline{\Omega})$ 16	$C^{l+\alpha}(\Omega)$ 160	$H^k(\overline{Q})$ 230, 246
$C^m(\overline{\Omega})$ 45	$L_2(\Omega)$ 160	$C^k(Q)$ 270
E^m 78	$W^l_2(\Omega)$ 160	$\tilde{C}^k(Q)$ 270
Q^k_ξ 104	$C^{k+\alpha}$ 160	C^n 279

Norms:

$\|u\|_{\bar{\Omega}_h}$ 17

$\|u\|_{\check{\Omega}_h}$ 17

$\|u\|_{D_h}$ 17

$\|u\|_{C,\tau} = \max_{\bar{\omega}_\tau} |u|$ 7

$\|\varphi\|_{C^m[0,1]}$ 45

(v, w)-scalar product   78, 279

$\|v\| = (v, v)^{1/2}$ 79, 279

$\|u\|_{C,h} = \max_{\bar{\omega}_h} |u|$ 94

$\|u\|_{C^{1+\alpha}(\bar{\Omega})}$ 160

$\|u\|_{L_2(\Omega)}$ 160

$\|u\|_{W_2^1(\Omega)}$ 160

$|u|$ 160

3. Net Regions and Boundaries

$\bar{\Omega}_h$ 17 $\check{\Omega}_h$ 17 D_h 17

$\bar{\omega}_\tau = \{t_j = j\tau, j = 0, 1, \ldots, M\}$ 46, 231

$\check{\omega}_\tau = \{t_{j+1/2} = (j + 1/2)\tau, j = 0, 1, \ldots, M - 1\}$ 46

$\mathring{\omega}_\tau = \{t_j = j\tau, j = 0, 1, \ldots, M - 1\}$ 62

$\omega_\tau = \{t_j = j\tau, j = 1, \ldots, M\}$ 79, 231

$\bar{\omega}_h = \{x_i = ih, i = 0, 1, \ldots, N\}$ 93

$\bar{\omega}_h = \{0 = x_0 < x_1 < \cdots < x_N = 1\}$-nonuniform net   105

$\check{\omega}_h = \{x_{i+1/2} = (i + 1/2)h, i = 0, 1, \ldots, N - 1\}$ 93

$\omega_h = \{x_i = ih, i = 1, \ldots, N - 1\}$ 93

$\check{\omega}_h = \{x_{i+1/2} = (x_i + x_{i+1})/2, i = 0, 1, \ldots, N - 1\}$ 105

Ω_h 163 Ω_h^{ir} 163 $\Omega_{h,y}^{ir}$ 177

$\Gamma_{h,x}$ 163 $\Omega_{h,x}^r$ 177 $\bar{Q}_h^\tau = \bar{\omega}_h \times \bar{\omega}_\tau$ 231

$\Gamma_{h,y}$ 163 $\Omega_{h,y}^r$ 177 $Q_h^\tau = \omega_h \times \omega_\tau$ 231

Ω_h^r 163 $\Omega_{h,x}^{ir}$ 177

4. Difference Expressions

Finite Differences

$u_{\hat{t}}(t) = \{u(t + \tau/2) - u(t - \tau/2)\}/\tau$ 46

$u_{\bar{t}}(t) = \{u(t + \tau/2) + u(t - \tau/2)\}/2$ 46

$u_t(t) = \{u(t + \tau) - u(t)\}/\tau$ 62

$u_{\hat{x}}(x) = \{u(x + h/2) - u(x - h/2)\}/h$ 94, 310

$u_{\bar{x}}(x) = \{u(x + h/2) + u(x - h/2)\}/2$ 98, 310

$u_{\hat{x}\hat{x}}(x) = \{u(x + h) - 2u(x) + u(x - h)\}/h^2$ 94, 312

$u_{\hat{y}\hat{y}}(x, y) = \{u(x, y + h) - 2u(x, y) + u(x, y - h)\}/h^2$ 163

Difference Operators

$D_x^{(m)}$ 168 J_n^x 178

$D_y^{(m)}$ 168 J_n^y 179

Bibliography

[1] Agoshkov, V. I.: On a variational form of Marchuk's integral identity. Novosibirsk: Computing Center of the USSR Academy of Sciences, 1977 (Russian).

[2] Albrecht, I. and Uhlmann, W.: Differenzenverfahren für die 1. Randwertaufgabe mit krummlinigen Rändern bei $\Delta u(x, y) = r(x, y, u)$. Z. Angew. Math. Mech. (1957), **37**, No. 5/6, 212–224.

[3] Alexeev, A. V.: Increasing the accuracy of approximate solutions of integral equations. Krasnoyarsk: Computing Center of the USSR Academy of Sciences, 1979 (Russian).

[4] Alibekov, H. A. and Sobolevskii, P. E.: On the stability of a difference scheme for parabolic equations. Soviet Math. Doklady (1977) **232**, No. 4, 737–740.

[5] Artemiev, S. S. and Demidov, G. V.: A stable Rosenbrock method for stiff initial-value problems with fourth-order accuracy. In: Some Problems of Computational and Applied Mathematics. Novosibirsk: Nauka, 1975, pp. 214–219 (Russian).

[6] Atkinson, F. V.: Discrete and Continuous Boundary Problems. New York, London: Academic Press, 1964.

[7] Aubin, I. P.: Behavior of the error on the approximate solution boundary-value problems for linear elliptic operators using Galërkin's and finite-difference methods. Ann. Sci. Norm. Pisa (1961), **21**, 599–637.

[8] Babuška, I.: The finite element method for domains with corners. J. Computing (1970), **6**, No. 3, 264–273.

[9] Babuška, I.: The finite element method with Lagrangian multipliers. Numer. Math. (1973), **21**, No. 16, 322–333.

[10] Babuška, I. and Aziz, A. K.: On the sector condition in the finite element method. SIAM J. Numer. Anal. (1976), **13**, No. 2, 214–226.

[11] Babuška, I., Práger, M., Vitásek, E.: Numerical Processes in Differential Equations. London: Wiley, 1966.

[12] Babuška, I., Rheinboldt, W., and Mesztenyi, C.: Self-adaptive refinements in the finite element method. Technical Report of the University of Maryland, 1975.

[13] Babuška, I. and Rosenzweig, M. B.: A finite element scheme for domains with corners. Numer. Math. (1972), **20**, No. 1, 1–21.

[14] Bakhvalov, N. S.: *Numerical Methods*, I. Moscow: Nauka, 1973 (Russian).

[15] Baker, C. T. H.: The approach to the limit for eigenvalues of integral equations. *SIAM J. Numer. Anal.* (1971), **8**, No. 1, 1–10.

[16] Baker, C. T. H.: Methods for Volterra equations of first kind. In: *Numerical Solution of Integral Equations*. Oxford: Clarendon Press, 1974, pp. 162–174.

[17] Bank, R. E. and Rose, D. J.: Extrapolated fast direct algorithms for elliptic boundary-value problems. In: *Algorithms and Complexity: New Directions and Recent Results*. New York: Academic Press, 1976, pp. 201–247.

[18] Bickley, W. G., Michaelson, S., and Osborne, M. R.: On finite-difference methods for the numerical solution of boundary-value problems. *Proc. Roy. Soc. London* (1961), ser. A, No. 262, 219–236.

[19] Bramble, J. H. and Hubbard, B. E.: On the formulation of finite difference analogues of the Dirichlet problem for Poisson's equation. *Numer. Math.* (1962), **4**, No. 4, 313–327.

[20] Bramble, J. H. and Schatz, A. H.: Rayleigh–Ritz–Galërkin methods for Dirichlet's problem using subspaces without boundary conditions. *Comm. Pure Appl. Math.* (1970), **23**, 653–675.

[21] Brezinski, C.: Etudes sur les ε- et ρ-algorithmes. *Numer. Math.* (1971), **17**, No. 2, 153–162.

[22] Brezinski, C.: Conditions d'application et de convergence de procedes d'extrapolation. *Numer. Math.* (1972), **20**, No. 1, 64–79.

[23] Bulirsch, R.: Bemerkungen zur Romberg-Integration. *Numer. Math.* (1964), **6**, No. 1, 6–16.

[24] Bulirsch, R. and Stoer J.: Fehlerabschätzungen und Extrapolation mit rationalen Funktionen bei Verfahren vom Richardson-Typus. *Numer. Math.* (1964), **6**, No. 5, 413–427.

[25] Bulirsch, R. and Stoer, J.: Numerical treatment of ordinary differential equations by extrapolation methods. *Numer. Math.* (1966), **8**, No. 1, 1–13.

[26] Cartan, H.: *Calcul Differentiel*. Hermann: Paris, 1967.

[27] Courant, R. and Hilbert, D.: *Methods of Mathematical Physics*, Vol. 1, New York: Interscience, 1953.

[28] Dahlquist, G.: A special stability problem for linear multistep methods. *BIT* (1963), **3**, 27–43.

[29] Dahlquist, G. and Lindberg, B.: On some implicit one-step methods for stiff differential equations. Stockholm: Royal Inst. of Tech., Dept. of Inf. Proc., TRITA-NA-7302.

[30] Davidenko, D. F.: On a method of constructing difference equations for solving the interior Dirichlet problem for the Poisson equation using the grid method. *Ukranian Math. J.* (1961), **13**, No. 4, 92–96.

[31] Demidov, G. V.: On a method of construction of stable schemes of higher order. *Numer. Meth. for the Mechanics of Continuous Media* (1970), **1**, No. 6, 60–69 (Russian).

[32] Descloux, J.: *Méthode des Éléments Finis*. Suisse: Lausanne, 1973.

[33] Enright, W. H. and Hull, T. E.: Test results on initial-value methods for nonstiff ordinary differential equations. *SIAM J. Numer. Anal.* (1976), **13**, No. 6, 944–961.

[34] Faddeev, D. K. and Faddeeva, V. N.: *Computational Methods of Linear Algebra*. New York: Dover, 1962.

[35] Falk, R. S. and King, J. T.: A penalty and extrapolation method for the stationary Stokes equations. *SIAM J. Numer. Anal.* (1976), **13**, No. 5, 814–829.

[36] Fikhtengol'ts, G. M.: *A Course in Differential and Integral Calculus*, Vol. 2. Moscow: Nauka, 1969 (Russian).

[37] Fikhtengol'ts, G. M.: *A Course in Differential and Integral Calculus*, Vol. 3. Moscow: Nauka, 1969 (Russian).

[38] Fedorova, O. A.: Variational difference scheme for the one-dimensional diffusion equation. *Mathematical Notes* (1975), **17**, No. 6, 893–898.

[39] Forsythe, G. and Wasow, W.: *Finite-Difference Methods for Partial Differential Equations.* New York, London: Wiley, 1961.

[40] Fox, L.: On the accuracy and precision of methods. In: *Numerical Solution of Ordinary and Partial Differential Equations.* Oxford: Pergamon Press, 1967, pp. 106–111, 205–312.

[41] Fox, L and Mayers, D. F.: *Computational Methods for Scientists and Engineers.* Oxford: Oxford University Press, 1968.

[42] Fryazinov, I. V.: Economical methods for solving multidimensional equations of parabolic type. *J. Computational Mathematics* (1969), **9**, No. 6, 1316–1326 (Russian).

[43] Fufaev, V. V.: The Dirichlet problem for domains with cusps. *Soviet Math. Doklady* (1960), **131**, No. 1, 37–39.

[44] Gerschgorin, S. A.: Fehlerabschätzung für das Differenzenverfahren zur Lösung partieller Differentialgleichungen. *Z. Angew. Math. Mech.* (1930), **10**, 373–382.

[45] Godunov, S. K.: *Equations of Mathematical Physics.* Moscow: Nauka, 1971 (Russian).

[46] Godunov, S. K. and Rjaben'kii, V. S.: *Introduction in the Theory of Difference Schemes.* Fizmatgiz: Moscow, 1962 (Russian).

[47] Gragg, W. B.: On extrapolation algorithms for initial-value problems. *SIAM J. Numer. Anal.* (1965), **2**, No. 3, 384–403.

[48] Hartman, P.: *Ordinary Differential Equations.* New York, London, Sydney: Wiley, 1964.

[49] Hunter, D. B.: The numerical evaluation of Cauchy principal values by Romberg integration. *Numer. Math.* (1973), **21**, No. 3, 185–191.

[50] Il'in, V. P.: *Difference Methods for Elliptic Equations.* Novosibirsk. Novosibirsk University, 1970 (Russian).

[51] Il'in, V. P.: *Numerical Methods for Electro-Optics Equations.* Novosibirsk: Nauka, 1974 (Russian).

[52] Ivanov, V. K.: On Ill-posed problems. *Math. Sbornik* (1963), **61**, No. 2, 211–223.

[53] Joyce, D. C.: A Survey of extrapolation methods in numerical analysis. *SIAM Review* (1971), **13**, No. 4, 435–490.

[54] Kantorovich, L. V. and Akimov, G. P.: *Functional Analysis.* New York: Academic Press, 1978.

[55] Kantorovich, L. V. and Krylov, V. I.: *Approximate Methods of Higher Analysis.* Moscow: *Govt. Publ. House*, 1949 (Russian).

[56] Keller, H. B.: Accurate difference methods for linear ordinary differential systems subject to linear constraints. *SIAM J. Numer. Anal.* (1969), **6**, No. 1, 8–30.

[57] Keller, H. B.: A new difference scheme for parabolic problems. In: *Numerical Solution of Partial Differential Equations*, Vol. 2. New York: Academic Press, 1971, pp. 327–350.

[58] Keller, H. B.: Accurate difference methods for nonlinear two-point boundary-value problems. *SIAM J. Numer. Anal.* (1974), **11**, No. 2, 305–320.

[59] Keller, H. B. and Cebeci, T.: Accurate numerical methods for boundary-layer flows. 2: Two-dimensional turbulent flows. *AIAA J.* (1972), **10**, No. 9, 1193–1199.

[60] Kellogg, B.: Singularities in interface problems. *Transactions of SYNSPASE* (1971), 351–400.

[61] Kobayashi, M.: On the numerical solution of the Volterra integral equations of the first kind by trapezoidal rule. *Rep. Stat. Appl. Res. JUSE* (1967), **14**, 65–78.

[62]   Kochergin, V. P., Klimok, V. I., and Shcherbakov, A. V.: On the computation of gradients of difference solutions. *Numer. Meth. for the Mechanics of Continuous Media* (1977), **8**, No. 3, 105–111 (Russian).

[63]   Kochergin, V. P. and Shcherbakov, A. V.: Analysis of difference schemes for elliptic equations with a small parameter. In: *Numerical Models of the Ocean*. Novosibirsk: Computing Center of the USSR Academy of Sciences, 1977, pp. 7–24 (Russian).

[64]   Kondratev, V. A.: Boundary-value problems for elliptic equations in domains with conic or angular points. *Transactions of the Moscow Math. Society* (1967), **16**, 109–192.

[65]   Krasnov, M. L.: *Integral Equations*. Moscow: Nauka, 1975 (Russian).

[66]   Krein, S. G.: *Linear Differential Equations in Banach Space*. New York: Academic Press, 1978.

[67]   Krylov, V. I., Bobkov, V. V., and Monastyrnyi, P. I.: *Computational Methods*, Vol. 1. Moscow: Nauka, 1976 (Russian).

[68]   Krylov, V. I., Bobkov, V. V., and Monastyrnyi, P. I.: *Computational Methods*, Vol. 2. Moscow: Nauka, 1977 (Russian).

[69]   Kurosh, A. G.: *A Course of Higher Algebra*. New York: Dover, 1972.

[70]   Kuznetsov, Yu. A. and Shidurov, V. V.: On the uniform convergence of difference schemes, I. In: *Computational Methods of Linear Algebra*. Novosibirsk: Computing Center of USSR Academy of Sciences, 1972, pp. 70–92 (Russian).

[71]   Ladyzenskaja, O. A., Solonnikov, V. A., and Ural'tseva, N. N.: *Linear and Quasilinear Equations of Parabolic type*. New York: Academic Press, 1970.

[72]   Ladyzenskaja, O. A. and Ural'tseva, N. N. *Linear and Quasilinear Equations of Elliptic Type*. New York: Academic Press, 1968.

[73]   Ladyzenskaja, O. A. and Ural'tseva, N. N.: *Linear and Quasilinear Equations of Elliptic Type*. Moscow: Nauka, 1973 (Russian).

[74]   Lattès, R. and Lions, J.-L.: *Methode de Quasireversibilitè et Applications*. Paris: Dunod, 1967.

[75]   Laurent, P.: Un théoréme de convergence pour le procédé d'extrapolation de Richardson. *C. R. Acad. Sci., Paris* (1963), **256**, 1435–1437.

[76]   Lavrentev, M. M.: *On Ill-Posed Problems of Mathematical Physics*. Novosibirsk: USSR Academy of Sciences, 1962 (Russian).

[77]   Lebedev, V. I.: On the Dirichlet and Neumann problems on triangular and hexagonal nets. *Dokl. Soviet Math. Doklady* (1961), **138**, No. 1, 33–36.

[78]   Lebedev, V. I.: On four-point difference schemes. *Soviet. Math. Doklady* (1962), **142**, No. 3, 526–529 (Russian).

[79]   Lindberg, B.: On smoothing and extrapolation for the trapezoidal rule. *BIT* (1971), **11**, 29–52.

[80]   Lindberg, B.: Error estimates and a step size strategy for the implicit midpoint rule. Stockholm: Royal Inst. of Tech., Dept. of Inf. Proc., Rept. N., 72.59, 1972.

[81]   Lindberg, B.: IMPEX–2. A procedure for solving systems of stiff differential equations. Stockholm: Royal Inst. of Tech., Dept. of Inf. Proc., TRITA-NA-7303, 1973.

[82]   Lindberg, B.: Optimal step size sequences and requirements for the local error for (stiff) differential equations. Univ. of Toronto, Dept. of Comput. Sci., Tech. Rept., No. 67, 1974.

[83]   Lins, P.: Numerical methods for Volterra integral equations of the first kind. *J. Computing* (1969), **12**, 393–397.

[84]   Lions, J. L.: *Quelques Méthodes de Résolution des Problèmes Aux Limites Non Linéaires*. Paris: Dunod–Gauthier-Villars, 1969.

[85]   Marchuk, G. I.: *Numerical Methods for Nuclear Reactors*. Moscow: Govt. Publ. House—Atomic Physics, 1961 (Russian).

[86]   Marchuk, G. I.: *Methods of Computational Mathematics*. Moscow: Nauka, 1980 (Russian).
[87]   Marchuk, G. I.: *Methods of Computational Mathematics*. Berlin and New York: Springer-Verlag, 1981.
[88]   Marchuk, G. I. and Kuznetsov, Yu. A.: Iterative methods and quadratic functions. In: *Methods of Computational Mathematics*. Novosibirsk: Nauka, 1975, pp. 4–143 (Russian).
[89]   Marchuk, G. I. and Lebedev, V. I.: *Numerical Methods in the Theory of Neutron Transport*. Moscow: Govt. Publ. House—Atomic Physics, 1971 (Russian).
[90]   Marchuk, G. I. and Shaidurov, V. V.: Increasing the accuracy of projective difference schemes. *Lecture Notes in Computer Science* (1974), **11**, 120–141.
[91]   Marchuk, G. I. and Shaidurov, V. V.: A variational method for increasing the accuracy of difference schemes. *Lecture Notes in Economics* (1976), **134**, 193–205.
[92]   Mayers, D. F.: The approach to the limit in ordinary differential equations. *J. Computing* (1964), **7**, 54–57.
[93]   Mickhlin, S. G.: *Variational Methods in Mathematical Physics*. New York: Academic Press, 1971.
[94]   Mikhlin, S. G.: *Linear Partial Differential Equations*, Moscow: Vyschaja Shkola, 1977 (Russian).
[95]   Mikeladze, S. E.: Numerische Integration der Gleichungen vom elliptischen und parabolischen Typus. *Izv. Akad. Nauk SSSR, Ser. Mat.* (1941), **5**, No. 1, 57–74 (Russian).
[96]   *Modern Numerical Methods for Ordinary Differential Equations*. Oxford: Clarendon Press, 1976.
[97]   Morozov, V. A.: On pseudosolutions. *J. Computational Mathematics* (1969), **9**, No. 6, 1387–1391 (Russian).
[98]   Morozov, V. A.: Regularization methods and systems of linear algebraic equations. In: *Computational Methods of Linear Algebra*. Novosibirsk: Computing Center of the USSR Academy of Sciences, 1972, pp. 62–69 (Russian).
[99]   Nitsche, J.: Ein Kriterium für die Quasi-Optimalität des Ritzschen Verfahrens. *Numer. Math.* (1968), **11**, No. 4, 346–348.
[100] Oganesyan, L. A., Rivkind, V. Ya., and Rukhovets, L. A.: Variational-difference methods for elliptic equations. Part I. In: *Differential Equations and Their Applications*, Vol. 5. Vilnius, 1973 (Russian).
[101] Oganesyan, L. A. and Rivkind, V. Ya.: Rukhovets, L. A. Variational-difference methods for elliptic equations. Part II. In: *Differential Equations and Their Applications*, Vol. 8. Vilnius, 1974 (Russian).
[102] Oganesyan, L. A. and Rukhovets, L. A.: Investigation of the rate of convergence of variational-difference schemes for elliptic second-order equations in two-dimensional domains with smooth boundary. *J. Computational Mathematics* (1969), **9**, No. 5, 1102–1120 (Russian).
[103] Ortega, J. and Rheinboldt, W.: *Iterative Solutions of Nonlinear Equations in Several Variables*. New York: Academic Press, 1970.
[104] Pereyra, V.: On improving an approximate solution of a functional equation. *Numer. Math.* (1966), **8**, No. 3, 376–391.
[105] Pereyra, V.: Accelerating the convergence of discretization algorithms. *SIAM J. Numer. Anal.* (1967), **4**, 508–553.
[106] Prikazchikov, V. G.: Uniform difference methods for Sturm–Liouville problems. *J. Computational Mathematics* (1969), **9**, No. 2, 315–336 (Russian).
[107] Richardson, L. F.: The approximate solution of physical problems involving differential equations using finite differences, with an application to the stress in a masonry dam. *Philos. Trans. Roy. Soc., London, Ser. A* (1910), **210**, 307–357.

[108]   Richardson, L. F.: The deferred approach to the limit. 1: The Single lattice.
        Philos. Trans. Roy. Soc., London, Ser. A (1927), **226**, 299–349.
[109]   Ryaben'ky, V. S. and Filippov, A. F.: *On the Stability of Difference Equations.*
        Moscow: GITTL, 1956 (Russian).
[110]   Romberg, W.: Vereinfachte numerische Integration. *Det. Kong. Norske
        Videnskabers Selskab Forhandlinger, Trondheim* (1955), **28**, No. 7, 30–36.
[111]   Rosenbrock, H. H.: Some general implicit methods for the numerical solution
        of differential equations. *J. Computing* (1963), **5**, No. 4, 329–330.
[112]   Samarsky, A. A.: Schemes of a higher order of accuracy for the multidi-
        mensional heat equation. *J. Computational Mathematics* (1963), **3**, No. 3,
        431–466 (Russian).
[113]   Samarsky, A. A.: *Introduction to the Analysis of Difference Schemes.* Moscow:
        Nauka, 1971 (Russian).
[114]   Samarsky, A. A. *Analysis of Difference Schemes.* Moscow: Nauka, 1977
        (Russian).
[115]   Samarsky, A. A. and Andreev, V. B.: *Difference Methods for Elliptic Equations.*
        Moscow: Nauka, 1976 (Russian).
[116]   Shcherbakov, A. V. and Kochergin, V. P.: The method of imbedding nets for
        the ocean dynamics problem. *Numer. Meth. for the Mechanics of Continuous
        Media* (1977), **8**, No. 2, 125–129 (Russian).
[117]   Shortley, G. and Weller, R.: The numerical solution of Laplace's equation.
        J. Appl. Phys. (1938), **9**, No. 5, 334–348.
[118]   Shampine, L. F., Watts, H. A., and Davenport, S. M.: Solving nonstiff ordinary
        differential equations—the state of the art. *SIAM Review* (1976), **18**, No. 3,
        376–411.
[119]   Shaidurov, V. V.: On a method for increasing the accuracy of difference
        solutions. *Numer. Meth. for the Mechanics of Continuous Media* (1972), **3**, No. 2.
        96–104 (Russian).
[120]   Shaidurov, V. V.: Extension by a parameter in the regularization method. In:
        Computational Methods of Linear Algebra. Novosibirsk: Computing Center of
        USSR Academy of Sciences, 1972, pp. 77–85 (Russian).
[121]   Shaidurov, V. V.: Regularization of systems with symmetric matrix. In:
        Computational Mathematics and Programming. Novosibirsk: Computing Center
        of the USSR Academy of Sciences, 1974, pp. 91–98 (Russian).
[122]   Shaidurov, V. V.: *Methods for Increasing the Accuracy of Approximate Solutions.*
        Novosibirsk: Novosibirsk University, 1978 (Russian).
[123]   Stetter, H. J.: *The Analysis of Discretization Methods for Ordinary Differential
        Equations.* Berlin: Springer-Verlag, 1973.
[124]   Stoer, J. and Bulirsch, R.: *Einführung in die Numerische Mathematik*, Band 2.
        Berlin, New York: Springer-Verlag, 1973.
[125]   Strang, G. and Fix, G.: *An Analysis of the Finite Element Method.* Englewood
        Cliffs, NJ.: Prentice-Hall, 1973.
[126]   Sultanova, I. A.: Effective estimates of the error in the method of nets for
        boundary-value problems for the Laplace and Poisson equations on the rect-
        angle and on special triangles. *J. Computational Mathematics* (1971), **11**, No. 5,
        1205–1218 (Russian).
[127]   Thatcher, R. W.: The use of infinite Grid refinements at singularities in the
        solution of Laplace's equation. *Numer. Math.* (1976), **25**, No. 3, 163–178.
[128]   Tikhonov, A. N.: On ill-posed problems of linear algebra and stable methods
        for their solution. *Soviet Math. Doklady* (1965), **163**, No. 3, 591–594.
[129]   Tikhonov, A. N. and Arsenin, V. Ja.: *Methods for Ill-posed Problems.* Moscow:
        Nauka, 1974 (Russian).
[130]   Urvantsev, A. L. and Shaidurov, V. V.: The correction method for one-di-
        mensional quasilinear diffusion equations. In: *Computational Mathematics and*

Programming. Novosibirsk: Computing Center of the USSR Academy of Sciences, 1974, pp. 81–90 (Russian).

[131] Valiullin, A. N.: *Difference Schemes for Problems of Mathematical Physics*. Novosibirsk: Novosibirsk University, 1970 (Russian).

[132] Vishik, M. I. and Lyusternik, L. A.: Regular decomposition and boundary layers for linear differential equation with a small parameter. *Uspekhi Mat. Nauk* (1957), **12**, Vyp. 5 (77), 3–122 (Russian).

[133] Vladimirov, V. S.: *Generalized Functions in Mathematical Physics*. Moscow: Nauka, 1976 (Russian).

[134] Voevodin, V. V.: *Linear Algebra*. Moscow: Nauka, 1974 (Russian).

[135] Volkov, E. A.: On a way of increasing the accuracy of the net method. *Soviet Math. Doklady* (1954), **96**, No. 4, 685–688.

[136] Volkov, E. A.: A method for improving the accuracy of grid solutions of the Poisson equation. *Computational Mathematics*, 1957, No. 1, 62–80 (Russian).

[137] Volkov, E. A.: Differentiability properties of solutions of boundary-value problems for the Laplace and Poisson equations on a rectangle. *Steklov Institute Publications* (1965), **77**, 89–112.

[138] Volkov, E. A.: Solving the Dirichlet problem by a correction method using higher-order differences, I. *J. Differential Equations* (1965), **1**, No. 7, 946–960.

[139] Volkov, E. A.: Solving the Dirichlet problem by a correction method using higher-order differences, II. *J. Differential Equations* (1965), **1**, No. 8, 1070–1084 (Russian).

[140] Volkov, E. A.: Approximate solutions of the Laplace and Poisson equations in weighted Hölder spaces. *Steklov Institute Publications* (1972), **128**, 76–112.

[141] Volkov, E. A.: On a regular composite net method for the Laplace equation on polygons. *Steklov Institute Publ.* (1976), **140**, 68–102.

[142] Whiteman, J. R. and Barnhill, R. E.: *Finite Element Methods for Elliptic Mixed Boundary-Value Problems with Singularities*. Brno: Proc. Conf. "Equadiff", 1972, pp. 261–267.

[143] Whittaker, E. T. and Watson, G. N.: *A Course of Modern Analysis*. Cambridge: Cambridge University Press, 1927.

[144] Wuytack, L.: A new technique for rational extrapolation to the limit. *Numer. Math.* (1971), **17**, No. 3, 215–221.

[145] Wynn, P.: On the convergence and stability of the epsilon algorithm. *SIAM J. Numer. Anal.* (1966), **3**, No. 1, 91–122.

[146] Yanenko, N. N.: *The Method of Fractional Steps*. Berlin, Heidelberg, New York: Springer-Verlag, 1971.

Index